STERLING
Test Prep

College Organic Chemistry

Practice Questions

6th edition

Copyright © Sterling Test Prep

Customer Satisfaction Guarantee

Your feedback is important because we strive to provide the highest quality prep materials. Email us questions, comments or suggestions.

info@sterling–prep.com

We reply to emails – check your spam folder

All rights reserved. This publication's content, including the text and graphic images or part thereof, may not be reproduced, downloaded, disseminated, published, converted to electronic media, or distributed by any means whatsoever without prior written consent from the publisher. Copyright infringement violates federal law and is subject to criminal and civil penalties.

6 5 4 3 2 1

ISBN-13: 978-1-9547256-9-0

Sterling Test Prep materials are available at quantity discounts.

Contact info@sterling–prep.com

Sterling Test Prep
6 Liberty Square #11
Boston, MA 02109

© 2023 Sterling Test Prep

Published by Sterling Test Prep

 Printed in the U.S.A.

Thousands of students use our study aids to achieve higher grades!

To achieve a high grade in college organic chemistry, you need to do well on your tests and final exam. This book helps you develop and apply knowledge to quickly choose the correct answers to questions typically tested in college organic chemistry courses. Solving targeted practice questions builds your understanding of fundamental concepts and is a more effective strategy than merely memorizing terms.

This book has over 690 practice questions covering organic chemistry topics. Chemistry instructors with years of teaching experience prepared this practice material to build your knowledge and skills crucial for success in a college chemistry course. Our editorial team reviewed and systematized the content for targeted preparation.

The detailed explanations describe why an answer is correct and – more important for your learning – why another attractive choice is wrong. They provide step-by-step solutions and teach the scientific foundations and details of essential organic chemistry topics. Read the explanations carefully to understand how they apply to the question and learn important organic chemistry principles and the relationships between them. With the practice material contained in this book, you will significantly improve your understanding, test scores, and your grade.

We wish you great success in your academic achievements and look forward to being an important part of your preparation!

College study aids

Cell and Molecular Biology Review

Organismal Biology Review

Cell and Molecular Biology Practice Questions

Organismal Biology Practice Questions

Physics Review (Part 1 and 2)

Physics Practice Questions (Vol. 1 and 2)

Organic Chemistry Practice Questions

United States History 101

American Government and Politics 101

Environmental Science 101

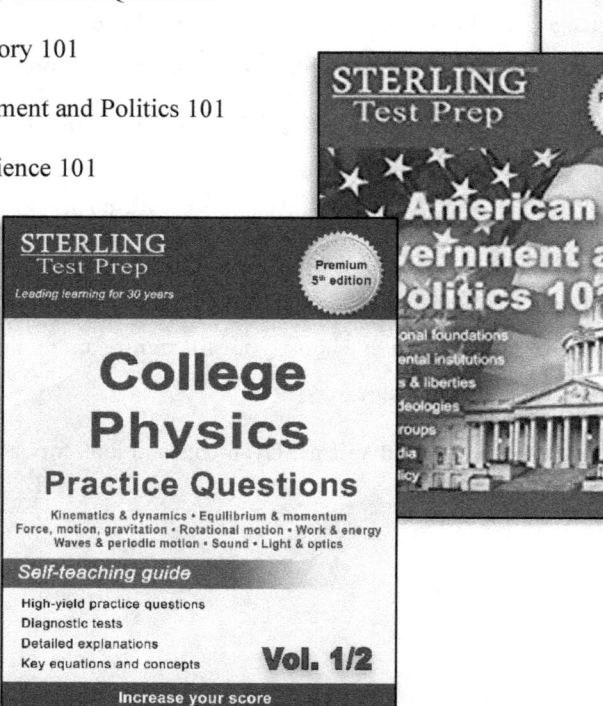

Visit our Amazon store

Table of Contents

Practice Questions 7
    Organic Chemistry Nomenclature 9
    Covalent Bond 21
    Stereochemistry 33
    Molecular Structure and Spectra 45
    Alkanes and Alkyl Halides 51
    Alkenes 61
    Alkynes 73
    Aromatic Compounds 83
    Alcohols 93
    Aldehydes and Ketones 103
    Carboxylic Acids 113
    COOH Derivatives 121
    Amines 131
    Amino Acids, Peptides, Proteins 139
    Lipids 149
    Carbohydrates 159
    Nucleic Acids 171

Answer Keys and Detailed Explanations 181
    Answer Keys 183
    Organic Chemistry Nomenclature 189
    Covalent Bond 199
    Stereochemistry 213
    Molecular Structure and Spectra 227
    Alkanes and Alkyl Halides 239
    Alkenes 251
    Alkynes 261
    Aromatic Compounds 271
    Alcohols 283
    Aldehydes and Ketones 293
    Carboxylic Acids 305
    COOH Derivatives 315
    Amines 325
    Amino Acids, Peptides, Proteins 335
    Lipids 353
    Carbohydrates 371
    Nucleic Acids 389

Glossary of Terms 405

Periodic Table of the Elements 432

Isomer Classification 433

College Level Examination Program (CLEP)

Biology Review	History of the United States I Review
Biology Practice Questions	History of the United States II Review
Chemistry Review	Western Civilization I Review
Chemistry Practice Questions	Western Civilization II Review
Introductory Business Law Review	Social Sciences and History Review
College Algebra Practice Questions	American Government Review
College Mathematics Practice Questions	Introductory Psychology Review

Visit our Amazon store

Topical Practice Questions

If you benefited from this book, we would appreciate if you left a review on Amazon, so others can learn from your input. Reviews help us understand our customers' needs and experiences while keeping our commitment to quality.

Organic Chemistry Nomenclature

1. Name the structure:

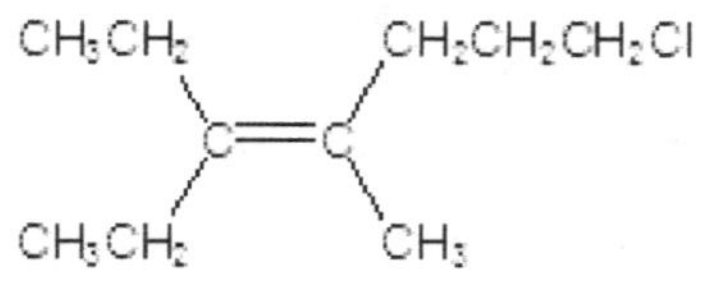

A. *cis*-7-chloro-3-ethyl-4-methyl-3-heptene
B. 1-chloro-3-pentenyl-2-pentene
C. 1-chloro-5-ethyl-4-methyl-3-heptene
D. 7-chloro-3-ethyl-4-methyl-3-heptene
E. *trans*-7-chloro-3-ethyl-4-methyl-3-heptene

2. What is the IUPAC name for the following compound?

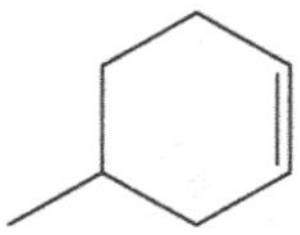

A. 1-methyl-4-cyclohexene
B. 1-methyl-3-cyclohexene
C. 4-methylcyclohexene
D. 5-methylcyclohexene
E. methylcyclohexene

3. Give the formula of the structure below:

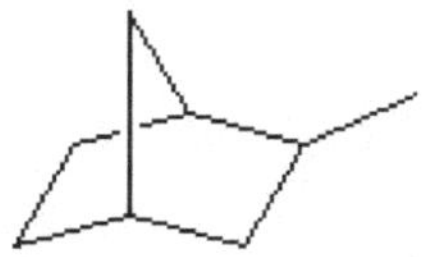

A. C_8H_{14}
B. C_8H_{12}
C. C_8H_{10}
D. C_8H_8
E. C_8H_{16}

4. Which structure is *para*-dibromobenzene?

I. 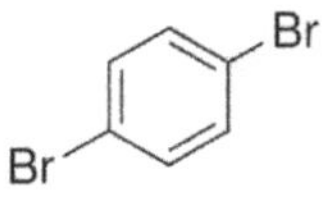

II. 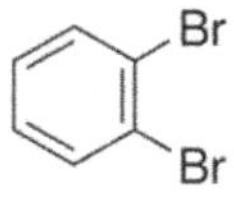

III. 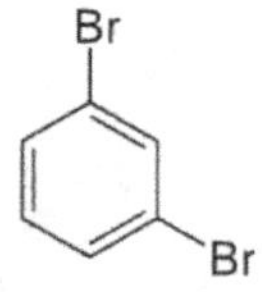

A. I only
B. II only
C. III only
D. I and II only
E. I, II and III

5. Ignoring geometric isomers, what is the IUPAC name for the following compound:

   CH₃–CH=CH–CH₃

   A. but-2-yne
   B. but-2-ene
   C. butene-3
   D. butene-2
   E. 2-butyl

6. What is the IUPAC name for the following structure:

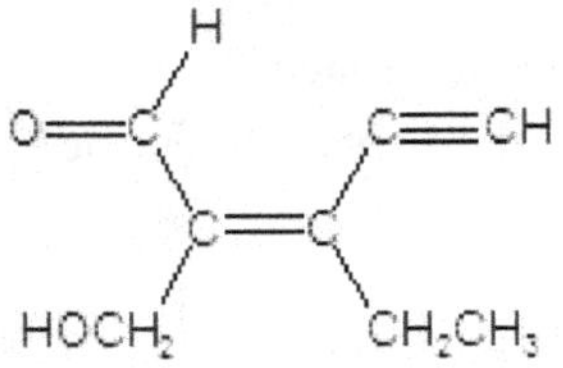

   A. (Z)-3-ethyl-5-hydroxymethyl-3-penten-1-ynal
   B. (E)-3-ethyl-5-hydroxymethyl-3-penten-1-ynal
   C. (Z)-3-ethyl-2-hydroxymethyl-2-penten-4-ynal
   D. (E)-3-ethyl-2-hydroxymethyl-2-penten-4-ynal
   E. 3-ethyl-5-hydroxymethyl-3-penten-1-ynal

7. Give the IUPAC name for the following structure:

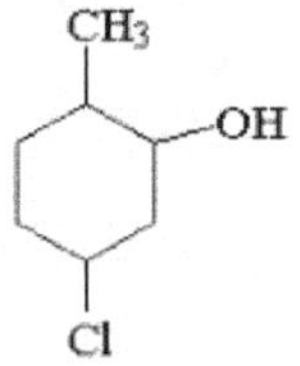

   A. 1-chloro-4-methylcyclohexanol
   B. 5-chloro-2-methylcyclohexanol
   C. 3-chloro-2-methylcyclohexanol
   D. 2-methyl-5-chlorocyclohexanol
   E. 2-methyl-3-chlorocyclohexanol

8. What is the name of the following compound?

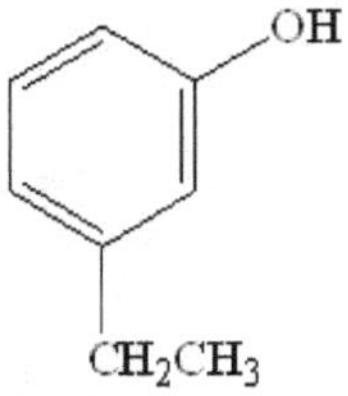

   A. p-ethylphenol
   B. m-ethylbenzene
   C. o-ethylphenol
   D. m-ethylphenol
   E. none of the above

9. Provide the IUPAC name of the compound:

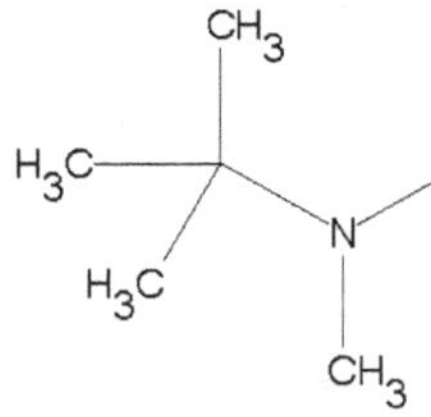

A. *N,N*,2-trimethyl-1-propanamine
B. *N,N*,2-trimethylpropanamine
C. *N,N*,1,1-tetramethylethanamine
D. *N,N*-dimethyl-2-butanamine
E. *N,N*,2-trimethyl-2-propanamine

10. Name the following structure:

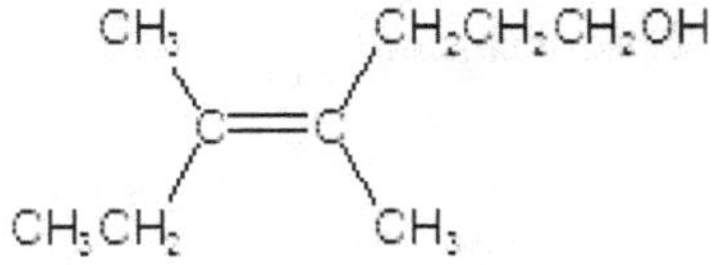

A. *cis*-3,4-dimethyl-3-hepten-7-ol
B. *trans*-4,5-dimethyl-4-hepten-1-ol
C. *cis*-4,5-dimethyl-4-hepten-1-ol
D. *trans*-3,4-dimethyl-3-hepten-7-ol
E. *trans*-4,5-dimethyl-4-heptenol

11. What is the systematic name for the following compound?

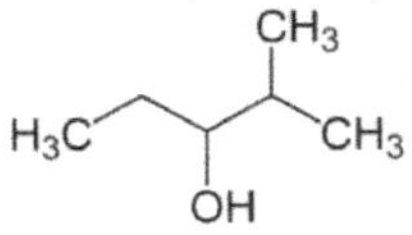

A. 3-methyl-2-pentanol
B. 4-methyl-3-pentanol
C. 2-methyl-3-pentenol
D. 2-methyl-3-pentanol
E. 3-methyl-3-pentanol

12. Which condensed structural formula below is the isopropyl group?

A.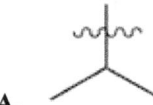

B.

C.

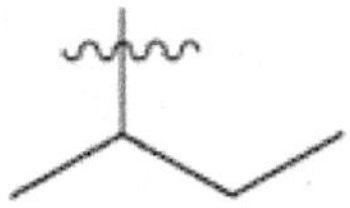

D.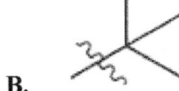

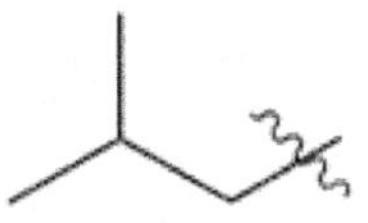

E.

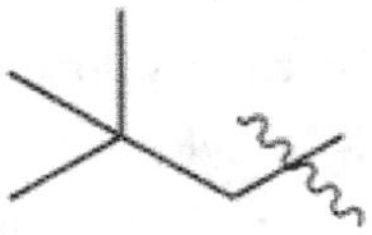

13. Which of the following compounds is named correctly?

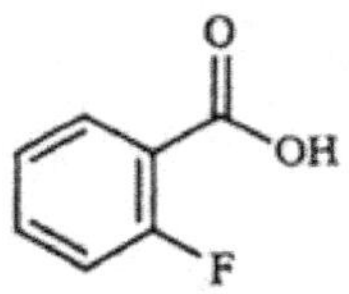

A. *meta*-fluorobenzoic acid

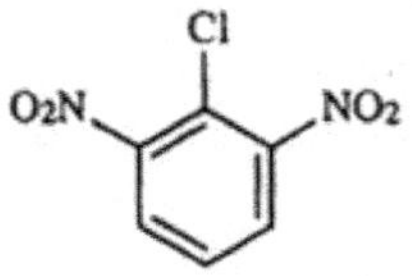

B. 2,5-dinitro-1-chlorobenzene

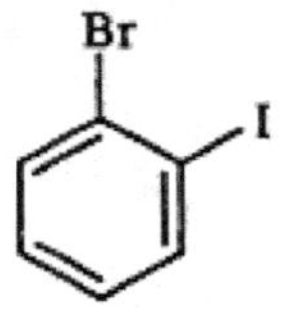

C. 2-iodo-1-bromobenzene

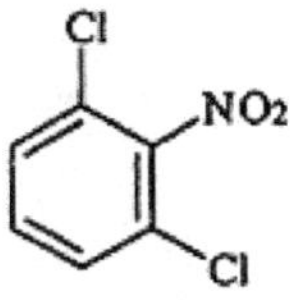

D. 1,3,dichloro-2-nitrobenzene

E. None of the above

14. Name the compound shown below.

A. *cis*-1,3-dichlorocyclohexane
B. *trans*-1,3-dichlorocyclohexane

C. *cis*-1,2-dichlorocyclohexane
D. *trans*-1,2-dichlorocyclohexane
E. *cis*-1,4-dichlorocyclohexane

15. What is the IUPAC name of the compound shown?

CH₃—CH—CH₂—CH—CH₃
    | |
    CH₃   OH

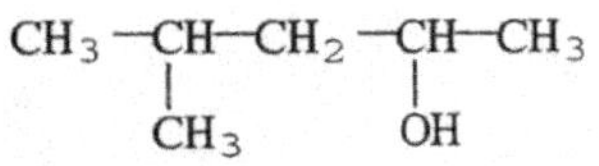

A. 2, 2-dimethyl-4-butanol
B. 4, 4-dimethyl-2-butanol

C. 2-methyl-4-pentanol
D. 4-methyl-2-pentanol
E. 2-isohexanol

16. What is the IUPAC name of the compound shown below?

A. (1*R*,4*S*)-1,4-dichloro-1-ethyl-4-methylcyclopentane
B. (1*R*,3*S*)-1,3-dichloro-1-ethyl-3-methylcyclopentane
C. (1*S*,3*S*)-1,3-dichloro-1-ethyl-3-methylcyclopentane
D. (1*R*,3*S*)-1,3-dichloro-1-methyl-3-ethylcyclopentane
E. (1*S*,3*R*)-1,3-dichloro-3-ethyl-1-methylcyclopentane

17. The compound CH$_3$(CH$_2$)$_5$CH$_3$ is:

   A. heptane
   B. hexene
   C. hexane
   D. pentane
   E. octene

18. Name the following structure:

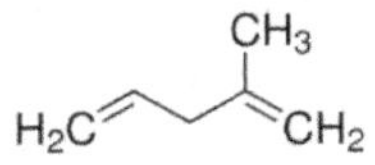

   A. 2-methylene-4-pentene
   B. 4-methylene-2-pentene
   C. 2-methyl-2,4-pentadiene
   D. 4-methyl-1,4-pentadiene
   E. 2-methyl-1,4-pentadiene

19. The name of the following alkyl group is ~CH$_2$CH$_2$CH$_3$:

   A. ethyl
   B. propyl
   C. isopropyl
   D. *sec*-butyl
   E. butyl

20. What is the complete systematic IUPAC name for the following compound?

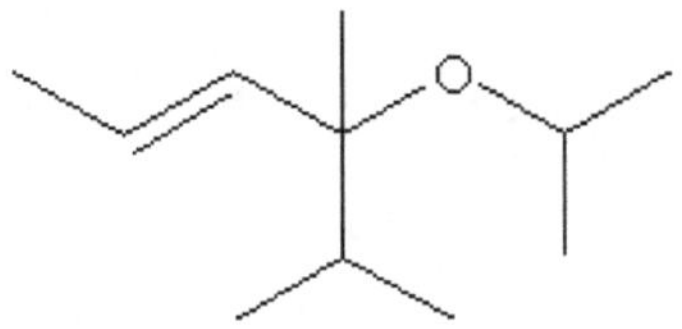

   A. isopropyl-(4-isopropyl-4-methylbut-2-enyl) ether
   B. (*E*)-4-isopropoxy-4,5-dimethylhex-2-ene
   C. 4-(1-methylethoxy)-4-isopropyl-4-methylpent-2-ene
   D. 4-isopropyl-2,4-dimethylhept-5-en-3-ol
   E. 4-isopropyl-4-methylbut-2-en-isopropyl ether

21. The IUPAC name for the compound H$_2$C=CH–CH=CH$_2$ is:

   A. 1,3-butadiene
   B. butane-1,3
   C. butene-2
   D. 1,3-dibutene
   E. 2-butyne

22. What is the IUPAC name for salicylic acid shown below:

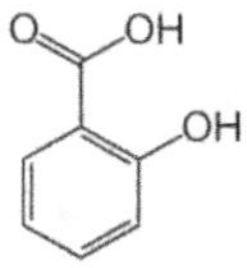

A. 2-hydroxybenzoic acid
B. α-hydroxybenzoic acid
C. 1-hydroxybenzoic acid
D. meta-hydroxybenzoic acid
E. 3-hydroxybenzoic acid

23. What is the common name for the simplest ketone, propanone?

A. acetal
B. acetone
C. carbanone
D. formalin
E. none of the above

24. The name of the compound shown below is:

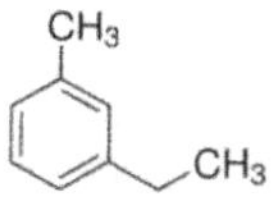

A. 3-ethyltoluene
B. 2-ethyltoluene
C. 1-ethyl-4-methylbenzene
D. 1-ethyl-2-methylbenzene
E. 4-ethyltoluene

25. A compound with the molecular formula C_3H_6 is:

A. butane
B. butyne
C. 2-methylpropane
D. cyclopropane
E. propane

26. What is the correct IUPAC name for the following compound?

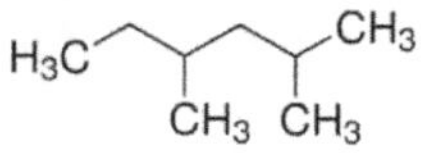

A. 2-ethyl-4-methylhexane
B. 2,4-dimethylhexane
C. 4-ethyl-2-methylhexane
D. 3,5-dimethylheptane
E. *sec*-butylpropane

27. Name the following structure according to IUPAC nomenclature:

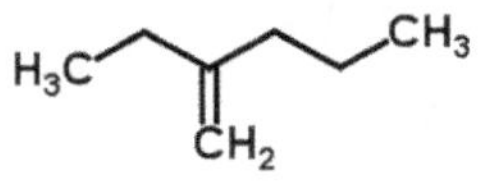

   A. 3-ethyl-3-hexene
   B. 4-methylenehexane
   C. 2-propyl-1-butene
   D. 2-ethyl-1-pentene
   E. ethyl-propylethene

28. What is the IUPAC name for the following structure?

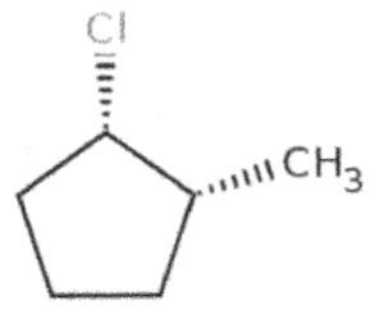

   A. *cis*-methylcyclohexane
   B. *cis*-1-chloro-2-methylcyclopentane
   C. *Z*-chloro-methylcyclohexane
   D. *cis*-2-chloro-2-methylcyclohexane
   E. *Z*-2-chloro-1-methylcyclohexane

29. The compound below is:

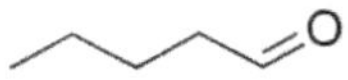

   A. pentanone
   B. pentanal
   C. butanaldehyde
   D. pentaketone
   E. pentanoic acid

30. Provide the common name of the compound:

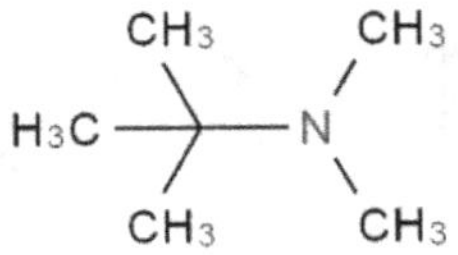

   A. neobutyldimethylamine
   B. *sec*-butyldimethylamine
   C. *tert*-butyldimethylamine
   D. isobutyldimethylamine
   E. *n*-butyldimethylamine

31. What is the correct IUPAC name for the following compound?

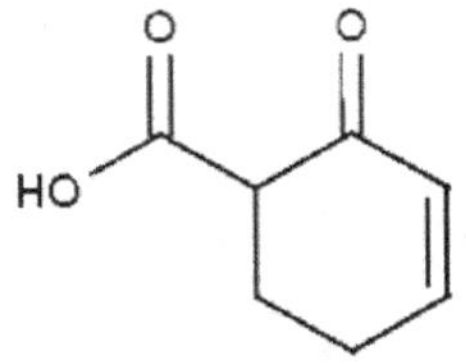

   A. 2-oxocyclohex-3-ene-1-carboxylic acid
   B. 5-formylcyclohex-2-enone-oic acid
   C. 2-formylcyclohex-5-enone
   D. 3-oxocyclohex-4-enoic acid
   E. 2-oxocyclohex-3-ene carbonate

32. What is the IUPAC name of the compound shown?

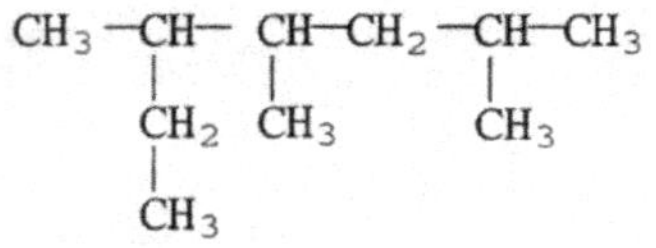

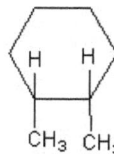

A. 3,4,6-trimethylheptane
B. 2,4,5-trimethylheptane
C. 3,5-dimethyl-2-ethylhexane
D. 2-ethyl-3,5-dimethylhexane
E. 5-ethyl-2,4-dimethylhexane

33. What is the IUPAC name for the following structure?

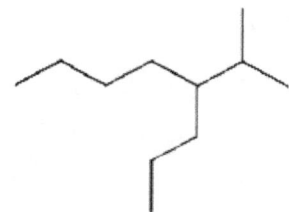

A. 5,6-dimethyl cyclohexane
B. *cis*-1,2-dimethyl cyclohexane
C. *trans*-1,2-dimethyl cyclohexane
D. 1,2-dimethyl cyclohexane
E. none of the above

34. Give the IUPAC name for the following structure:

A. 4-isopropyloctane
B. 5-isopropyloctane
C. 3-ethyl-2-methylheptane
D. 2-methyl-3-ethylheptane
E. 2-methyl-3-propylheptane

35. Which name is NOT correct for IUPAC nomenclature?

A. 2,2-dimethylbutane
B. 2,3-dimethylpentane
C. 2,3,3-trimethylbutane
D. 2,3,4-trimethylpentane
E. All are correct names

36. Which compound has the common name s*ec*-butylamine?

A. 1-butanamine
B. 2-butanamine
C. *N*-methyl-1-propanamine
D. *N*-methyl-2-propanamine
E. *N*-ethylethanamine

37. Which of the following is *cis*-2,3-dichloro-2-butene?

A. 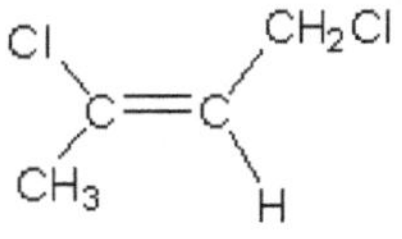 [structure: Cl and CH₃ on left carbon, CH₂Cl and H on right carbon]

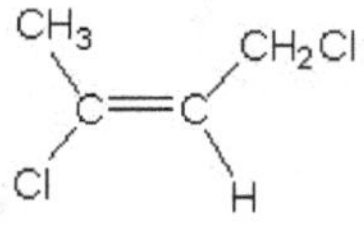

B. 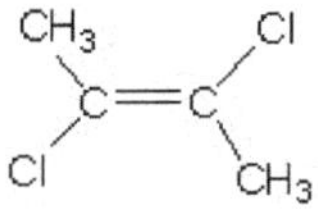 [structure: CH₃ and Cl on left carbon, Cl and CH₃ on right carbon]

C. [structure: CH₃ and Cl on left carbon, CH₂Cl and H on right carbon]

D. 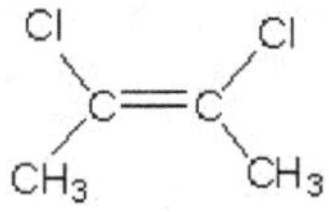 [structure: Cl and CH₃ on left carbon, Cl and CH₃ on right carbon]

E. None of the above

38. Which of the following is the IUPAC name for this compound?

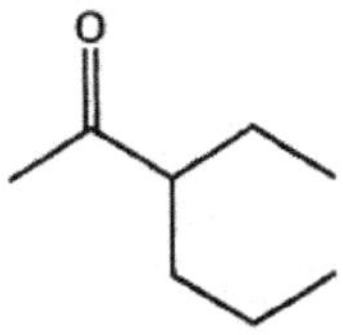

A. 3-ethylhexan-2-one
B. 4-ethylhexan-5-one
C. 3-propylpentan-2-one
D. 3-propylpentan-4-one
E. hexana-3-one

39. Name the structure shown below:

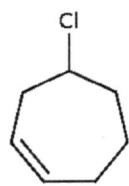

A. 1-chloro-3-cycloheptene
B. 4-chlorocycloheptane
C. 4-chlorocyclohexene
D. 1-chloro-3-cyclohexene
E. 4-chlorocycloheptene

40. Provide the IUPAC name of the compound:

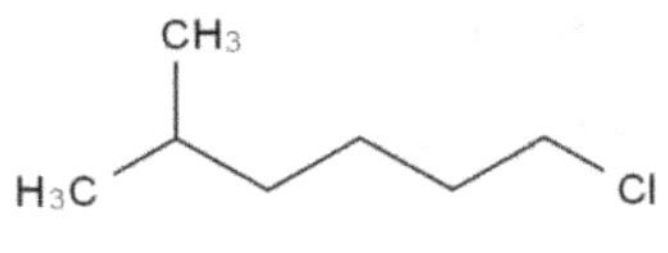

A. 1,1-dimethyl-5-chloropentane
B. 6-chloro-2-methylhexane
C. 1-chloro-5-methylhexane
D. 2-methyl-chloro-heptane
E. 1,1-dichloro-5-methylpentane

41. Ethyl propanoate is a(n):

A. ester
B. alcohol
C. aldehyde
D. carboxyl alcohol
E. amide

42. Provide the common name of the compound:

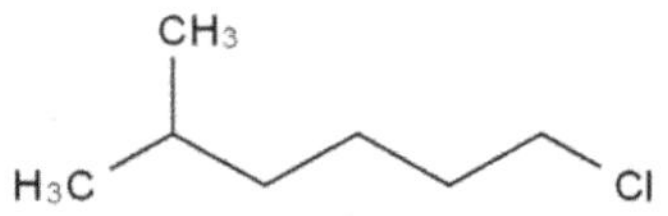

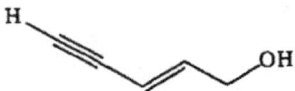

A. isoheptyl chloride
B. *tert*-heptyl chloride
C. neoheptyl chloride
D. *sec*-heptyl chloride
E. n-heptyl chloride

43. What is the IUPAC name of the molecule shown?

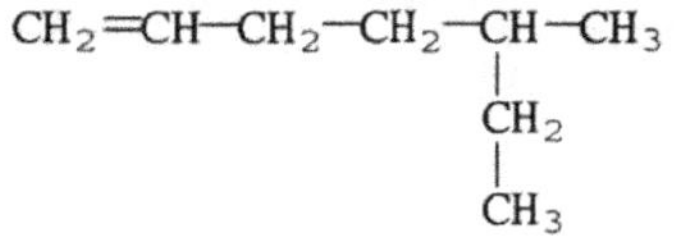

$$CH_2=CH-CH_2-CH_2-CH(CH_2CH_3)-CH_3$$

A. 2-ethyl-5-hexene
B. 5-ethyl-1-hexene
C. 5-methyl-1-heptene
D. 3-methyl-6-heptene
E. octane

44. Which of the following is NOT a proper condensed structural formula for an alkane?

A. $CH_3CHCH_3CH_2CH_3$
B. $CH_3CH_2CH_2CH_2CH_3$
C. $CH_3CH_3CH_3$
D. $CH_3CH_2CH_2CH_3$
E. CH_3CH_3

45. Which of the following is the IUPAC name for this compound?

A. (Z)-pent-3-en-1-yn-5-ol
B. (E)-pent-3-en-1-yn-5-ol
C. (Z)-pent-2-en-4-yn-1-ol
D. (E)-pent-2-en-4-yn-1-ol
E. Cis-pent-2-en-4-yn-1-ol

46. What is the name of the following compound: $H_3CCH_2CH_2CH_2CH_2CONH_2$

A. 1-hexanamide
B. hexanamide
C. hexanamine
D. hexamine
E. hexketoneamine

Notes for active learning

Notes for active learning

Covalent Bond

1. Which of the following is a benzylic cation?

   A.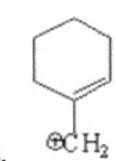

   B. ⊕CH₂ (on cyclohexene)

   C.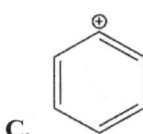

   D.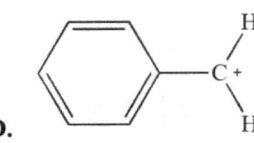

   E. 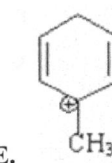 CH₃

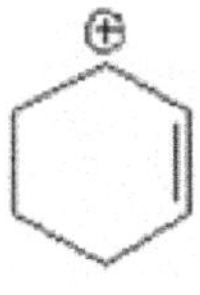

2. Which orbitals overlap to create the H–C bond in CH₃⁺?

   A. *s–p*
   B. *s–sp²*
   C. *sp³–sp³*
   D. *sp²–sp³*
   E. *s–sp³*

3. Which of the following structures, including formal charges, is correct for diazomethane, CH₂N₂?

   A. H₂C=N⁺=N⁻
   B. H₂C=N⁺≡N⁻
   C. H2C=N⁻=N⁺
   D. ⁻CH₂–N≡N:
   E. H₂C–N⁺³≡N⁻³

4. What chemical reaction did German chemist Friedrich Wohler use to synthesize urea for the first time?

   A. Heating ammonium cyanide
   B. Combining the elements carbon, hydrogen, oxygen, and nitrogen
   C. Evaporating urine
   D. Heating ammonium cyanate
   E. None of the above

5. Which of the following is an allylic cation?

A.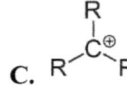

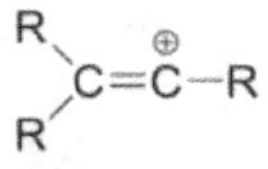

B.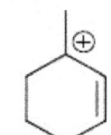

C. R–C⁺(R)–R

D.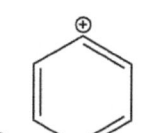

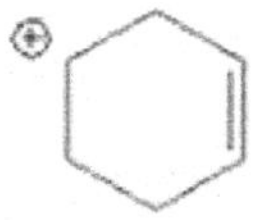

E. CH₃C⁺HCH₃

6. Which of the following best approximates the C–C–C bond angle of propene?

A. 90°
B. 109°
C. 120°
D. 150°
E. 180°

7. What determines the polarity of a covalent bond?

A. The difference in the number of protons
B. The difference in the number of valence electrons
C. The difference in the atomic size
D. The difference in the electronegativity
E. None of the above

8. What is the molecular geometry of an open-chain noncyclical hydrocarbon with the generic molecular formula C_nH_{2n-2}?

A. trigonal pyramidal
B. trigonal planar
C. tetrahedral
D. linear
E. none of the above

9. Identify the most stable carbocation:

A. H₂C=CH⊕

B. 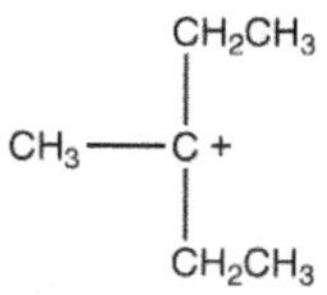 CH₃–C⁺(CH₂CH₃)(CH₂CH₃)

C. CH₃–CH=CH–CH₃ (with ⊕)

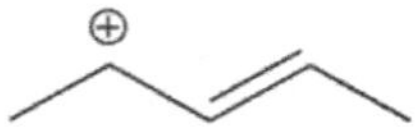

D. phenyl cation (⊕ on benzene ring)

E. 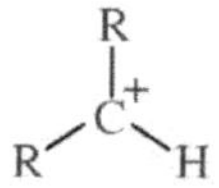 R–C⁺(R)–H

10. Consider the interaction of two hydrogen 1s atomic orbitals of the same phase. Which of the statements below is NOT a correct description of this interaction?

   A. The molecular orbital formed is cylindrically symmetric
   B. The molecular orbital formed has a node between the atoms
   C. The molecular orbital formed is lower in energy than a hydrogen 1s atomic orbital
   D. A *sigma* bonding molecular orbital is formed
   E. A maximum of two electrons may occupy the molecular orbital formed

11. In *trans*-hept-4-en-2-yne, the shortest carbon-carbon bond is between carbons:

   A. 1 and 2
   B. 2 and 3
   C. 3 and 4
   D. 4 and 5
   E. 5 and 6

12. What two atomic orbitals (or hybrid atomic orbitals) overlap to form the C=C π bond in ethylene?

   A. $C\ sp^2 + C\ p$
   B. $C\ sp^2 + C\ sp^2$
   C. $C\ sp^3 + C\ sp^2$
   D. $C\ sp^3 + C\ sp^3$
   E. $C\ p + C\ p$

13. Which of the following is the most stable resonance contributor to acetic acid?

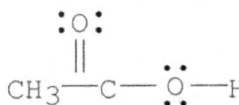

A.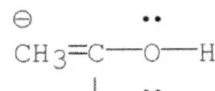

C.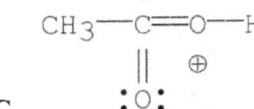

B.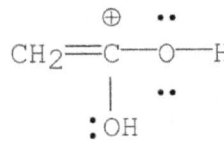

D.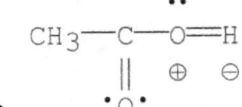

E.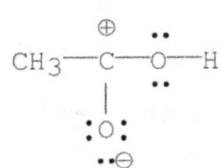

14. The compound methylamine, CH_3NH_2, contains a C–N bond. In this bond, which of the following best describes the charge on the nitrogen atom?

A. uncharged
B. slightly negative
C. +1
D. slightly positive
E. −1

15. Which of the following molecules is the most polar?

A. acetaldehyde
B. acetic acid
C. ethane
D. ethylene
E. benzene

16. A molecule of acetylene (C_2H_2) has a [ ] geometry and a molecular dipole moment that is [ ].

A. bent, zero
B. linear, nonzero
C. tetrahedral, nonzero
D. bent, nonzero
E. linear, zero

17. Which of the following is NOT a resonance structure of the species shown?

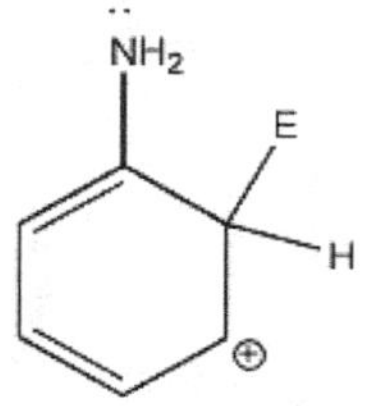

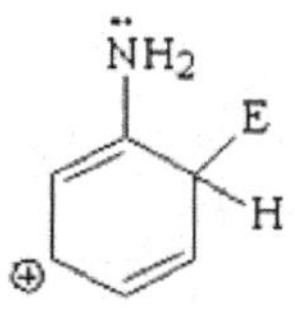

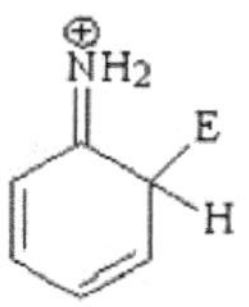

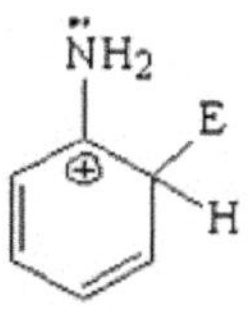

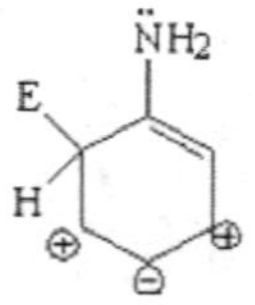

E. All are resonance forms

18. In an aqueous environment, which bond requires the most energy to break?

A. hydrogen
B. dipole-dipole
C. *sigma*
D. ionic
E. *pi*

19. The nitrogen atom of trimethylamine is [ ] hybridized which is reflected in the C–N–C bond angle of [ ]:

   A. *sp*, 180°
   B. *sp²*, 120°
   C. *sp²*, 108°
   D. *sp³*, 120°
   E. *sp³*, 108°

20. Which of the following statements concerning the cyclic molecule shown is NOT true?

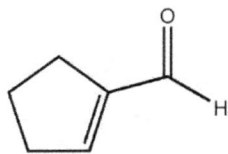

   A. It contains a π molecular orbital formed by the overlap of carbon *p* atomic orbitals
   B. It contains a σ molecular orbital formed by the overlap of carbon *sp³* hybrid atomic orbitals
   C. It contains a σ molecular orbital formed by the overlap of carbon *sp²* hybrid atomic orbitals
   D. It contains a π molecular orbital formed by the overlap of a carbon *p* atomic orbital with an oxygen *p* atomic orbital
   E. It contains a σ molecular orbital formed by the overlap of a carbon *p* atomic orbital with an oxygen *sp³* atomic orbital

21. Draw a structural formula for cyclohexane, a cyclic saturated hydrocarbon (C_6H_{12}). How many π bonds are in a cyclohexane molecule?

   A. 3
   B. 4
   C. 0
   D. 2
   E. 6

22. The N–H bond in the ammonium ion, NH_4^+, is formed by the overlap of which two orbitals?

   A. *sp²–s*
   B. *sp²–sp²*
   C. *sp³–sp²*
   D. *sp³–sp³*
   E. *sp³–s*

23. Among the hydrogen halides, the strongest bond is in [ ], and the longest bond is in [ ]?

   A. HI … HI
   B. HI … HF
   C. HF … HI
   D. HF … HF
   E. HCl … HBr

24. Which of the following is the most stable carbocation?

A.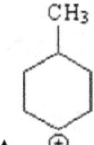

B.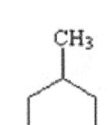

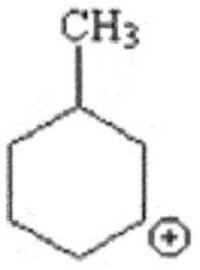

C.

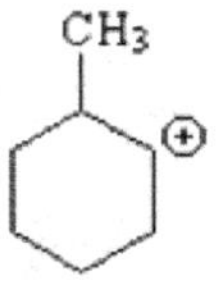

D.

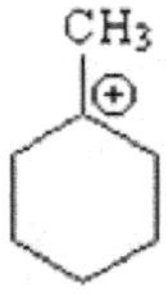

E.

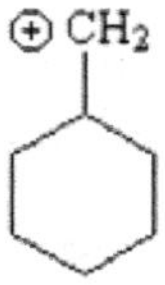

25. Which of the following is the least stable carbocation? (Use Ph = phenyl group)

A. PhH$_2$C$^+$
B. CH$_3$CH$_2$CH$_2$$^+$
C. Ph$_3$C$^+$
D. Ph$_2$HC$^+$
E. CH$_3$CH$_2$C$^+$HCH$_3$

26. The energy of an *sp*^{3} hybridized orbital for a carbon atom is:

A. lower in energy than the 2*s* and 2*p* atomic orbitals
B. higher in energy than the 2*s* and 2*p* atomic orbitals
C. higher in energy than the 2*p* atomic orbital but lower in energy than the 2*s* atomic orbital
D. higher in energy than the 2*s* atomic orbital but lower in energy than the 2*p* atomic orbital
E. equal in energy to the 2*p* atomic orbital

27. Which of the following pairs are resonance structures?

A. 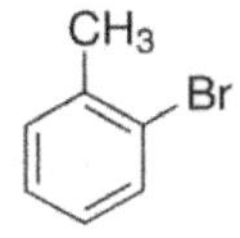 (2-bromotoluene and 3-bromotoluene) and

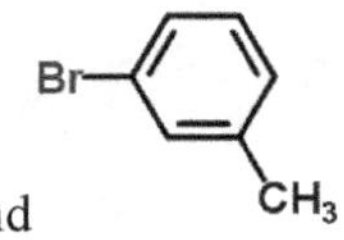

B. 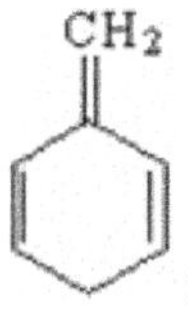 (methylenecyclohexadiene and toluene) and

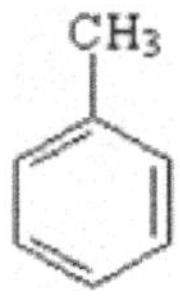

C. 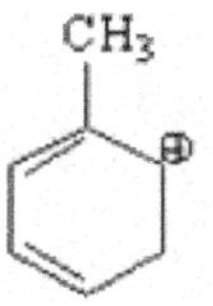 (methylcyclohexadienyl cation) and (methylcyclohexadienyl cation)

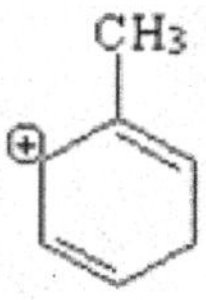

D. 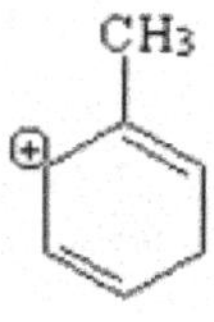 (methylcyclohexadienyl cation) and (methylcyclopentadiene)

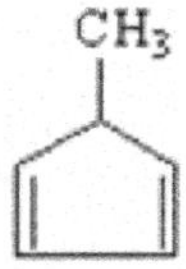

E. 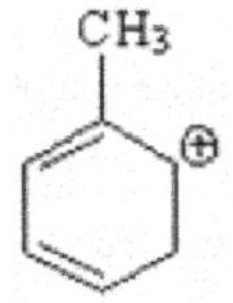 (methylcyclohexadienyl cation) and (methylcyclohexadienyl cation)

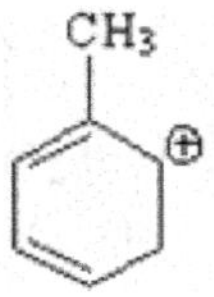

28. Due to electron delocalization, the carbon-oxygen bond in acetamide, CH₃CONH₂:

A. is longer than the carbon-oxygen bond of dimethyl ether, (CH₃)₂O
B. is longer than the carbon-oxygen bond of acetone, (CH₃)₂CO
C. is nonpolar
D. has more double bond character than the carbon-oxygen bond of acetone, (CH₃)₂CO
E. is formed by overlapping sp^3 orbitals

29. Which of the compounds listed below is linear?

A. 1,3,5-heptatriene
B. acetylene
C. 2-butyne
D. dichloromethane
E. 1,3-hexadiene

30. How many single and double bonds are in the benzene molecule (not including the C–H bonds)?

A. 6 single, 2 double
B. 5 single, 2 double
C. 6 single, 0 double
D. 5 single, 0 double
E. 6 single, 3 double

31. Determine the number of *pi* bonds in CH_3CN:

A. 0
B. 1
C. 2
D. 3
E. 4

32. Give the hybridization, shape, and bond angle for carbon in ethene:

A. sp^3, tetrahedral, 120°
B. sp^3, tetrahedral, 109.5°
C. sp^2, trigonal planar, 120°
D. sp^2, trigonal planar, 109.5°
E. none of the above

33. C=C, C=O, C=N, and N=N bonds are observed in many organic compounds. However, C=S, C=P, C=Si, and other similar bonds are not often found. What is the most probable explanation for this observation?

A. the comparative sizes of $3p$ atomic orbitals make effective overlap less likely than between two $2p$ orbitals
B. S, P, and Si do not undergo hybridization of orbitals
C. S, P and Si do not form π bonds, lacking occupied *p* orbitals in their ground state electron configurations
D. carbon does not combine with elements found below the second row of the periodic table
E. none of the above

34. How many σ bonds are in cyclohexane (C_6H_{12}), a saturated cyclic hydrocarbon?

A. 12
B. 14
C. 16
D. 18
E. 20

35. Triethylamine [$(CH_3CH_2)_3N$] is a molecule in which the nitrogen atom is [ ] hybridized, and the C–N–C bond angle is approximately [ ].

A. sp^3 ... < 107°
B. sp^3 ... 109.5°
C. sp^2 ... > 109.5°
D. sp^2 ... < 109.5°
E. sp ... 109.5°

36. How many distinct and degenerate *p* orbitals exist in the second electron shell, where n = 2?

A. 3
B. 2
C. 1
D. 0
E. 4

37. Which of the following compounds exhibits the greatest dipole moment?

A. (1S,2S)-1,2-dichloro-1,2-diphenylethane
B. 1,2-dichlorobutane
C. (1R,2S)-1,2-dichloro-1,2-diphenylethane
D. (E)-1,2-dichlorobutene
E. (Z)-1,2-dibromobutene

38. Draw a structural formula for benzene. How many σ bonds are in the molecule?

A. 14
B. 18
C. 6
D. 12
E. 20

39. Which is the formal charge of nitrogen in NH$_4$?

A. –2
B. –1
C. 0
D. +1
E. +2

40. The nitrogen's lone pair in pyrrolidine is occupying which type of orbital?

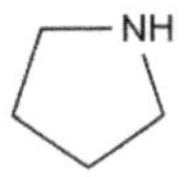

A. s
B. sp^3
C. sp^2
D. sp
E. p

41. Acetone is a common solvent used in organic chemistry laboratories. Which of the following statements is/are correct regarding acetone?

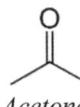
Acetone

I. One atom is sp^3 hybridized and tetrahedral
II. One atom is sp^2 hybridized and trigonal planar
III. The carbonyl carbon contains an unshared pair of electrons

A. I only
B. II only
C. I and II only
D. II and III only
E. I, II and III

42. Which of the following molecules represents the most stable carbocation?

A. 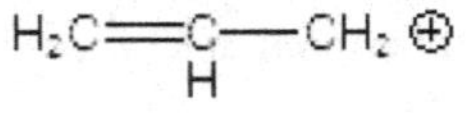 H₂C=C(H)—CH₂⁺

B. H₂C=C(CH₃)—CH₂⁺

C. CH₃CH=C(H)—CH₂⁺

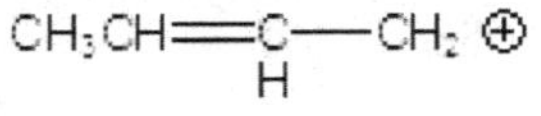

D. 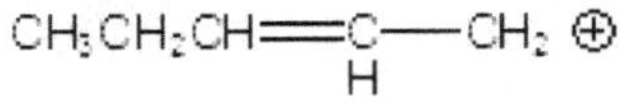 CH₃CH₂CH=C(H)—CH₂⁺

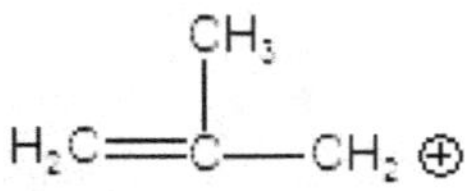

E. 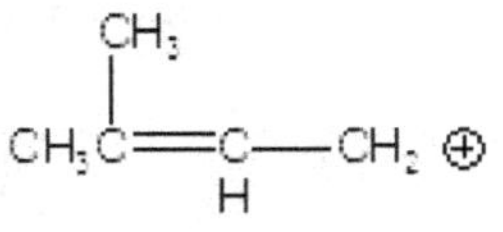 CH₃C(CH₃)=C(H)—CH₂⁺

43. Which of the following pairs are resonance structures?

A. 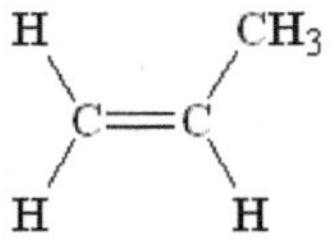 (H)(H)C=C(CH₃)(H) and cyclopropane (triangle)

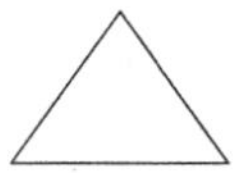

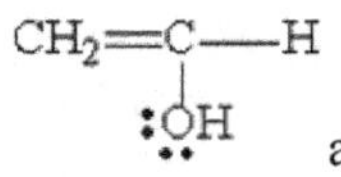

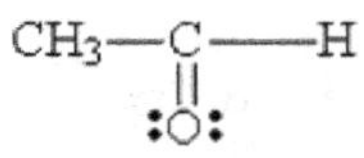

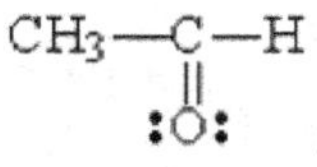

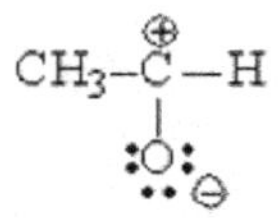

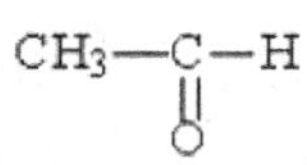

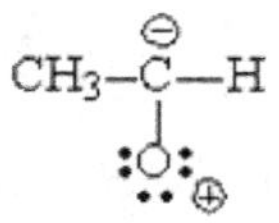

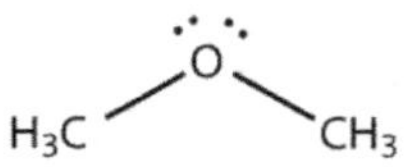

B. CH₂=C(H)—ÖH and CH₃—C(=Ö)—H

C. CH₃—C(=Ö)—H and CH₃—C⁺(H)—Ö:⁻

D. CH₃—C(=O)—H and CH₃—C⁻(H)—Ö:⁺

E. H₃C—O—CH₃ and H₃C—OH

44. Which of the following is closest to the C–O–C bond angle in CH₃–O–CH₃?

A. 109.5°
B. 90°
C. 180°
D. 120°
E. 140°

45. Identify the number of carbon atoms for each hybridization in the molecule?

O=CH–CH$_2$–CH=C=C=CH$_2$

	sp	sp^2	sp^3
A.	1	4	1
B.	2	3	1
C.	0	3	3
D.	1	3	2
E.	2	3	2

46. Identify the correctly drawn arrows:

A. 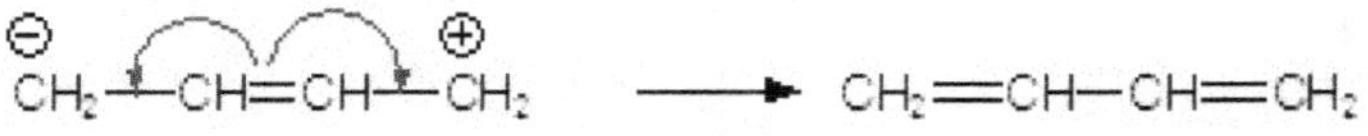 $CH_2^{\ominus}$—CH=CH—$CH_2^{\oplus}$ ⟶ CH_2=CH—CH=CH_2

B. $CH_2^{\ominus}$—CH=CH—$CH_2^{\oplus}$ ⟶ CH_2=CH—CH=CH_2

C. $CH_2^{\ominus}$—CH=CH—$CH_2^{\oplus}$ ⟶ CH_2=CH—CH=CH_2

D.  $CH_2^{\ominus}$—CH=CH—$CH_2^{\oplus}$ ⟶ CH_2=CH—CH=CH_2

E. $CH_2^{\ominus}$—CH=CH—$CH_2^{\oplus}$ ⟶ CH_2=CH—CH=CH_2

47. Triethylamine, (CH$_3$CH$_2$)$_3$N, is a molecule in which the nitrogen atom is [ ] hybridized and the molecular shape is [ ].

A. *sp^3* ... trigonal pyramidal
B. *sp^3* ... tetrahedral
C. *sp^2* ... tetrahedral
D. *sp^2* ... trigonal planar
E. *sp* ... bent

48. Which of the following is the most stable cation?

A. CH$_3$C$^{\oplus}$=CH$_2$

B. 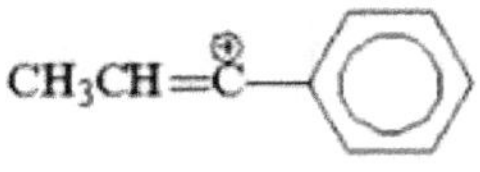 CH$_3$CH=C$^{\oplus}$—(phenyl)

C. CH$_3$—C$^{\oplus}$(CH$_3$)—(phenyl)

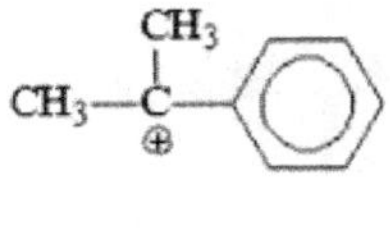

D. 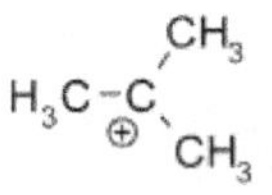 H$_3$C—C$^{\oplus}$(CH$_3$)(CH$_3$)

E. 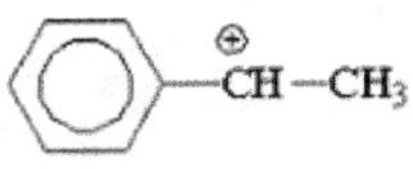 (phenyl)—C$^{\oplus}$H—CH$_3$

Notes for active learning

Stereochemistry

1. Butene, C₄H₈, is a hydrocarbon with one double bond. How many isomers are there of butene?

    A. two
    B. three
    C. four
    D. five
    E. no isomers

2. Which of the following compounds has an asymmetric center?

    A. 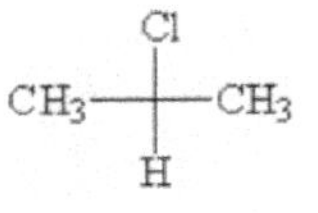 CH₃—C(Cl)(H)—CH₃

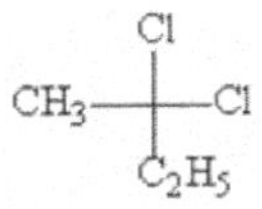

    B. CH₃—C(Cl)(Cl)—C₂H₅

    C. Cl—C(Cl)(H)—CH₃

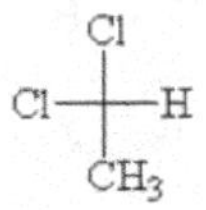

    D. 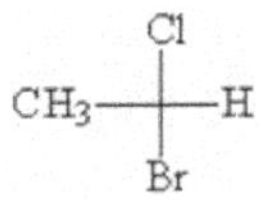 CH₃—C(Cl)(H)—Br

    E. 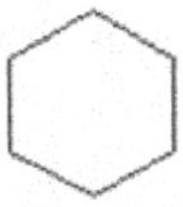 (cyclohexane)

3. How many stereoisomers are possible for the structure below?

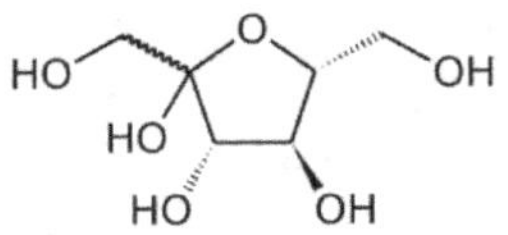

    A. 2
    B. 4
    C. 8
    D. 16
    E. 32

4. *Cis-trans* isomerism occurs when:

    A. each carbon in an alkene double bond has two different substituent groups
    B. the carbons in the *para* position of an aromatic ring have the same substituent groups
    C. a branched alkane has a halogen added to two adjacent carbon atoms
    D. an alkene is hydrated according to Markovnikov's Rule
    E. hydrogen is added to both carbon atoms in a double bond

5. What is the relationship between the following molecules?

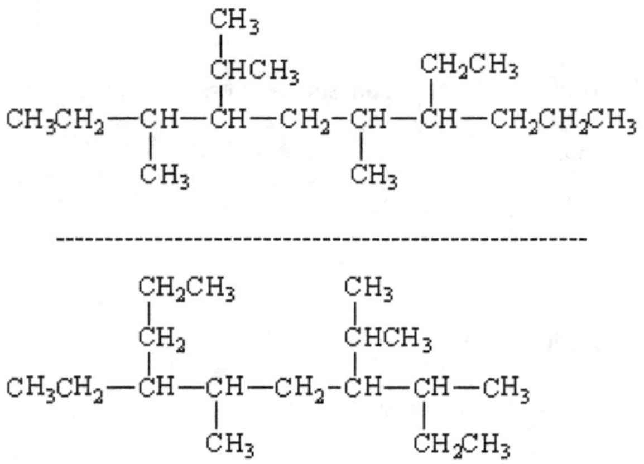

```
              CH3                      CH2CH3
              |                        |
              CHCH3                    CH
CH3CH2—CH—CH—CH2—CH—CH—CH2CH2CH3
       |                  |
       CH3                CH3
```

```
              CH2CH3              CH3
              |                   |
              CH2                 CHCH3
CH3CH2—CH—CH—CH2-CH—CH—CH3
       |                   |
       CH3                 CH2CH3
```

A. different molecules
B. enantiomers
C. identical
D. isomers
E. diastereomers

6. What is the relationship between the structures shown below?

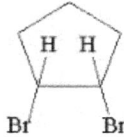

A. geometric isomers
B. conformational isomers
C. constitutional isomers
D. diastereomers
E. enantiomers

7. What is the relationship between the following compounds?

A. conformational isomers
B. diastereomers
C. superimposable without bond rotation
D. constitutional isomers
E. enantiomers

8. Which of the following is NOT true of enantiomers?

A. Enantiomers have the same chemical reactivity with non-chiral reagents
B. Enantiomers have the same density
C. Enantiomers have the same melting point
D. Enantiomers have the same boiling point
E. Enantiomers have the same direction of specific rotation

9. When two compounds consist of the same number and kind of atoms but differ in molecular structure are:

    A. hydrocarbons
    B. isomers
    C. homologs
    D. isotopes
    E. allotropes

10. Identify the relationship between the compounds:

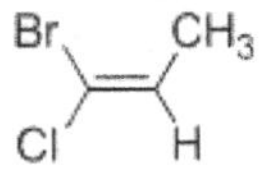

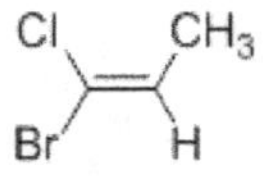

    A. constitutional isomers
    B. configurational isomers
    C. identical
    D. conformational isomers
    E. none of the above

11. Which of the following carbons in the molecule below are chiral carbons?

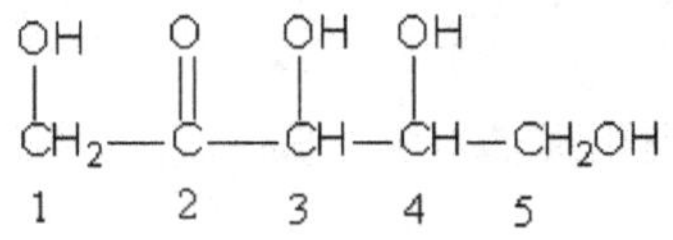

    A. carbons 1 and 5
    B. carbons 3 and 4
    C. carbons 2, 3 and 4
    D. all carbons are chiral
    E. none of the carbon atoms are chiral

12. How many different stereoisomers can the following compound have?

    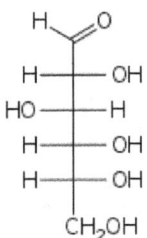

    D-glucose

    A. 2
    B. 4
    C. 8
    D. 16
    E. 32

13. CH₃–CH₂–O–H and CH₃–O–CH₃ are a pair of:

A. isomers
B. epimers
C. anomers
D. allotropes
E. geometric isomers

14. What is the relationship between the two structures shown below?

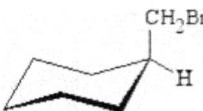 and

A. conformational isomers
B. configurational isomers
C. enantiomers
D. constitutional isomers
E. not isomers

15. Among the butane conformers, which occur at energy minima on a graph of potential energy versus dihedral angle?

A. *anti*
B. eclipsed
C. *gauche*
D. eclipsed and *gauche*
E. *gauche* and *anti*

16. How many isomers exist for dibromobenzene, C₆H₄Br₂?

A. 0
B. 1
C. 2
D. 3
E. 4

17. How many isomers are there of butane, C₄H₁₀?

A. 3
B. 4
C. 1 (no isomers)
D. 2
E. 5

18. What is the relationship between the following compounds?

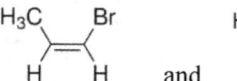 and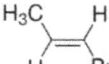

A. constitutional isomers
B. structural isomers
C. geometric isomers
D. conformational isomers
E. positional isomers

19. Which of the following compounds are isomers?

   I. CH$_3$CH$_2$OCH$_3$
   II. CH$_3$CH$_2$CH$_2$OH
   III. CH$_2$COHCH$_2$CH$_3$
   IV. CH$_3$CH$_2$OCH$_2$CH$_3$

   A. I and II only
   B. II and III only
   C. III and IV only
   D. I, II and III only
   E. I, II, III and IV

20. The cause of *cis-trans* isomerism is:

   A. short length of the double bond
   B. strength of the double bond
   C. lack of rotation of the double bond
   D. stability of the double bond
   E. vibration of the double bond

21. Which of the following compounds is an isomer of CH$_3$CH$_2$CH$_2$CH$_2$OH?

   A. CH$_3$CH$_2$CH$_2$OH
   B. CH$_3$(OH)CHCH$_3$
   C. CH$_3$CH$_2$CH$_2$CHO
   D. CH$_3$CH$_2$(OH)CHCH$_3$
   E. CH$_3$OH

22. If two of the hydrogen atoms in ethylene, H$_2$C=CH$_2$, are replaced by one chlorine atom and one fluorine atom to form chlorofluoroethene, C$_2$H$_2$ClF, how many different chlorofluoro-ethene isomers are there?

   A. 4
   B. 3
   C. 2
   D. 1
   E. 6

23. In the Fischer projection below, what are the configurations of the two asymmetric centers?

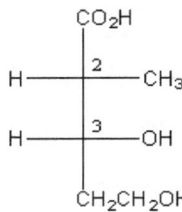

   A. 2*S*, 3*S*
   B. 2*S*, 3*R*
   C. 2*R*, 3*S*
   D. 2*R*, 3*R*
   E. Cannot be determined

24. Which of the following terms best describes the pair of compounds shown?

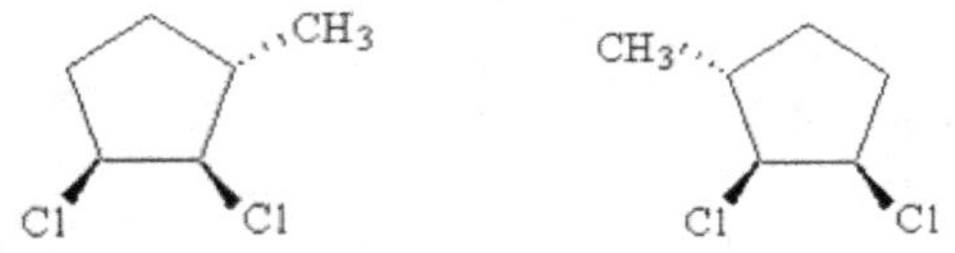

A. same molecule

B. conformational isomers

C. enantiomers

D. diastereomers

E. configurational isomers

25. If two of the hydrogen atoms in ethylene, $H_2C{=}CH_2$, are replaced by two chlorine atoms to form dichloroethylene, how many different dichloroethylene isomers are there?

A. 3

B. 4

C. 1

D. 2

E. 6

26. The enantiomer of the compound below is:

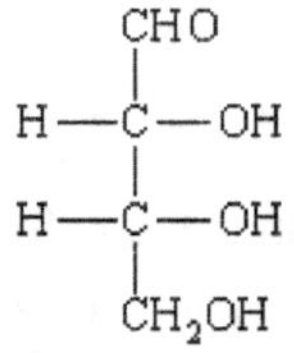

A.

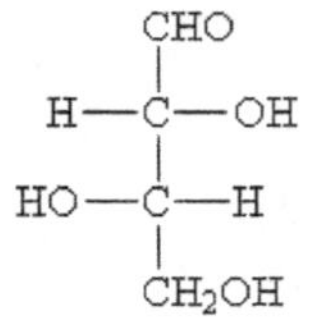

C.

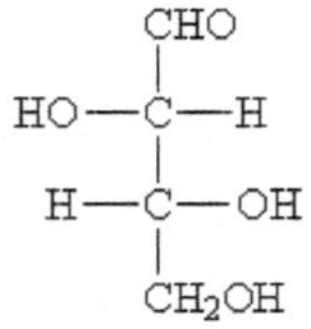

B.

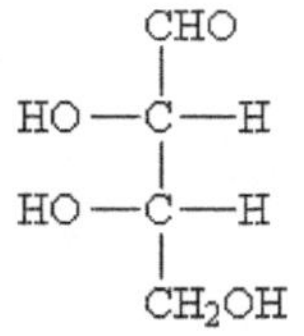

D.

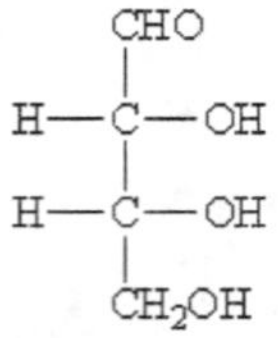

E. none of the above

27. Which of the statements correctly describes an achiral molecule?

    A. The molecule has an enantiomer
    B. The molecule may be a *meso* form
    C. The molecule has a non superimposable mirror image
    D. The molecule exhibits optical activity when it interacts with plane-polarized light
    E. None of the above

28. Which of the following statements does NOT correctly describe *cis*-1,2-dimethylcyclopentane?

    A. Its diastereomer is *trans*-1,2-dimethylcyclopentane
    B. It contains two asymmetric carbons
    C. It is achiral
    D. It is a *meso* compound
    E. It has an enantiomer

29. How many structural isomers of $C_4H_8Cl_2$ exhibit optical activity?

    A. 0
    B. 1
    C. 2
    D. 3
    E. 4

30. How many chiral carbon atoms are in this structure?

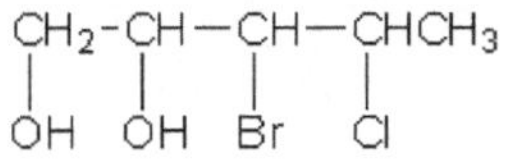

    $$CH_2\text{-}CH\text{-}CH\text{-}CHCH_3$$
    $$\;\;\;|\;\;\;\;\;|\;\;\;\;\;|\;\;\;\;\;|$$
    $$\;\;OH\;\;OH\;\;Br\;\;Cl$$

    A. 6
    B. 5
    C. 3
    D. 4
    E. 2

31. Which of the following is a true statement?

    A. A mixture of achiral compounds is optically inactive
    B. Molecules that possess a single chirality center of the *S* configuration are levorotatory
    C. Achiral molecules are *meso*
    D. Chiral molecules possess a plane of symmetry
    E. Molecules that possess 2 or more chiral centers are chiral

32. Which of the following best describes the geometry of the carbon-carbon double bond in the alkene below?

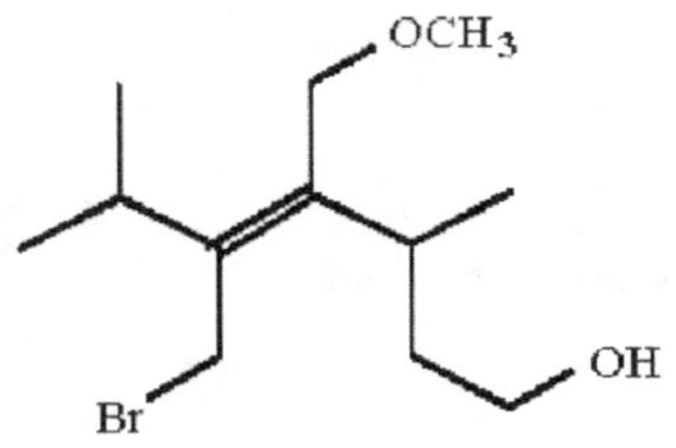

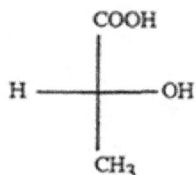

A. E
B. Z
C. cis
D. R
E. S

33. Which of the following compounds share the same absolute configuration?

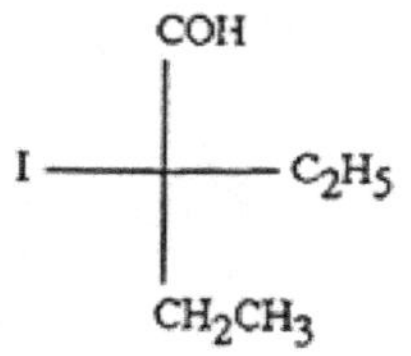

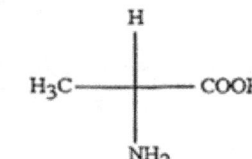

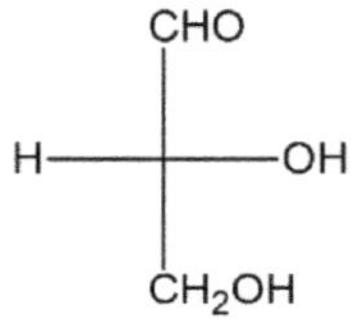

A. I and III only
B. I and II only
C. I and IV only
D. II, III and IV only
E. II and IV only

34. What is the correct IUPAC name for the following structure?

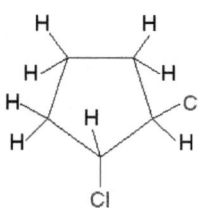

A. trans-1,2-dichlorocyclopentane
B. 1,2-dichlorocyclopentane
C. cis-1,2-dichlorocyclopentane
D. trans-dichlorocyclopentane
E. Z-dichlorocyclopentane

35. Which of the following compounds is NOT chiral?

   A. 1,2-dichlorobutane
   B. 1,4-dibromobutane
   C. 2,3-dibromobutane
   D. 1,3-dibromobutane
   E. 1-bromo-2-chlorobutane

36. The specific rotation of a pure enantiomeric substance is –6.30°. What is the percentage of this enantiomer in a mixture with an observed specific rotation of –3.15°?

   A. 75%
   B. 80%
   C. 25%
   D. 50%
   E. 0%

37. Which of the following is a result of the reaction below?

   (S)-3-bromo-3-methylhexane + HCN

   A. loss of optical activity
   B. mutarotation
   C. retention of optical activity
   D. inversion of absolute configuration
   E. epimerization

38. If two chlorine atoms replace two of the hydrogen atoms in butane, how many different dichlorobutane constitutional isomers can there be?

   A. 4
   B. 3
   C. 2
   D. 1
   E. 6

39. How many asymmetric centers are present in the compound shown below?

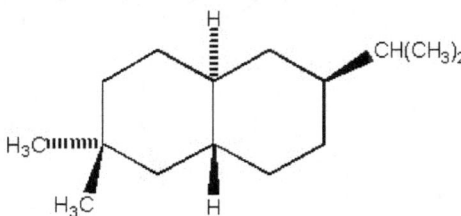

   A. 1
   B. 2
   C. 3
   D. 4
   E. 5

40. What is the relationship between the structures shown below?

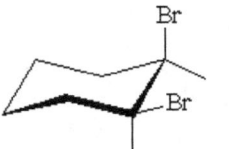

A. identical compounds
B. configurational isomers
C. diastereomers
D. enantiomers
E. constitutional isomers

41. Which of the following statement(s) for the compound *meso*-tartaric acid is/are true?

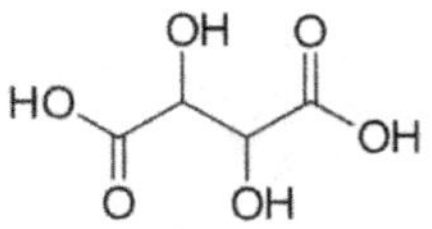

I. achiral
II. the polarimeter reads a zero deflection
III. racemic mixture

A. I only
B. II only
C. III only
D. I and II only
E. II and III only

42. Which of the following molecules are identical?

I.   II. 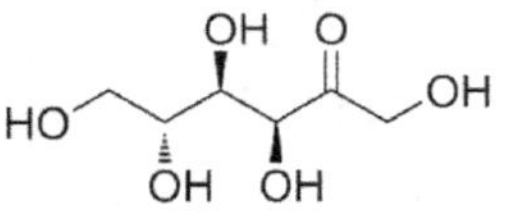   III.

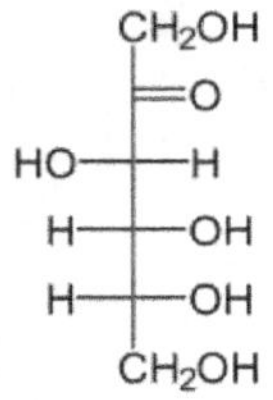

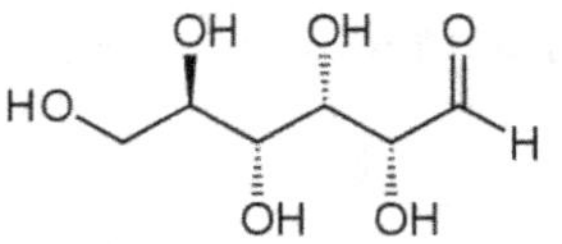

A. I and II
B. I and III
C. II and III
D. I, II and III
E. None of the above

43. Which of the following molecules contains a chiral carbon?

A. 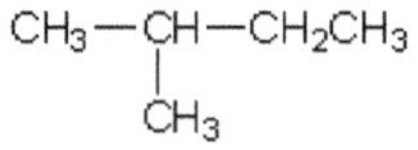 CH₃—CH(CH₃)—CH₂CH₃

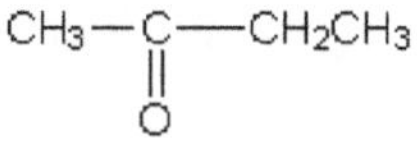

B. 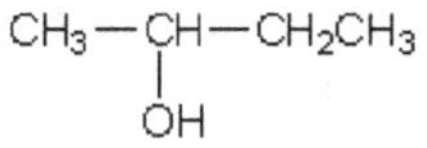 CH₃—CH(OH)—CH₂CH₃

C. CH₃—C(=O)—CH₂CH₃

D. 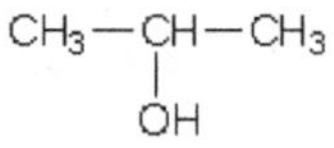 CH₃—CH(OH)—CH₃

E. none of the above

44. What is the relationship between the following compounds?

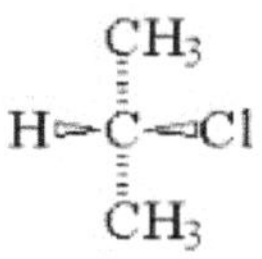

(structures shown)

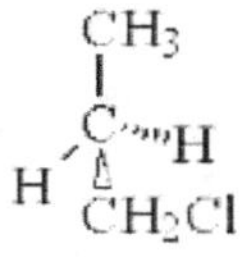

A. conformational isomers
B. constitutional isomers
C. diastereomers
D. enantiomers
E. identical

45. Which of the following statements about *cis-* / *trans-*isomers is NOT correct?

A. Conversion between *cis–* and *trans-*isomers occurs by rotation around the double bond
B. In the *trans-*isomer, the groups of interest, are on opposite sides across the double bond
C. In the *cis-*isomer, the reference groups, are on the same side of the double bond
D. There are no *cis-* / *tran-*isomers in alkynes
E. There are no *cis* / *trans* isomers in alkanes

46. Which of the following best describes the geometry about the carbon-carbon double bond in the alkene below?

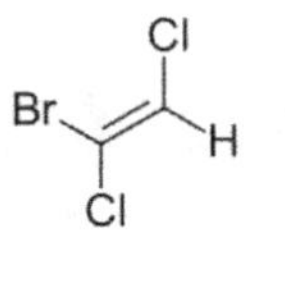

A. E
B. Z
C. cis
D. S
E. R

47. Which of the following is a structural isomer of 2-methylbutane?

A. *n*-propane
B. 2-methylpropane
C. *n*-butane
D. *n*-pentane
E. *n*-heptane

Notes for active learning

Molecular Structure and Spectra

1. In the proton NMR, in what region of the spectrum does one typically observe hydrogens bound to the aromatic ring?

A. 1.0-1.5 ppm
B. 2.0-3.0 ppm
C. 4.5-5.5 ppm
D. 6.0-8.0 ppm
E. 8.0-9.0 ppm

2. ^{1}H nuclei located near electronegative atoms are [ ] relative to ^{1}H nuclei not near them.

A. coupled
B. split
C. shielded
D. deshielded
E. none of the above

3. Which compound is expected to show intense IR absorption at 1,715 cm^{-1}?

A. $(CH_3)_2CHNH_2$
B. hex-1-yne
C. 2-methylhexane
D. $(CH_3)_2CHCO_2H$
E. $CH_3CH_2CH_2COH_2CH_3$

4. Free-radical chlorination of propane gives two isomeric monochlorides: 1-chloropropane and 2-chloropropane. How many NMR signals does each of these compounds display, respectively?

A. 3, 2
B. 3, 3
C. 2, 2
D. 2, 3
E. None of the above

5. Which of the following transitions is usually observed in the UV spectra of ketones?

A. n to π^*
B. n to π
C. σ to n
D. σ to σ^*
E. n to σ^*

6. What is the relative area of each peak in a quartet spin-spin splitting pattern?

A. 1:1:1:1
B. 1:4:4:1
C. 1:2:1
D. 1:2:2:1
E. 1:3:3:1

7. Which NMR signal represents the most deshielded proton?

A. δ 2.0
B. δ 3.8
C. δ 6.5
D. δ 7.3
E. δ 4.5

8. The mass spectrum of alcohols often fails to exhibit detectable M peaks, instead showing relatively large [ ] peaks.

A. M+1
B. M+2
C. M–16
D. M–17
E. M–18

9. In ^{1}H NMR, protons on the α-carbon of amines typically resonate between:

A. 0.5 and 1.0 ppm
B. 2.0 and 3.0 ppm
C. 3.0 and 4.0 ppm
D. 6.0 and 7.0 ppm
E. 9.0 and 10.0 ppm

10. A researcher recorded the NMR spectra of each of the following compounds. Ignoring chemical shifts, which possesses a spectrum significantly different from the others?

A. 1,1,2-tribromobutane
B. bromobutane
C. 3,3-dibromoheptane
D. dibutyl ether
E. the spectra are similar

11. Which compound(s) show(s) intense IR absorption at 1680 cm^{-1}?

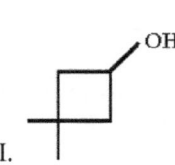 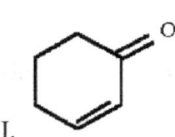
I.   II.   III.

A. I only
B. II only
C. III only
D. II and III only
E. I, II and III

12. What is the relationship between H$_a$ and H$_b$ in the following structure?

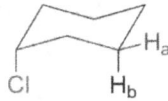

A. diastereotopic
B. enantiotopic
C. homotopic
D. isotopic
E. conformers

13. Which of the following compounds gives the greatest number of proton NMR peaks?

    A. 3,3-dichloropentane
    B. 4,4-dichloroheptane
    C. 1-chlorobutane
    D. 1,4-dichlorobutane
    E. dichloromethane

14. Where would one expect to find the ^{1}H NMR signal for the carboxyl group's hydrogen in propanoic acid?

    A. δ 4.1-5.6 ppm
    B. δ 10-13 ppm
    C. δ 8-9 ppm
    D. δ 6.1-7.8 ppm
    E. δ 9.5-10 ppm

15. While the carbonyl stretching frequency for simple aldehydes, ketones, and carboxylic acids is about 1710 cm^{-1}, the carbonyl stretching frequency for acid chlorides is about:

    A. 1700 cm^{-1}
    B. 1735 cm^{-1}
    C. 1800 cm^{-1}
    D. 1660 cm^{-1}
    E. 2200 cm^{-1}

16. A clear liquid is subjected to infrared spectroscopy and produces a spectrum with a prominent, broad peak at approximately 3000 cm^{-1} and a sharp peak at 1710 cm^{-1}, as well as several smaller peaks between 1420 cm^{-1} and 940 cm^{-1}. This substance is most likely:

    A. ketone
    B. carboxylic acid
    C. alcohol
    D. aldehyde
    E. anhydride

17. Which of the following laboratory techniques is used primarily as a compound identification procedure?

    A. extraction
    B. NMR spectroscopy
    C. crystallization
    D. distillation
    E. none of the above

18. While the carbonyl stretching frequency for simple aldehydes, ketones, and carboxylic acids is about 1710 cm^{-1}, the carbonyl stretching frequency for esters is about:

    A. 1660 cm^{-1}
    B. 1700 cm^{-1}
    C. 1735 cm^{-1}
    D. 1800 cm^{-1}
    E. 2200 cm^{-1}

19. A compound with nine carbon atoms produces a single NMR signal. Which is a structural formula for the compound?

A. $(CH_3)_3CCCl_2C(CH_3)_3$

B. $(CH_3)_2CHCH_2CH_2CH(CH_3)CH_2CH_3$

C. $(CH_3)_2CHCH_2(CH_2)_4CH_3$

D. $CH_3(CH_2)_7CH_3$

E. None of the above

20. In mass spectrometry plots, the relative abundance is the unit along the *y*-axis in a mass spectrum. What are the units on the *x*-axis?

A. mass / charge (m/z)

B. mass (m)

C. molecular weight (amu)

D. frequency (v)

E. wavelength (λ)

21. How many nuclear spin states are allowed for the ^1H nucleus?

A. 1

B. 2

C. 3

D. 4

E. 10

22. Infrared spectroscopy provides a scientist with information about:

A. molecular weight

B. distribution of protons

C. functional group

D. conjugation

E. polarity

23. In the mass spectrum of 3,3-dimethyl-2-butanone, the base peak occurs at *m/z*:

A. 43

B. 58

C. 84

D. 85

E. 100

24. The protons marked H_a and H_b in the molecule below are:

A. diastereotopic

B. heterotopic

C. homotopic

D. enantiotopic

E. endotopic

25. Which of the following compounds is NOT IR active?

A. Cl_2

B. CO

C. $CH_3CH_2CH_2OH$

D. CH_3Br

E. HCN

26. Which sequence correctly ranks the regions of the electromagnetic spectrum in order of increasing energy?

       1) infrared   2) ultraviolet   3) radio wave

A. 3 < 1 < 2   C. 2 < 1 < 3
B. 3 < 2 < 1   D. 1 < 3 < 2
                                            E. 1 < 2 < 3

27. In the UV-visible spectrum of (E)-1,3,5-hexatriene, the lowest energy absorption corresponds to a:

A. π to σ* transition   C. σ to π transition
B. σ to σ* transition   D. σ to π* transition
                                          E. π to π* transition

28. What type of spectroscopy would be the LEAST useful in distinguishing dimethyl ether from bromoethane?

A. UV spectroscopy   C. IR spectroscopy
B. mass spectrometry   D. ^{1}H NMR spectroscopy
                                          E. ^{13}C NMR spectroscopy

29. What 1H NMR spectral data is expected for the compound shown?

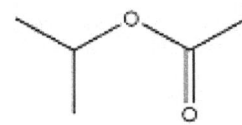

A. 4.9 (1H, sextet), 4.3 (3H, s), 3.0 (6H, d)   C. 4.3 (1H, septet), 3.3 (3H, s), 1.2 (6H, d)
B. 3.6 (3H, s), 2.8 (3H, septet), 1.2 (6H, d)   D. 3.8 (1H, septet), 2.2 (3H, s), 1.0 (6H, d)
                                                            E. 2.8 (1H, septet), 2.2 (3H, s), 1.0 (6H, d)

30. Which of the following compounds absorbs the longest wavelength of UV-visible light?

A. (Z)-1,3-hexadiene   C. (Z)-but-2-ene
B. hex-1-ene   D. (E)-but-2-ene
                                            E. (E)-1,3,5-hexatriene

31. Which of the following compounds generates only one signal on its ^{1}H NMR spectrum?

A. tert-butyl alcohol   C. toluene
B. 1,2-dibromoethane   D. methanol
                                          E. phenol

32. Methyl salicylate and aspirin exhibit a strong absorption for IR at 1,735 cm^{-1}. This absorption indicates:

A. ester
B. aromatic ring
C. alcohol
D. phenol
E. amine

33. Absorption of what type of electromagnetic radiation results in electronic transitions?

A. X-rays
B. radio waves
C. microwaves
D. ultraviolet light
E. infrared light

34. Which of the following is NOT true?

A. NMR spectroscopy utilizes magnetic fields
B. IR spectroscopy utilizes light of wavelength 200-400 nm
C. UV spectroscopy utilizes transitions in electron states
D. Mass spectrometry utilizes fragmentation of the sample
E. IR spectroscopy is useful in the absorption region of 1500 to 3500 cm^{-1}

35. The IR absorption at 1710 cm^{-1} indicates which of the following functional groups?

A. carbon-carbon double bond
B. ketone
C. alcohol
D. ether
E. sulfide

36. A compound with a broad, deep IR absorption at 3300 cm^{-1} indicates the presence of:

A. acyl halide
B. alcohol
C. alkene
D. ketone
E. alkyne

Alkanes and Alkyl Halides

1. A nucleophile is:

   A. an oxidizing agent
   B. electron deficient
   C. a Lewis base
   D. a Lewis acid
   E. none of the above

2. Identify the number of tertiary carbons in the following structure:

   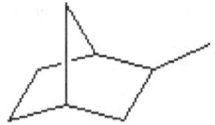

   A. 4
   B. 5
   C. 2
   D. 3
   E. 6

3. If the concentration of ⁻OH doubles in a reaction with bromopropane, then the reaction rate:

   A. quadruples
   B. doubles
   C. remains the same
   D. is halved
   E. none of the above

4. The rate of an S_N1 reaction depends on:

   A. the concentration of the nucleophile and electrophile
   B. neither the concentration of the nucleophile nor the electrophile
   C. the concentration of the nucleophile only
   D. the concentration of the electrophile only
   E. none of the above

5. Which of the following properties is NOT characteristic of alkanes?

   A. They are tasteless and colorless
   B. They are nontoxic
   C. Their melting points increase with molecular weight
   D. They are generally less dense than water
   E. They have strong hydrogen bonds

6. Predict the most likely mechanism for the reaction shown below:

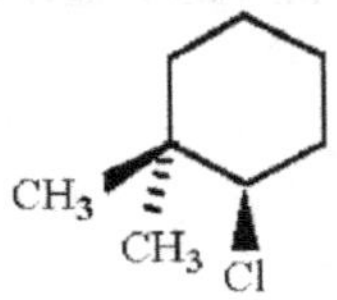

[structure: cyclohexane with CH₃, CH₃, Cl substituents] + NaOCH₃ / CH₃OH → ?

A. E₂
B. E₁
C. S_N2
D. S_N1
E. Cannot be determined

7. An alkyl halide forms a more stable carbocation than the carbocation formed from isopropyl bromide. Which of the following alkyl halides forms the most stable carbocation?

A. *n*-propyl chloride
B. *tert*-butyl chloride
C. methyl chloride
D. ethyl chloride
E. propyl chloride

8. Which of the following reactions is most likely to proceed by an S_N2 mechanism?

A. *t*-butyl iodide with ethanol
B. 2-bromo-3-methyl pentane with methanol
C. 2-bromo-2-methyl pentane with HCl
D. 1-bromopropane with NaOH
E. None of the above

9. Which of the following molecules can rotate freely around its carbon-carbon bond?

A. acetylene
B. cyclopropane
C. ethane
D. ethylene
E. none of the above

10. The rate of an S_N2 reaction depends on:

A. neither the concentration of the nucleophile nor the substrate
B. the concentration of the nucleophile and the substrate
C. the concentration of the substrate only
D. the concentration of the nucleophile only
E. none of the above

11. The heat of combustion of propane is –530 kcal/mol. A reasonable approximation for the heat of combustion for heptane is:

A. –432 kcal/mol
B. –682 kcal/mol
C. –865 kcal/mol
D. –1158 kcal/mol
E. None of the above

12. Which compound is least soluble in water?

A.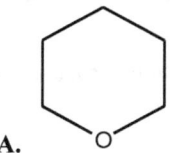

B.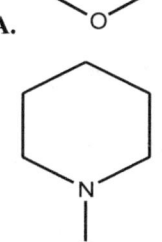

C.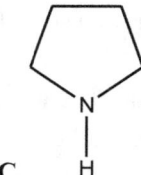

D.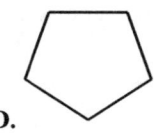

E.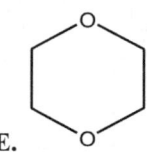

13. Which of the following statements best describes the mechanism of the unimolecular elimination of *tert*-butyl chloride with ethanol?

A. The reaction is a concerted single-step process
B. The reaction involves homolytic cleavage of the C–Cl bond
C. The rate-determining step is the formation of $(CH_3)_3C\cdot$
D. The rate-determining step is the formation of $(CH_3)_3C^+$
E. None of the above

14. Which of the following best describes the process of an S_N1 reaction in which the leaving group is on a chiral carbon atom?

A. inversion of stereochemistry
B. double inversion of stereochemistry
C. racemic mixture
D. retention of stereochemistry
E. none of the above

15. Which of the following is true regarding S$_N$1 and S$_N$2 reactions?

    A. The rates of S$_N$1 reactions depend mainly on steric factors, while the rates of S$_N$2 reactions depend mostly on electronic factors
    B. S$_N$1 reactions proceed more readily with a tertiary alkyl halide substrate, while S$_N$2 reactions proceed more readily with a primary alkyl halide substrate
    C. S$_N$1 and S$_N$2 reactions produce rearrangement products
    D. S$_N$1 reactions proceed *via* a carbocation intermediate, while S$_N$2 reactions proceed *via* a carbocation intermediate under certain conditions
    E. S$_N$1 reactions proceed more readily with a polar aprotic solvent, while S$_N$2 reactions proceed more readily with a polar protic solvent

16. Which of the compounds listed below has the lowest boiling point?

    A. 3-methylheptane
    B. 2,4-dimethylhexane
    C. octane
    D. 2,2,4-trimethylpentane
    E. decane

17. Which of the following is NOT a feature of S$_N$2 reactions?

    A. single-step mechanism
    B. pentacoordinate transition state
    C. bimolecular kinetics
    D. carbocation intermediate
    E. nucleophilic substitution

18. Which conformer is at a local energy minimum on the potential energy diagram in the chair-chair interconversion of cyclohexane?

    A. boat
    B. twist-boat
    C. half-chair
    D. planar
    E. fully eclipsed

19. If propane was reacted with Cl$_2$ in the presence of UV light, what products form, and what are their approximate percentages?

    A. Propyl chloride yields 100%
    B. Propyl chloride yields 42%, and isopropyl chloride yields 58%
    C. Propyl chloride yields 75%, and isopropyl chloride yields 25%
    D. Propyl chloride yields 90%, and isopropyl chloride yields 10%
    E. None of the above

20. What is true about an S_N1 reaction?

    I. A carbocation intermediate is formed
    II. The rate determining step is bimolecular
    III. The mechanism has two steps

- A. I only
- B. II only
- C. I and II only
- D. I and III only
- E. I, II and III

21. The most stable conformational isomer of 1,2-dibromoethane is:

- A. eclipsed, *anti*
- B. staggered, *gauche*
- C. staggered, *anti*
- D. eclipsed, *gauche*
- E. staggered, eclipsed

22. What is the most likely mechanism for the reaction between 1-bromobutane and sodium cyanide?

- A. E_1
- B. E_2
- C. S_N1
- D. S_N2
- E. S_N1 and S_N2

23. Which of the following alkyl halogens reacts the fastest with NaOH?

- A. *t*-butyl bromide
- B. *t*-butyl iodide
- C. *t*-butyl fluoride
- D. *t*-butyl chloride
- E. *t*-butyl cyanide

24. Acetate can react with a tertiary alkyl chloride to form an ester. The reaction occurs more rapidly in water than in dimethylsulfoxide (DMSO) because water stabilizes the following:

- A. intermediate racemates
- B. configuration inversion
- C. carbocation intermediate
- D. acetate
- E. anion intermediate

25. Ethers can be formed from ethyl bromide in a reaction whereby the incoming ⁻OR group represents a(n):

- A. substrate
- B. electrophile
- C. nucleophile
- D. leaving group
- E. solvent

26. Which of the following compounds can most easily undergo E₁ and S_N2 reactions?

   A. (CH₃CH₂CH₂)₂CHBr
   B. (CH₃CH₂CH₂)₃CBr
   C. (CH₃CH₂CH₂)₃CCH₂Cl
   D. (CH₃CH₂CH₂)₂CHCN
   E. CH₃CH₂CH₂CH₂Br

27. Which of the following molecules has the lowest boiling point?

   A. *cis*-2-pentene
   B. 2-pentyne
   C. pentane
   D. neopentane
   E. 3-pentanol

28. Which of the following is the most stable conformer of *trans*-1-isopropyl-3-methylcyclohexane?

   A. methyl and isopropyl are axial
   B. methyl and isopropyl are equatorial
   C. methyl is axial, and isopropyl is equatorial
   D. methyl is equatorial, and isopropyl is axial
   E. none of the above

29. Which of the following compounds readily undergoes E₁, S_N1 and E₂ reactions, but not S_N2 reactions?

   A. (CH₃CH₂CH₂)₃CBr
   B. CH₃CH₂CH₂CH₃
   C. (CH₃CH₂)₃COH
   D. CH₃CH₂CH₂CH₂Br
   E. none of the above

30. The complete combustion of one mole of nonane in oxygen would produce [ ] moles of CO_2 and [ ] moles of H_2O?

   A. 9 … 10
   B. 9 … 9
   C. 9 … 4.5
   D. 4.5 … 4.5
   E. 4.5 … 9

31. Halogenation of alkanes proceeds by the mechanism shown below:

   I. Br₂ + hν → 2 Br•
   II. Br• + RH → HBr + R•
   III. R• + Br₂ → RBr + Br•

Which of these steps involve chain propagation?

   A. I only
   B. III only
   C. I and II only
   D. II and III only
   E. I and III only

32. What statement(s) is/are true about an S_N2 reaction?

　　I. A carbocation intermediate is formed
　　II. The rate-determining step is bimolecular
　　III. The mechanism has two steps

- **A.** I only
- **B.** II only
- **C.** I and III only
- **D.** II and III only
- **E.** I, II and III

33. Which of the following compounds undergo(es) a substitution reaction?

　　I. C_2H_6　　II. C_2H_2　　III. C_2H_4

- **A.** I only
- **B.** II only
- **C.** I and II only
- **D.** II and III only
- **E.** I, II and III

34. Which of the following undergoes bimolecular nucleophilic substitution at the fastest rate?

- **A.** 1-chloro-2,2-diethylcyclopentane
- **B.** 1-chlorocyclopentane
- **C.** 1-chlorocyclopentene
- **D.** 1-chloro-1-ethylcyclopentane
- **E.** *tert*-butylchloride

35. Which of the following is the best leaving group?

- **A.** Cl^-
- **B.** NH_2^-
- **C.** ^-OH
- **D.** Br^-
- **E.** F^-

36. Which of the following is an example of the termination step for a free-radical chain reaction?

   A. Cl–Cl + hv → 2 Cl·

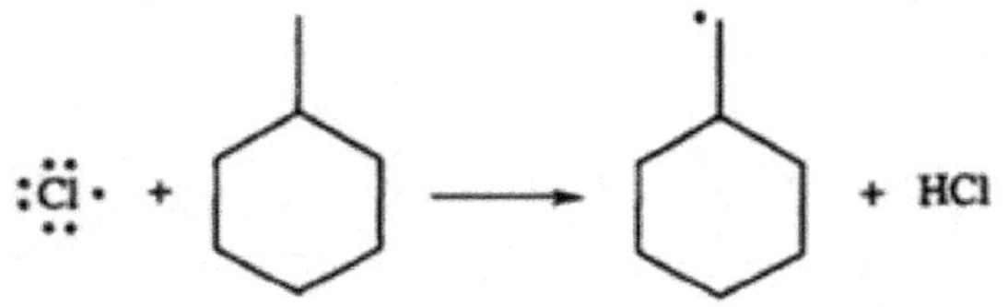

   B. :Cl· + (cyclohexane) → (cyclohexyl radical at position 1) + HCl

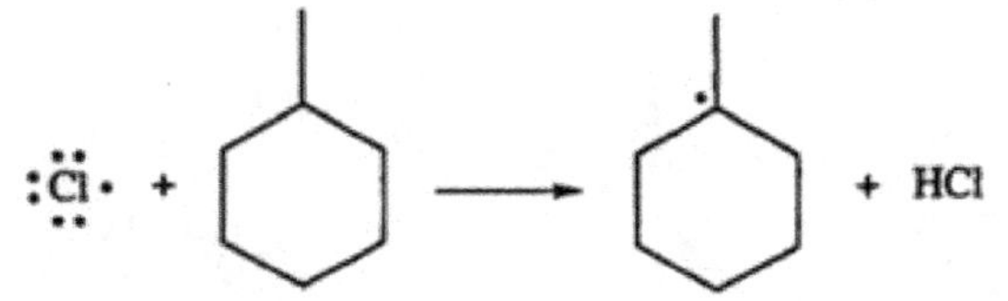

   C. :Cl· + (methylcyclohexane) → (radical) + HCl

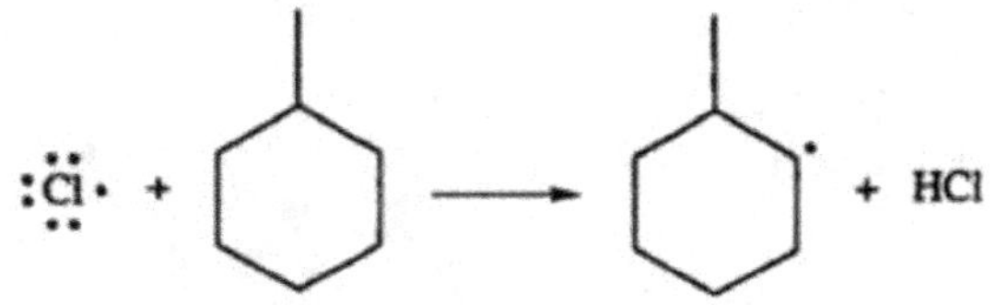

   D. :Cl· + (methylcyclohexane) → (radical) + HCl

   E. Cl· + Cl· → Cl$_2$

Notes for active learning

Notes for active learning

Alkenes

1. (*E*)- and (*Z*)-hex-3-ene can be subjected to a hydroboration-oxidation sequence. How are the products from these two reactions related?

    A. The products of the two isomers are diastereomers
    B. The products of the two isomers are constitutional isomers
    C. The (*E*)- and (*Z*)-isomers generate the same products in the same amounts
    D. The (*E*)- and (*Z*)-isomers generate the same products but in differing amounts
    E. The products of the two isomers are not structurally related

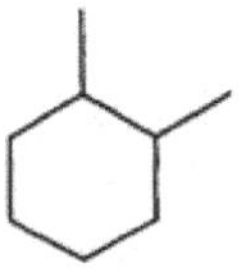

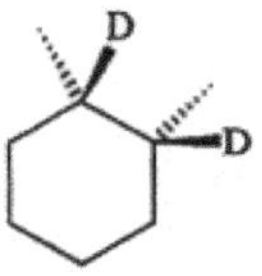

2. What is the major product of this reaction?

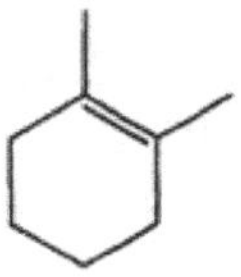

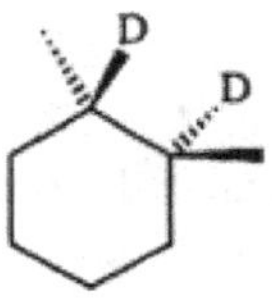

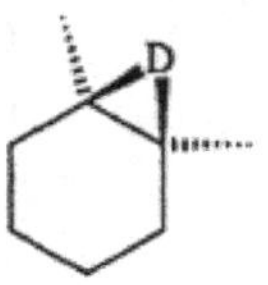

    E. None of the above

3. Give the best product for the following reaction:

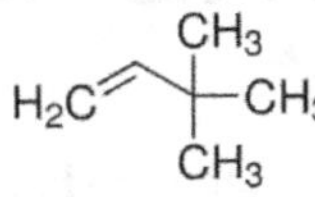

H₂C=C(CH₃)(C(CH₃)₃) + 1) Hg(O₂CCF₃)₂, CH₃CH₂OH; 2) NaBH₄ → ?

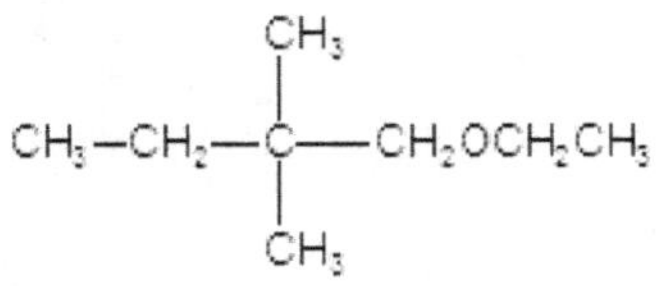

A. 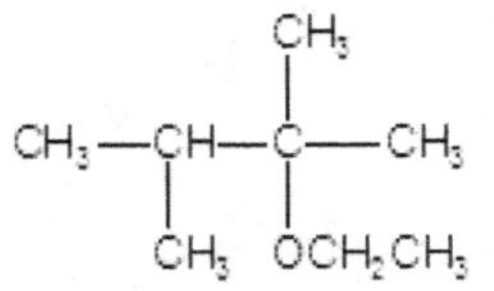 CH₃—CH(CH₃)—C(CH₃)(OCH₂CH₃)—CH₃

B. CH₃—CH₂—C(CH₃)₂—CH₂OCH₂CH₃

C. CH₂—CH₂—C(CH₃)(OCH₂CH₃)—CH₃

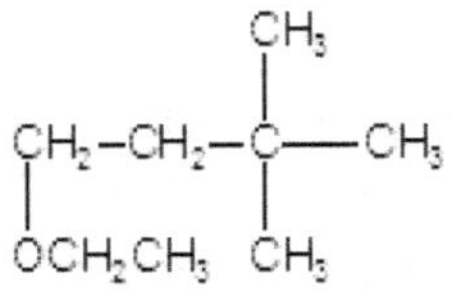

D. 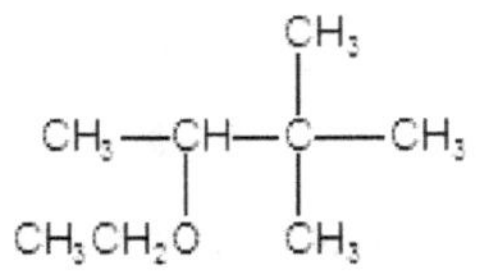 CH₃—CH(OCH₂CH₃)—C(CH₃)₂—CH₃

E. 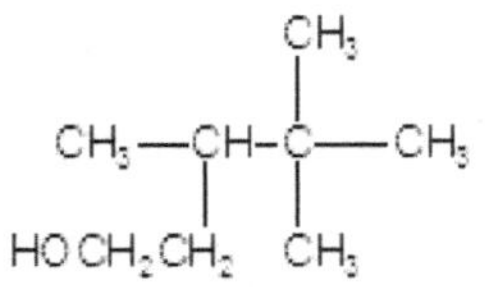 CH₃—CH(CH₂CH₂OH)—C(CH₃)₂—CH₃

4. Which of the following correctly ranks the halides in order of increasing rate for addition to 3-hexene in a nonpolar aprotic solvent?

A. HI < HBr < HCl
B. HBr < HCl < HI
C. HCl < HI < HBr
D. HCl < HBr < HI
E. None of the above

5. The carbon–carbon single bond in 1,3-butadiene has a bond length that is shorter than a carbon–carbon single bond in an alkane. This is a result of:

A. overlap of two sp^3 orbitals
B. overlap of one sp^2 and one sp^3 orbital
C. partial double-bond character due to the σ electrons
D. overlap of two sp^2 orbitals
E. none of the above

6. What is the name of the major organic product of the following reaction?

(CH₃)₂C=C(CH₃)₂ + H⁺/H₂O → ?

A. 2,3-dimethyl-2-butanol
B. 2,3-dimethyl-1-butanol
C. 3,3-dimethyl-1-butanol
D. 3,3-dimethyl-2-butanol
E. 4-methyl-2-pentanol

7. The rate law for the addition of HBr to many simple alkenes may be approximated as rate = k[alkene]·[HBr]. This rate law indicates the following EXCEPT that the reaction:

   A. occurs in a single step involving one HBr molecule and one alkene molecule
   B. involves one HBr and one alkene molecule in the rate-determining step and may have many steps
   C. is first order in HBr
   D. is second order overall
   E. is first order in the alkene

8. Which reactant is used to convert propene to 1,2-dichloropropane?

   A. HCl
   B. NaCl
   C. H_2
   D. Cl_2
   E. BrCl

9. Using Zaitsev's rule, choose the most stable alkene among the following:

   A. 1-methylcyclohexene
   B. 3-methylcyclohexene
   C. 4-methylcyclohexene
   D. 3,4-dimethylcyclohexene
   E. they are of equal stability

10. Which of the following has the lowest heat of hydrogenation per mole of H_2 absorbed?

    A. 1,2-hexadiene
    B. 1,3-hexadiene
    C. 1,3,5-heptatriene
    D. 1,5-hexadiene
    E. 1,2,3-heptatriene

11. Which of the following is the best solvent for the addition of HCl to 3-hexene?

    A. 3-hexene
    B. CH_3CH_3
    C. CH_3OH
    D. H_2O
    E. None of the above

12. What is the major product of the following reaction?

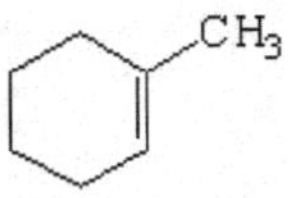

[1-methylcyclohexene] + H⁺/H₂O → ?

A. 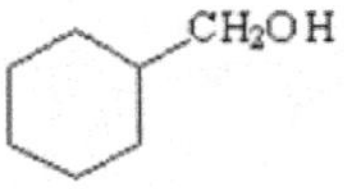 cyclohexyl-CH₂OH

B. 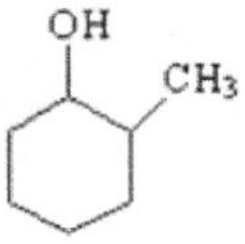 1-hydroxy-2-methylcyclohexane (OH and CH₃ on adjacent carbons)

C. 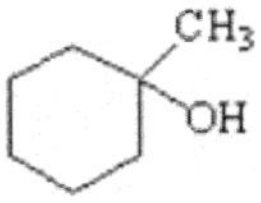 1-methylcyclohexan-1-ol (CH₃ and OH on same carbon)

D. 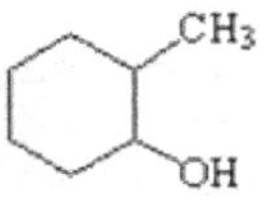 1-methyl-2-hydroxycyclohexane

E. 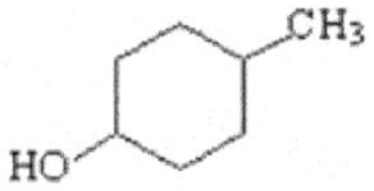 4-methylcyclohexan-1-ol

13. Which of the following is NOT an example of a conjugated system?

A. 1,2-butadiene
B. cyclobutadiene
C. benzene
D. 1,3-cyclohexadiene
E. 2,4-pentadiene

14. Which of the following would show the LEAST regioselectivity for HBr addition?

A. $(CH_3)_2C=C(CH_3)CH_2CH_3$
B. $H_2C=C(CH_3)CH_2CH_3$
C. $CH_2HC=C(CH_3)CH_2CH_3$
D. $(CH_3)_2C=CHCH_2CH_3$
E. $CH_2=CHCH_3$

15. Which statement is true in the oxymercuration-reduction of an alkene?

A. *Anti*-Markovnikov orientation and *anti*-addition occur
B. *Anti*-Markovnikov orientation and *syn*-addition occur
C. Markovnikov orientation and *anti*-addition occur
D. Markovnikov orientation and *syn*-addition occur
E. Zaitsev orientation and *anti*-addition occur

16. Which of the following compounds is/are geometric isomers?

   I. Isobutene
   II. (*E*)-2-butene
   III. *cis*-2-butene
   IV. *trans*-2-butene

   A. I and II only
   B. II and III only
   C. III and IV only
   D. II, III and IV only
   E. I, II, III and IV

17. Consider the following alcohol A.

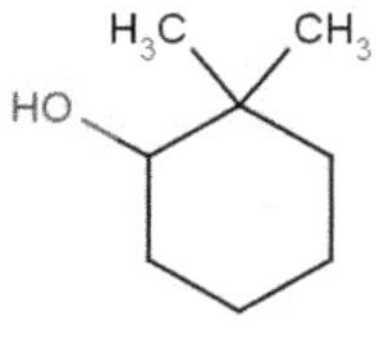

    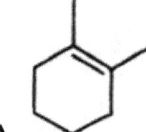

    The major product resulting from the dehydration of A is:

    A.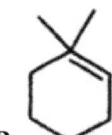

    B.

    C.

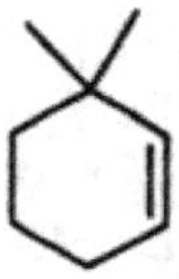

    D.

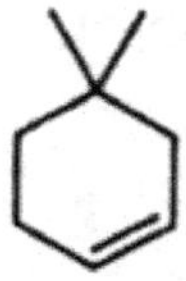

    E. None of the above

18. What is the major product of the following reaction?

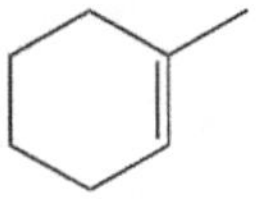

1-methylcyclohexene + Br₂ / H₂O → ?

A. 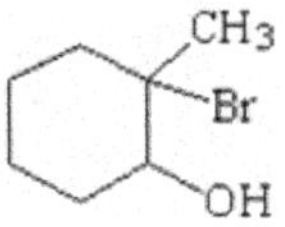 cyclohexane with CH₃, Br, OH (Br on same C as CH₃... actually Br adjacent, OH on CH₃-bearing C) — structure: CH₃ and Br on one carbon, OH on adjacent — (drawn: CH₃ top, Br on same carbon, OH on adjacent)

A. [cyclohexane with CH₃ and Br on one carbon, OH on adjacent carbon]

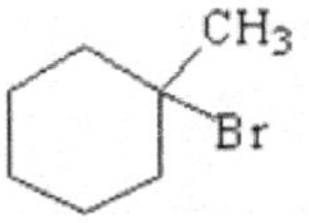

B. 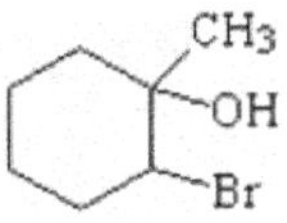 [cyclohexane with CH₃ and OH on one carbon, Br on adjacent carbon]

C. [cyclohexane with CH₃ and Br on same carbon]

D. 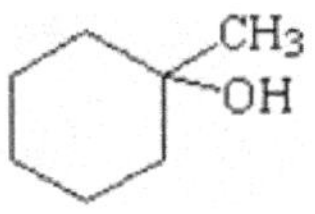 [cyclohexane with CH₃ and OH on same carbon]

E. 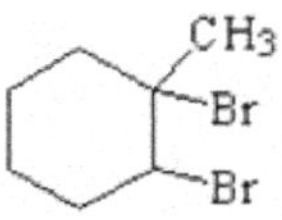 [cyclohexane with CH₃ and Br on one carbon, Br on adjacent carbon]

19. Which of the following reactions does NOT proceed through a bromonium ion intermediate?

A. $CH_3CH=CH_2 + Br_2 + H_2O \rightarrow CH_3CH_2OHCH_2Br + HBr$
B. $CH_2=CHCH_2CH_2CH=CH_2 + Br_2 \rightarrow CH_2BrCHBrCH_2CH_2CHBrCH_2Br$
C. $CH_3CH_2CH=CH_2 + Br_2 \rightarrow CH_3CH_2CHBrCH_2Br$
D. $CH_3CH_2CH=CH_2 + HBr \rightarrow CH_3CH_2CHBrCH_3$
E. None of the above

20. When reacted with HBr, *cis*-3-methyl-2-hexene most likely undergoes:

A. *anti*-Markovnikov *syn*- and *anti*-addition
B. *anti*-Markovnikov *syn*-addition
C. Markovnikov *syn*- and *anti*-addition
D. Markovnikov *syn*-addition
E. none of the above

21. The following are examples of addition reactions of alkenes, EXCEPT:

A. hydrogenation
B. hydration
C. oxidation
D. bromination
E. ozonolysis

22. Which of the following reactions will NOT occur?

    Reaction 1: butene + NBS → 2-bromobutane
    Reaction 2: 2-methylbutane + 8 O$_2$ + heat → 5 CO$_2$ + 6 H$_2$O
    Reaction 3: 2-methylbutane + Br$_2$ + hv → 2-bromo-2-methylbutane
    Reaction 4: 2-methyl-2-butene + Br$_2$ + CCl$_4$ → 2,3-dibromo-2-methylbutane (+ enantiomer)

    A. Reaction 1
    B. Reaction 2
    C. Reaction 3
    D. Reaction 4
    E. Reactions 2 and 4

23. Give the possibilities in the structure for a compound with a formula of C$_6$H$_{10}$:

    A. no rings; no double bonds; no triple bonds
    B. one double bond; or one ring
    C. two rings; two double bonds; one double bond and one ring; or one triple bond
    D. three rings; three double bonds; two double bonds and one ring; one ring and two double bonds; one triple bond and one ring; or one double bond and one ring
    E. benzene

24. Which of the following is/are the most stable diene?

    A.

    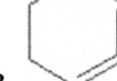
    B.

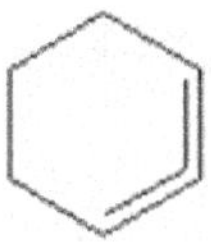

    C.

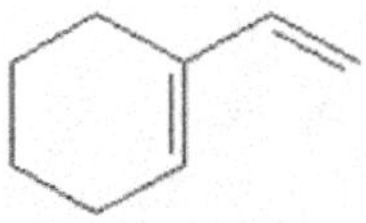

    D.

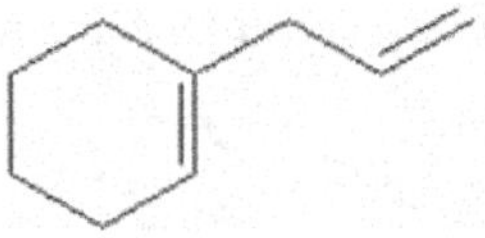

    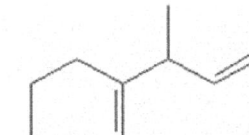
    E.

College Organic Chemistry Practice Questions with Detailed Explanations

25. Products A, B, and C of the following reaction are, respectively:

2-methyl-2-butene + HBr → product A
2-methyl-2-butene + HBr / H_2O_2 → product B
2-methyl-2-butene + H_2O / H^+ / heat → product C

Product A	Product B	Product C
A. 2-bromo-2-methylbutane	2-bromo-3-methylbutane	2-methyl-2-butanol
B. 3-methyl-2-bromobutene	3-methyl-2-bromobutane	2-methyl-2-butanol
C. 2-bromo-2-methylbutane	2-bromo-2-methylbutane	3-methyl-2-butanol
D. 1-bromo-2-methyl-2-butene	2-bromo-2-methylbutane	2-methyl-2-butane
E. None of the above		

26. Which of the following is vinyl chloride?

A. CH_2=CHCl

B. CH_2=CHCH$_2$Cl

C. [structure: cyclohexanone with Cl at α-position]

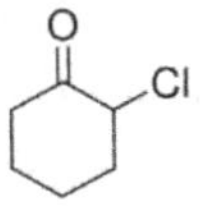

D. [structure: benzene ring with CH(Cl)(CH$_3$) substituent]

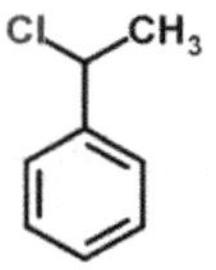

E. CH_3CH_2Cl

27. What reagent(s) is/are needed to accomplish the following transformation?

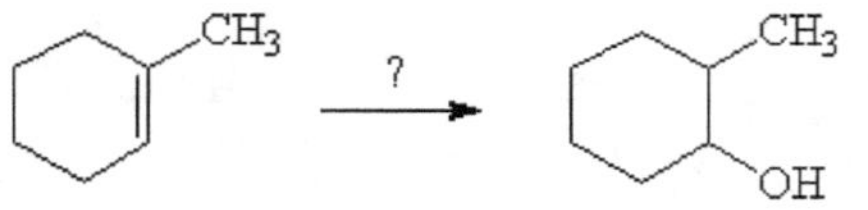

[structure: 1-methylcyclohexene] → [structure: 2-methylcyclohexanol with OH trans to CH$_3$]

A. BH$_3$ / THF
B. ⁻OH
C. H_2O / H_2O_2
D. H_2O / H^+
E. 1) BH$_3$ / THF; 2) ⁻OH, H_2O_2, H_2O

28. The reaction below can be classified as a(n):

cis-pent-2-ene → pentane

A. tautomerization
B. elimination
C. oxidation
D. reduction
E. substitution

Alkenes

29. The major product of the following reaction will likely be the result of a(n) [ ] mechanism?

    2-bromobutane + *tert*-butyl alkoxide

A. E_2

B. E_1

C. S_N2

D. S_N1

E. S_N1 and E_1

30. What is the major product from the following reaction?

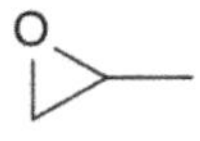

epoxide + HBr → ?

A. 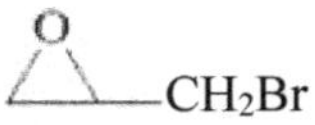 epoxide—CH$_2$Br

B. 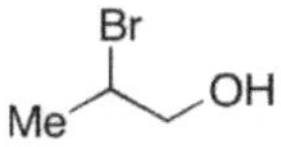 Me-C(Br)-CH$_2$OH

C. HO-CH(CH$_3$)-CH$_2$Br

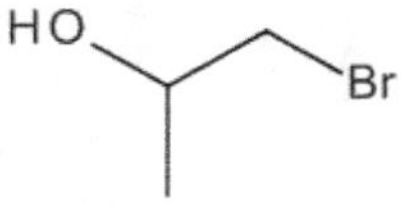

D. 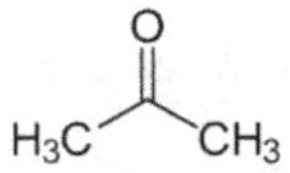 H$_3$C-C(=O)-CH$_3$

E.  Br-CH$_2$-CH$_2$-CH$_2$-OH

31. When CH_3–CH=CH_2 is reacted in water with a catalytic amount of acid, a new compound is formed. What might be the product of this reaction?

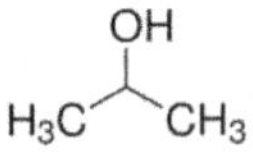

A. 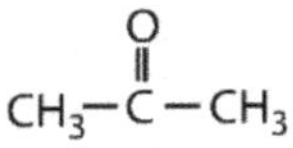 $CH_3-\overset{O}{\underset{\|}{C}}-CH_3$

B. $H_3C-CH(OH)-CH_3$

C. HO-CH$_2$-CH(OH)-CH$_3$

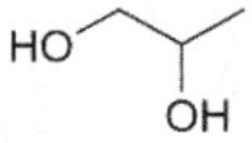

D. CH_3–CH–CH_2

E. 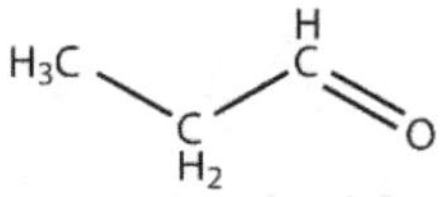 H_3C-CH_2-CHO

32. What is the product of the following reaction?

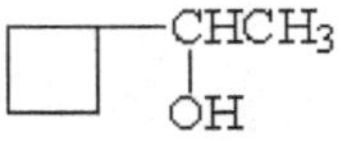

cyclobutyl-CH(OH)CH₃ + H₂SO₄/Δ → ?

A. 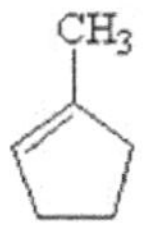 1-methylcyclopentene (with CH₃)

B. 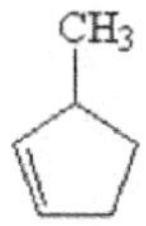 3-methylcyclopentene (with CH₃)

C. 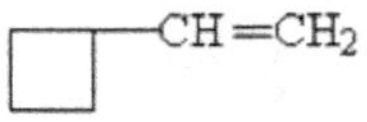 cyclobutyl-CH=CH₂

D. 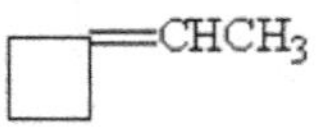 cyclobutylidene=CHCH₃

E. 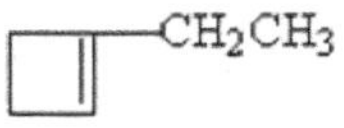 cyclobutene with -CH₂CH₃ substituent

33. Alkenes are more acidic than alkanes. What is the best explanation of this property?

A. The *sp²* hybridized orbitals in alkenes stabilize the negative charge generated when a proton is abstracted
B. The *sp²* hybridized orbitals in alkenes destabilize the negative charge generated when a proton is abstracted
C. The *sp³* hybridized orbitals in alkenes stabilize the negative charge generated when a proton is abstracted
D. The *sp³* hybridized orbitals in alkenes destabilize the negative charge generated when a proton is abstracted
E. The alkene is small and less sterically hindered compared to the alkane.

34. What is the major product formed from the reaction of 2-bromo-2-methylpentane with sodium ethoxide?

A. 2-methylpent-2-ene
B. 2-methylpent-3-ene
C. 2-methyl-2-methoxypentane
D. 2-methylpentene
E. 1-methylpentene

35. What reagents can best be used to accomplish the following transformation?

A. 1) Hg(OAc)₂, H₂O / THF; 2) NaBH₄
B. 1) Hg(O₂CCF₃)₂, CH₃OH; 2) NaBH₄
C. 1) BH₃·THF; 2) HO⁻, H₂O₂
D. H⁺, H₂O
E. NaOH, H₂O

36. Heating a(n) [ ] results in a Cope elimination.

A. amine oxide
B. imine
C. enamine
D. oxime
E. enol

70

Notes for active learning

Notes for active learning

Alkynes

1. What is the major organic product when 3-heptyne is subjected to excess hydrogen and a platinum catalyst?

   A. (*E*)-3-heptene
   B. (*Z*)-3-heptene
   C. (*Z*)-2-heptene
   D. 2-heptyne
   E. heptane

2. In the addition of hydrogen bromide to alkynes in the absence of peroxides, which of the following species is proposed as an intermediate?

   A. carbene
   B. vinyl radical
   C. vinyl cation
   D. vinyl anion
   E. none of the above

3. Which of the following molecular formulas correspond(s) to an acyclic alkyne?

   I. C_9H_{20} II. C_9H_{18} III. C_9H_{16}

   A. I only
   B. II only
   C. III only
   D. I and II only
   E. I, II and III

4. What is the term for a family of unsaturated hydrocarbon compounds with a triple bond?

   A. alkynes
   B. arenes
   C. alkanes
   D. alkenes
   E. none of the above

5. Which of the following molecular formulas correspond(s) to an alkyne?

   I. $C_{10}H_{18}$ II. $C_{10}H_{20}$ III. $C_{10}H_{22}$

   A. I only
   B. II only
   C. III only
   D. I and II only
   E. I, II and III

6. What is the product from the reaction of one mole of acetylene and two moles of hydrogen gas using a platinum catalyst?

   A. propene
   B. propane
   C. ethene
   D. ethane
   E. none of the above

7. Which of the following does NOT accurately describe the physical properties of an alkyne?

   A. Less dense than water
   B. Insoluble in most organic solvents
   C. Relatively nonpolar
   D. Nearly insoluble in water
   E. Boiling point nearly the same as an alkane with a similar carbon skeleton

8. What are the two products from the complete combustion of an alkyne?

   A. CO_2 and H_2O
   B. CO_2 and H_2
   C. CO and H_2O
   D. CO and H_2
   E. None of the above

9. The compound propyne consists of how many carbon atoms and how many hydrogen atoms?

   A. 3C,2H
   B. 3C,6H
   C. 3C,4H
   D. 2C,2H
   E. None of the above

10. Given that 1-butyne has a boiling point of 8.1 °C, what is the phase of propyne at room temperature and 1 atm pressure?

    A. solid
    B. supercritical fluid
    C. gas
    D. liquid
    E. vapor

11. How many moles of hydrogen are required to convert a mole of pentyne to pentane?

    A. 0
    B. 1
    C. 2
    D. 3
    E. 4

12. What is the product of the reaction of one mole of acetylene and one mole of bromine vapor?

    A. 1,1,2,2-tetrabromoethane
    B. 1,1,2,2-tetrabromoethene
    C. 1,2-dibromoethane
    D. 1,2-dibromoethene
    E. None of the above

13. What reagents are used to convert 1-hexyne into 2-hexanone?

    A. 1) Si_2BH; 2) H_2O_2, NaOH
    B. Hg^{2+}, H_2SO_4, H_2O
    C. 1) O_3; 2) $(CH_3)_2S$
    D. 1) CH_3MgBr; 2) CO_2
    E. 1) H_2, Ni; 2) $Na_2Cr_2O_7$, H_2SO_4

14. What is the major product of this reaction?

$$H_3C-C\equiv C-CH_3 + Na\ (s),\ NH_3\ (l) \rightarrow\ ?$$

A. 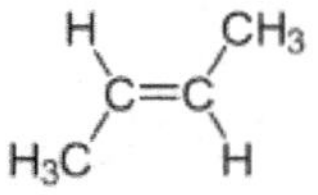 (trans-2-butene: H and CH₃ on one carbon, H₃C and H on the other)

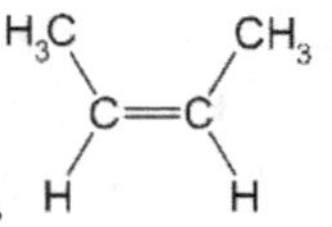

B.  (structure of pentane/butane skeletal)

C. (cis-2-butene: H₃C and CH₃ on same side)

D. 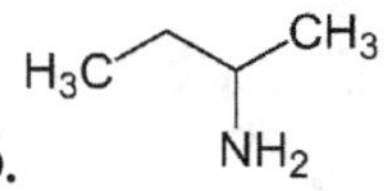 H₃C–CH(NH₂)–CH₃ with CH₃

E.  H₃C–CH₂–CH₂–NH₂

15. What is the major product of the following acid/catalyzed hydration reaction?

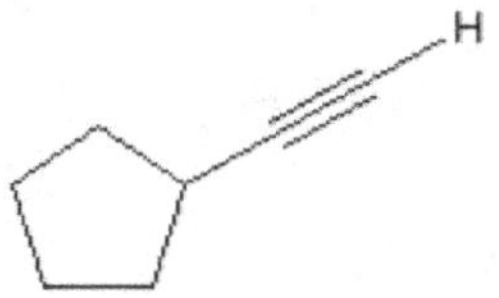

(cyclopentyl–C≡C–H) + $H_2O\,/\,H_2SO_4$, $HgSO_4 \rightarrow\ ?$

A.  cyclopentyl–CH=CH₂

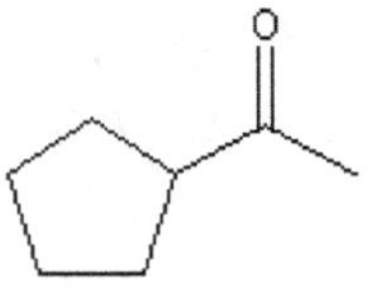

B. 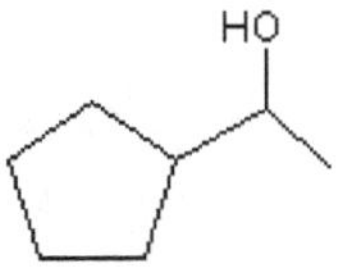 cyclopentyl–CH(OH)–CH₃

C. cyclopentyl–C(=O)–CH₃

D. 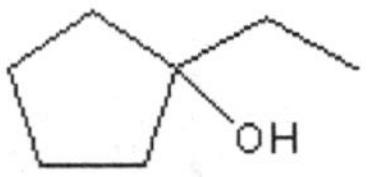 1-ethyl-1-hydroxycyclopentane

E. 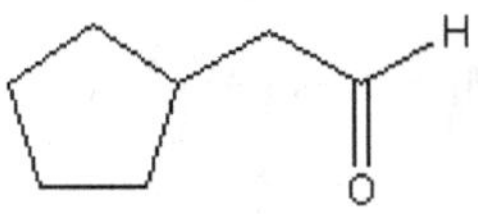 cyclopentyl–CH₂–CHO

16. When 2,2-dibromobutane is heated to 200 °C with molten KOH, what is the major organic product?

A. but-1-yne
B. but-2-yne

C. 1-bromobut-1-yne
D. 1-bromobut-2-yne
E. but-1-ene

17. What class of organic product results when 1-heptyne is treated with a mixture of mercuric acetate [Hg(OAc)$_2$] in aqueous sulfuric acid (H$_2$SO$_4$), followed by sodium borohydride (NaBH$_4$)?

A. diol
B. ketone
C. aldehyde
D. alcohol
E. carboxylic acid

18. Which of the following describes the reaction below?

H$_3$C—≡—CH$_3$ + H$_2$, Pd, CaCo$_3$, quinolone, hexane → cis-CH$_3$CH=CHCH$_3$

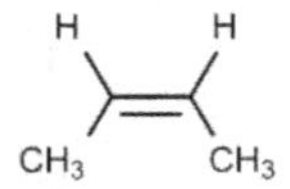

A. oxidation
B. reduction
C. substitution
D. catalytic hydration
E. elimination

19. What is the general molecular formula for the alkyne class of compounds?

I. C$_n$H$_{2n+2}$ II. C$_n$H$_{2n}$ III. C$_n$H$_{2n-2}$

A. I only
B. II only
C. III only
D. II and III only
E. I, II and III

20. Among the following compounds, which acids are stronger than ammonia?

I. water   II. ethane   III. butyne   IV. but-2-yne

A. I and II only
B. II only
C. I and III only
D. II and III only
E. I, II, III and IV

21. In reducing alkynes using sodium in liquid ammonia, which of the species below is NOT an intermediate in the commonly accepted mechanism?

A. vinyl anion
B. vinyl cation
C. anion
D. vinyl radical
E. all are intermediates

22. For isomers with the formula C$_{10}$H$_{16}$, which of the following structural features are NOT possible?

A. 2 rings and 1 double bond
B. 2 double bonds and 1 ring
C. 2 triple bonds
D. 1 ring and 1 triple bond
E. 3 double bonds

23. What is the product from the reaction of one mole of acetylene and one mole of hydrogen gas using a platinum catalyst?

    A. propane
    B. propene
    C. ethane
    D. ethene
    E. 2-butene

24. Which of the species below is less basic than an acetylide?

    I. CH_3Li II. CH_3MgBr III. CH_3ONa

    A. I only
    B. II only
    C. III only
    D. I and II only
    E. I, II and III

25. Which is the most stable product for the reaction below:

    [structure] + 1) BH_3 / THF; 2) ^-OH, H_2O_2, H_2O

    A. 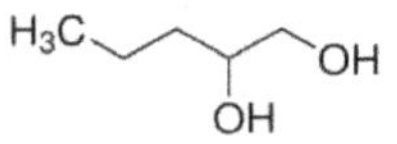 [structure with H3C, OH, OH]
    B. $CH_3CH_2CH_2CH=CHOH$
    C. 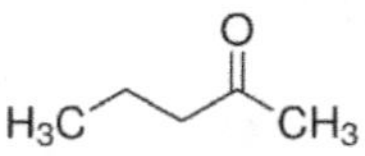 [ketone structure H3C-CO-CH3 region]
    D. 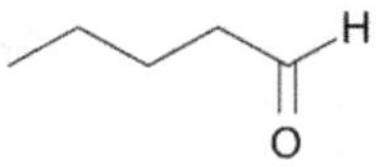 [aldehyde structure]
    E. $CH_3CH_2CH_2(OH)C=CH_2$

26. The compound 1-butyne contains:

    A. a ring structure
    B. a triple bond
    C. a double bond
    D. all single bonds
    E. a bromine atom

27. The *pi* bond of an alkyne is [ ] and [ ] than the *pi* bond of an alkene.

    A. longer; stronger
    B. longer; weaker
    C. shorter; stronger
    D. shorter; weaker
    E. none of the above

28. What is the major product of this reaction?

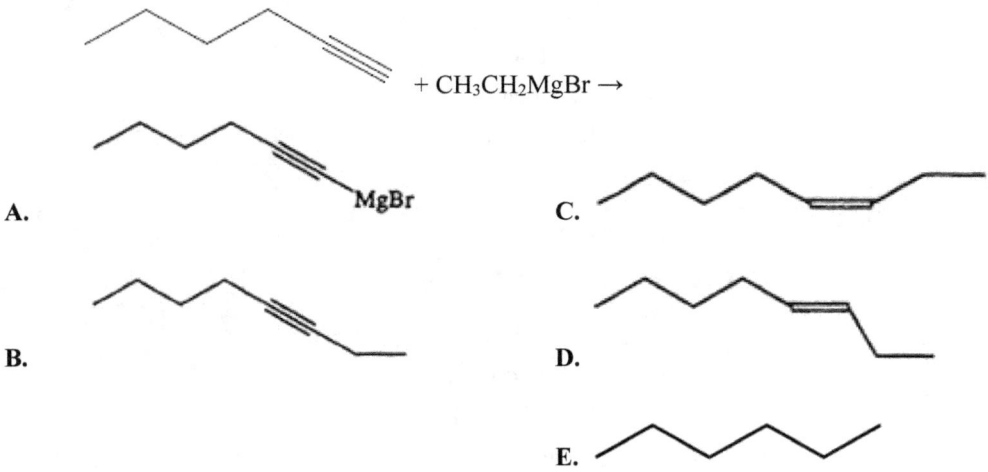

29. Which of the following is the product for the reaction?

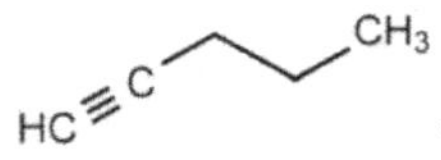

HC≡C—CH(CH₃)—CH₂CH₃ (approximate) + H₂O + HgSO₄ / H₂SO₄ →

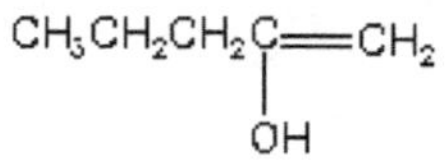

A. CH₃CH₂CH₂C(OH)=CH₂

B. CH₃CH₂CH₂CH=CHOH

C. 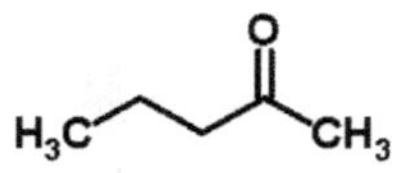 H₃C—C(=O)—CH₂—CH₃ (approx, shown as H₃C–CO–CH₃ type)

D. CH₃CH₂CH₂CH₂CHO

E. CH₃CH₂CH₂CH(OH)CH₂OH

30. What is the product from the reaction of one mole of acetylene and two moles of bromine vapor?

A. 1,1,2,2-tetrabromoethene
B. 1,1,2,2-tetrabromoethane
C. 1,2-dibromoethene
D. 1,2-dibromoethane
E. none of the above

31. Which of the alkyne addition reactions below involve(s) an enol intermediate?

   I. hydroboration/oxidation
   II. treatment with HgSO₄ in dilute H₂SO₄
   III. hydrogenation

A. I only
B. II only
C. III only
D. I and II only
E. I and III only

78

32. Which of the following is the final and major product of this reaction?

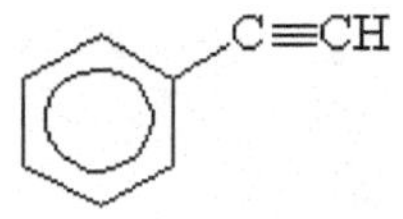

Ph-C≡CH + H₂O, H₂SO₄/HgSO₄ → ?

A. 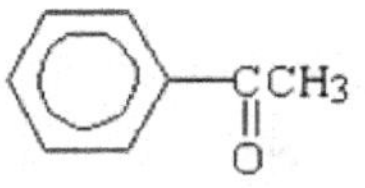 Ph-C(=O)-CH₃

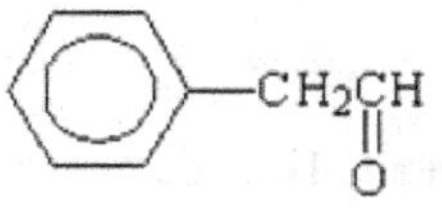

B. 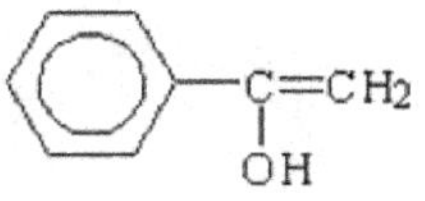 Ph-C(OH)=CH₂

C. Ph-CH₂-CH=O

D. 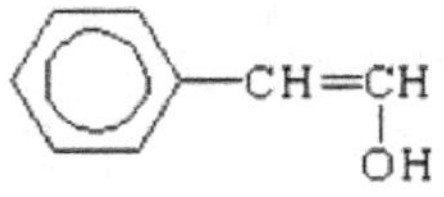 Ph-CH=CH-OH

E. 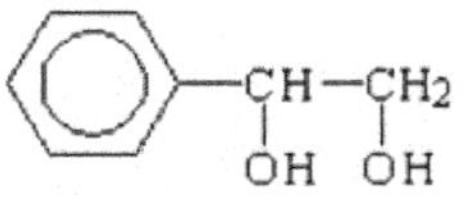 Ph-CH(OH)-CH₂(OH)

33. If the compound C_5H_7NO contains 1 ring, how many *pi* bonds are there in this compound?

A. 0
B. 1
C. 2
D. 3
E. 4

34. Which of the following is the major product of this reaction?

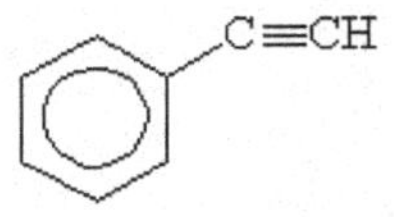

Ph-C≡CH + 1) BH₃, THF + 2) ⁻OH, H₂O₂, H₂O →

A. 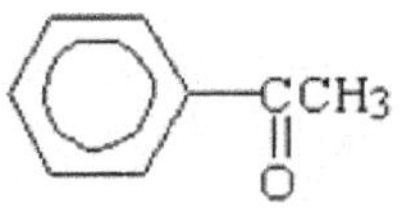 Ph-C(=O)-CH₃

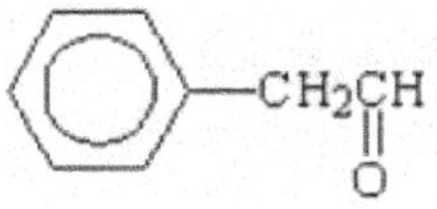

B. 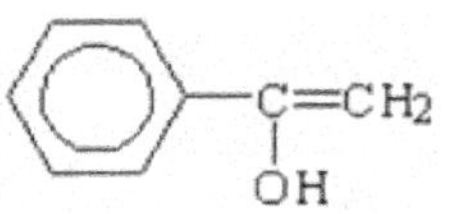 Ph-C(OH)=CH₂

C. Ph-CH₂-CH=O

D. 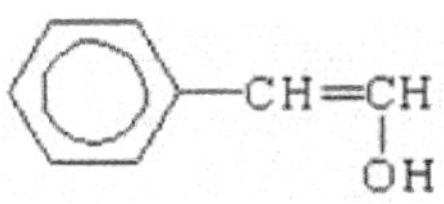 Ph-CH=CH-OH

E. 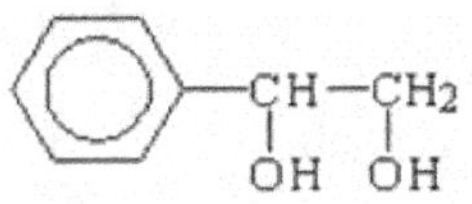 Ph-CH(OH)-CH₂(OH)

35. Which of the following statements correctly describes the general reactivity of alkynes?

A. Unlike alkenes, alkynes fail to undergo electrophilic addition reactions
B. Alkynes are generally more reactive than alkenes
C. An alkyne is an electron-rich molecule and therefore reacts as a nucleophile
D. The σ bonds of alkynes are higher in energy than the π bonds, and thus are more reactive
E. Alkynes react as electrophiles, whereas alkenes react as nucleophiles

36. What is the major product when C_2H_4 undergoes an addition reaction with 1 mole equivalent of Br_2 in CCl_4?

A. $C_2H_2 + H_2$
B. $C_2H_4Br_2$
C. $C_2HBr + HBr$
D. C_2H_3Br
E. None of the above

Notes for active learning

Notes for active learning

Aromatic Compounds

1. What is the major product of this reaction?

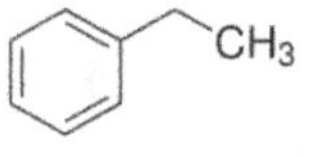

   (PhCH₃ derivative: benzyl-CH₃) + C(CH₃)₃Br + FeBr → ?

   A. 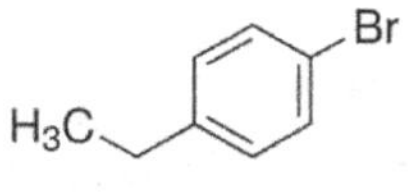 H₃C—(benzene)—Br (para)

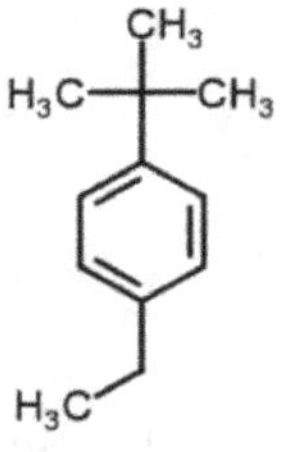

   B. 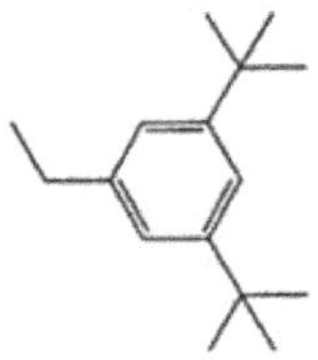 3,5-di-tert-butyl ethylbenzene

   C. para-tert-butyl ethylbenzene (H₃C—(benzene)—C(CH₃)₃)

   D. 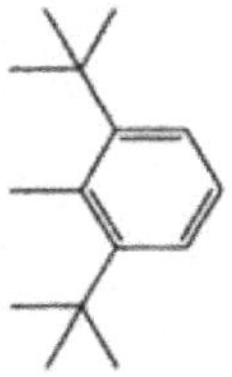 2,6-di-tert-butyl methylbenzene

   E. 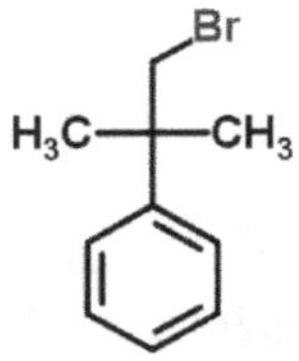 (CH₃)₂C(CH₂Br)—phenyl

2. The following are common reactions of benzene, EXCEPT:

   A. nitration
   B. hydrogenation
   C. chlorination
   D. bromination
   E. sulfonation

3. How many pairs of degenerate π molecular orbitals are in benzene?

   A. 6
   B. 5
   C. 4
   D. 3
   E. 2

4. What is the effect of an ammonium substituent on electrophilic aromatic substitution?

    A. *ortho/para*-directing with activation   **C.** *meta*-directing with activation

    B. *ortho/para*-directing with deactivation   **D.** *meta*-directing with deactivation

                                                            E. neither directing nor activating

5. Which of the following compounds undergoes Friedel-Crafts alkylation with $(CH_3)_3CCl$, $AlCl_3$ most rapidly?

    A. toluene   **C.** acetophenone

    B. iodobenzene   **D.** benzenesulfonic acid

                                                                  E. cyanobenzene

6. Which sequence correctly ranks the following aromatic rings in order of increasing rate of reactivity in an electrophilic aromatic substitution reaction?

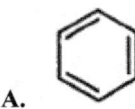

    A. II < I < III   **C.** III < I < II

    B. II < III < I   **D.** I < II < III

                                                                      E. I < III < II

7. Of the following, which reacts most readily with $Br_2/FeBr_3$ in an electrophilic aromatic substitution?

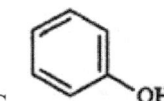

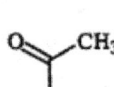

                                                                      E. The compounds have the same reactivity

8. In the molecular orbital representation of benzene, how many π molecular orbitals are present?

    A. 1   **C.** 4

    B. 2   **D.** 6

                                                                    E. 8

9. Which of the following structures is aromatic?

A.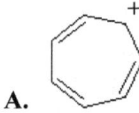

B.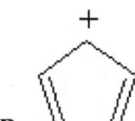

C.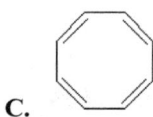

D.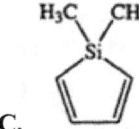

E.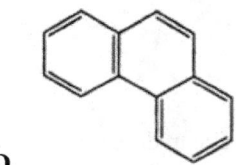

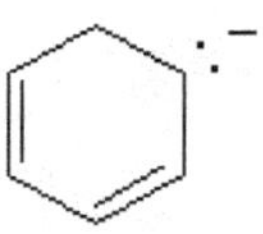

10. In the electrophilic aromatic substitution of phenol, substituents add predominantly in which position(s)?

   I. *ortho* to the hydroxyl group
   II. *meta* to the hydroxyl group
   III. *para* to the hydroxyl group

   A. I only
   B. II only
   C. III only
   D. I and III only
   E. I, II and III

11. Which of the following is NOT aromatic?

A.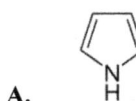

B.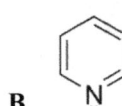

C. (silacyclopentadiene with Si(CH₃)₂)

D. (phenanthrene)

E. All of the above

12. Which of these molecules is aromatic?

A.

B. 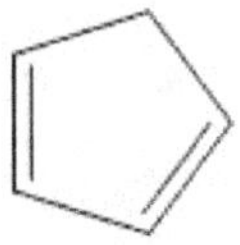 (cyclopentadiene)

C.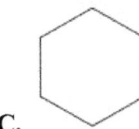

D. 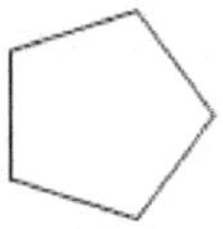 (cyclopentane)

E. None of the above

13. What is the most likely regiochemistry of this electrophilic aromatic substitution reaction?

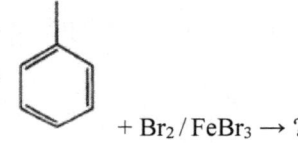

+ Br$_2$ / FeBr$_3$ → ?

A. 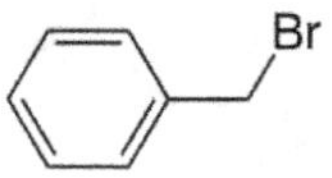 (benzyl bromide, PhCH$_2$Br)

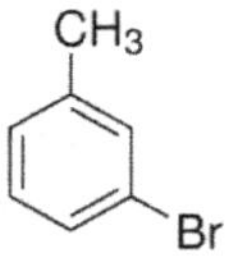

B. 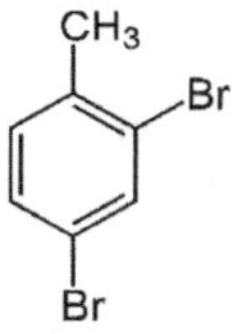 (2,4-dibromotoluene)

C. (3-bromotoluene)

D. 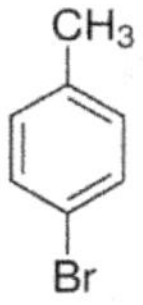 (4-bromotoluene)

E. 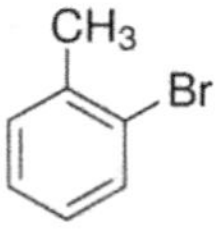 (2-bromotoluene)

14. A compound is a six-carbon cyclic hydrocarbon. It is inert to bromine in water and bromine in dichloromethane, yet it decolorizes bromine in carbon tetrachloride when a small quantity of FeBr$_3$ is added. Which of the following is the identity of the compound?

A. 1,4-cyclohexadiene
B. 1,3-cyclohexadiene
C. benzene
D. cyclohexane
E. cyclohexyne

15. While electron-withdrawing groups (such as ~NO₂ and ~CO₂R) are *meta*-directing with regard to electrophilic aromatic substitution reactions, they are *ortho-* / *para*-directing in nucleophilic aromatic substitution reactions. This observation would best be explained by using which concept?

   A. tautomerism
   B. aromaticity
   C. hydrogen bonding
   D. resonance
   E. equilibration

16. Derivatives of the compound shown below are currently being examined for their effectiveness in treating drug addiction and metabolic syndrome. Which sequence ranks the aromatic rings of this compound in order of increasing reactivity (slowest to fastest reacting) in an electrophilic aromatic substitution reaction?

   A. 1 < 2 < 3
   B. 2 < 3 < 1
   C. 3 < 2 < 1
   D. 3 < 1 < 2
   E. 2 < 1 < 3

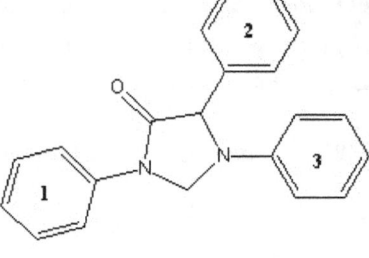

17. The major aromatic product of the following reaction is:

   A. methyl ketone substitutes in the *ortho* / *para* position
   B. methyl ketone substitutes in the *meta* position
   C. methyl ketone replacing the bromine
   D. formation of phenol
   E. formation of benzene

18. In electrophilic aromatic substitution reactions, an extremely reactive electrophile is typically used because the aromatic ring is:

   A. a poor electrophile
   B. nonpolar
   C. reactive
   D. a poor nucleophile
   E. unstable

19. What is the effect of each of ~Cl substituents on electrophilic aromatic substitution?

   A. *meta*-directing with deactivation
   B. *meta*-directing with activation
   C. *ortho/para*-directing with deactivation
   D. *ortho*-directing with activation
   E. *para*-directing with activation

20. 1,3-cyclopentadiene reacts with sodium metal at low temperatures according to:

What is the best explanation for this observation?

A. Rehybridization of the saturated carbon atoms provides additional product stability
B. Sodium metal is highly selective for cycloalkenes
C. Aromaticity stabilizes the carbocation
D. The reactant is more unstable at reduced temperatures
E. Aromaticity stabilizes the anion

21. What is the major product of this electrophile aromatic substitution (EAS) reaction?

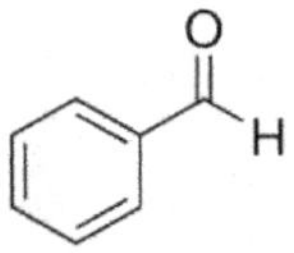

benzaldehyde + $HNO_3 / H_2SO_4 \rightarrow$?

A. 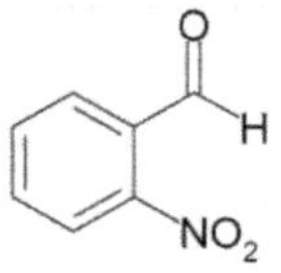 2-nitrobenzaldehyde
B. 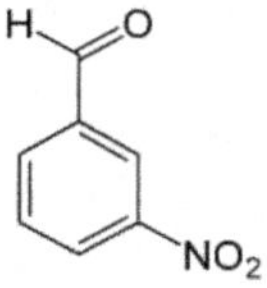 3-nitrobenzaldehyde
C. 3-nitrobenzoic acid
D. 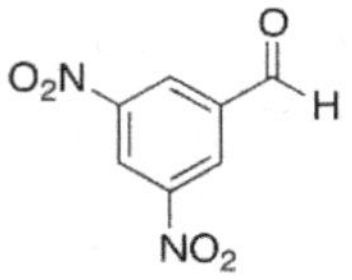 3,5-dinitrobenzaldehyde
E. 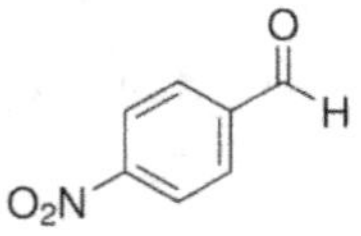 4-nitrobenzaldehyde

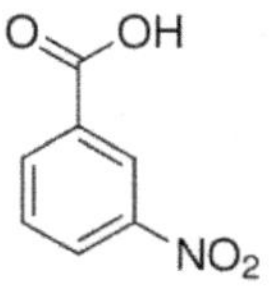

22. Which of the following molecules reacts the slowest in electrophilic nitration?

A. 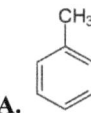 Toluene

B. Aniline

C. 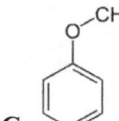 Anisole (methoxybenzene)

D. 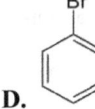 Bromobenzene

E. 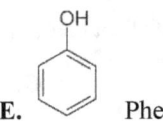 Phenol

23. In electrophilic aromatic substitution, the aromatic ring acts as a(n):

A. leaving group
B. dienophile
C. nucleophile
D. electrophile
E. spectator

24. What is the effect of ~F substituents on electrophilic aromatic substitution?

A. *meta*-directing with activation
B. *meta*-directing with deactivation
C. *ortho-* / *para*-directing with activation
D. *ortho-* / *para*-directing with deactivation
E. *ortho*-directing with activation

25. Which of the following compounds is least susceptible to electrophilic aromatic substitution?

A. *p*-H$_3$CCH$_2$O–C$_6$H$_4$–O–CH$_2$CH$_3$
B. *p*-O$_2$N–C$_6$H$_4$–NH–CH$_3$
C. *p*-Cl–C$_6$H$_4$–NH$_3^+$
D. *p*-CH$_3$CH$_2$–C$_6$H$_4$–CH$_2$CH$_3$
E. benzene

26. Which of the following statements is NOT correct about benzene?

A. The carbon-carbon bond lengths are the same
B. The carbon-hydrogen bond lengths are the same
C. All of the carbon atoms are *sp* hybridized
D. It has delocalized electrons
E. All twelve atoms lie in the same plane

27. Which reaction is NOT characteristic of aromatic compounds?

A. addition
B. halogenation
C. nitration
D. sulfonation
E. acylation

28. The reason why complete hydrogenation of benzene to cyclohexane requires H_2, rhodium (Rh) catalyst, and 1,000 psi pressure at 100 °C is because:

A. the double bonds in benzene have the same reactivity as *pi* bonds of non-aromatic alkenes
B. hydrogenation produces an aromatic compound
C. the double bonds in benzene are more reactive than a typical alkene
D. the double bonds in benzene are less reactive than a typical alkene
E. none of the above

29. Which of the following reactions is NOT an electrophilic aromatic substitution reaction?

A. $CH_3C_6H_5 + C_6H_5CH_2CH_2Cl / AlCl_3$
B. $CH_3C_6H_5 + Br_2 / FeBr_3$
C. $CH_3C_6H_5 + CH_3CH_2CH_2COCl / AlCl_3$
D. $CH_3C_6H_5 + H_2, Rh / C$
E. $C_6H_6 + HSO_3 / H_2SO_4$

30. Which steps may be used to synthesize 1-chloro-4-nitrobenzene, starting from benzene?

A. 1) Na / NH_3; 2) $Cl_2 / FeCl_3$
B. 1) HNO_3 / H_2SO_4; 2) $Cl_2 / FeCl_3$
C. 1) HCl / H_2O; 2) HNO_3 / H_2SO_4
D. 1) $Cl_2 / FeCl_3$; 2) HNO_3 / H_2SO_4
E. Cl_2 / CCl_4

31. If bromobenzene is treated with sulfur trioxide (SO_3) and concentrated H_2SO_4, what is/are the major product(s)?

A. *ortho*- and *para*-bromobenzenesulfonic acid
B. benzene
C. benzenesulfonic acid
D. *meta*-bromobenzenesulfonic acid
E. toluene

32. Rank the following three molecules in increasing order according to the rate they react with $Br_2/FeBr_3$.

I. 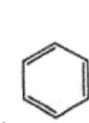 II. 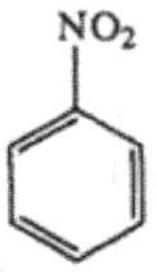 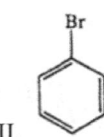 III.

A. II < III < I
B. II < I < III
C. I < II < III
D. I < III < II
E. III < II < I

33. Which of the following is true about the benzene molecule?

    A. It is a saturated hydrocarbon
    B. The *pi* electrons of the ring move around the ring and have resonance
    C. It is a hydrocarbon with the molecular formula of C_nH_{2n+2}
    D. It contains heterocyclic oxygen
    E. Attachments to the ring can exhibit *cis/trans* isomerism

34. Which of the following statements is supported by the table below?

	Solubility (g/L H_2O)	Melting point (°C)
para-nitrophenol	10	112
meta-nitrophenol	2.7	98
ortho-nitrophenol	0.8	47

    A. *Meta*- and *para*-nitrophenol form intramolecular hydrogen bonds
    B. *Ortho*-nitrophenol does not form intermolecular hydrogen bonds
    C. *Ortho*-nitrophenol has the greatest intramolecular hydrogen bonding
    D. *Para*-nitrophenol has the weakest intermolecular hydrogen bonding
    E. *Ortho*-nitrophenol has the weakest intramolecular hydrogen bonding

35. Which of the molecules shown below is NOT an *aromatic* compound?

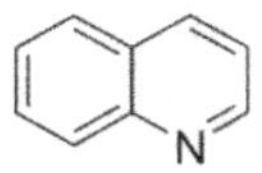

    A. 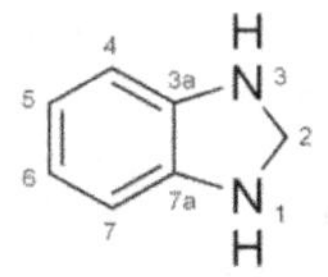 Benzimidazoline
    B.  Thiophene
    C. Quinoline
    D. 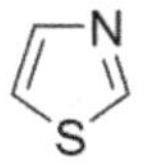 Thiazole
    E. 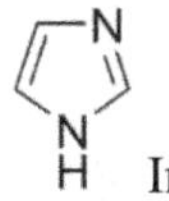 Imidazole

36. What is the degree of unsaturation for benzene?

    A. 1
    B. 2
    C. 3
    D. 4
    E. 5

Notes for active learning

Alcohols

1. Which of the following alcohols has the highest boiling point?

   A. 2-methyl-1-propanol
   B. hexanol
   C. ethanol
   D. propanol
   E. methanol

2. When phenol acts as an acid, a [ ] ion is produced.

   A. phenolic acid
   B. benzol
   C. phenyl
   D. benzyl
   E. phenoxide

3. When alcohol reacts with phosphoric acid, the product is a:

   A. pyrophosphate
   B. phosphate anion
   C. phosphate ester
   D. phosphate salt
   E. none of the above

4. Which formula is alcohol?

   A. 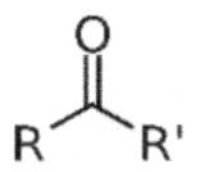 R–CO–R'
   B. 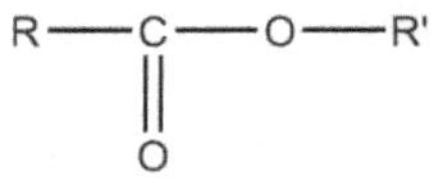 R–C(=O)–O–R'
   C. 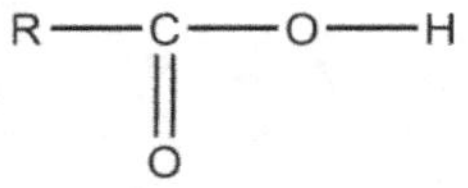 R–C(–O–H)=O
   D. 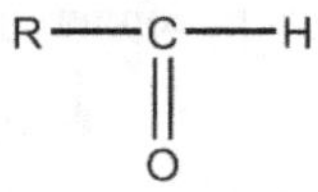 R–C(=O)–H
   E. R–O–H

5. The alcohol and carboxylic acid required to form propyl ethanoate are [ ] and [ ]:

   A. 1-propanol … ethanoic acid
   B. propanol … propanoic acid
   C. ethanol … propionic acid
   D. methanol … propionic acid
   E. 2-propanol … ethanoic acid

6. Which compound has the highest boiling point?

A. $CH_3CH_2CH_2CH_2OH$

B. $CH_3CH_2CH_2CH_3$

C. $CH_3CH_2CH_2CH_2CH_2OH$

D. $CH_3CH_2CH_2CH_2CH_3$

E. $CH_3CH_2CH_2OH$

7. The functional group, ~OH, is in which of these types of organic compounds?

A. amines

B. alcohols

C. alkanes

D. alkenes

E. ethers

8. What is the major product of this reaction?

Ph-CH(OH)-CH$_3$ (Ph-CH$_2$CH$_2$OH) + $Na_2Cr_2O_7 / H_2SO_4$ → ?

A. 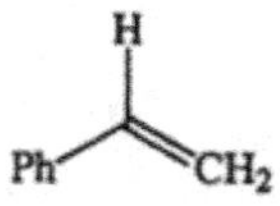 Ph-CH=CH$_2$ (with H)

B.  Ph-CH$_2$CH$_2$-OSO$_3$H

C. 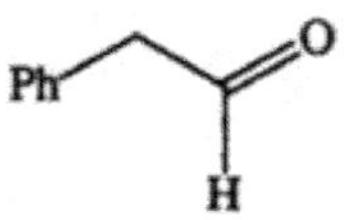 Ph-CH$_2$-CHO

D. 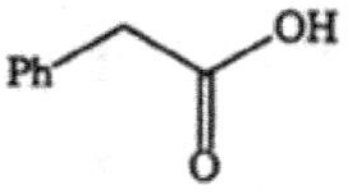 Ph-CH$_2$-COOH

E. None of the above

9. Which of the following reagents is best to convert methyl alcohol to methyl chloride?

A. Cl$^-$

B. $SOCl_2$

C. Cl_2/CCl_4

D. $Cl_2/h\nu$

E. NaCl

10. The compound below has which functional groups?

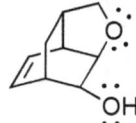

A. ether, alkene and alcohol

B. ester, alkene and alcohol

C. aromatic, alcohol and ether

D. aromatic, alcohol and ester

E. ether and ester

11. Which compound is NOT an unsaturated compound?

    A. $H_2C=CH-Cl$
    B. CH_3-CH_2-O-H
    C. $H_2C=CH_2$
    D. $CH_3-CH=CH_2$
    E. $H_2C=CH-O-H$

12. Compounds of the type R_3C-OH are referred to as [ ] alcohols.

    A. primary
    B. secondary
    C. tertiary
    D. quaternary
    E. none of the above

13. Compounds with the ~OH group attached to a saturated alkane-like carbon are:

    A. ethers
    B. hydroxyls
    C. alcohols
    D. alkyl halides
    E. phenols

14. When (S)-2-heptanol is subjected to $SOCl_2$/pyridine, the compound is transformed into:

    A. (R/S)-2-chloroheptane
    B. (R)-2-chloroheptane
    C. (S)-2-chloroheptane
    D. 2-heptone
    E. heptanoic acid

15. The ester prepared by heating 1-pentanol with acetic acid in the presence of an acidic catalyst is:

    A. 1-pentyl acetate
    B. acetyl 1-pentanoate
    C. acetic pentanoate
    D. pentanoic acetate
    E. acetyl pentanol

16. Which of the following has the highest boiling point?

    A. ethyl methyl ether
    B. dihexyl ether
    C. dimethyl ether
    D. diethyl ether
    E. dipropyl ether

17. Based on the properties of the attached functional group, which compound interacts most strongly with water, thus making it the most soluble compound?

    A. CH_3-CH_2-S-H
    B. CH_3-CH_2-I
    C. CH_3-CH_2-F
    D. CH_3-CH_2-Cl
    E. CH_3-CH_2-O-H

18. What is the product of the following reaction?

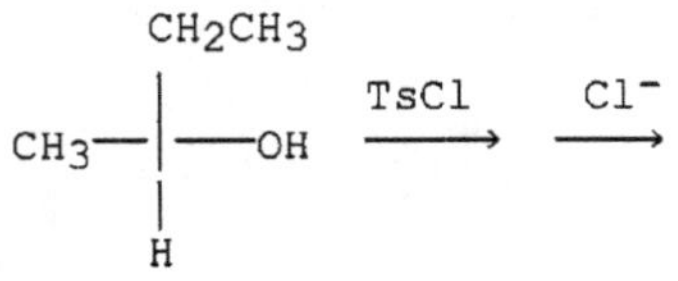

CH₃—C(CH₂CH₃)(OH)(H) →TsCl→ →Cl⁻→

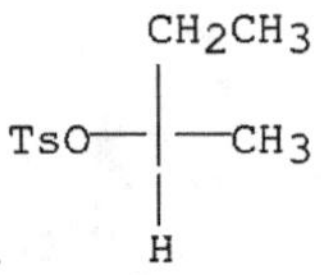

A. TsO—C(CH₂CH₃)(CH₃)(H)

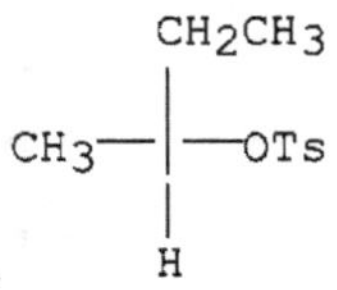

B. 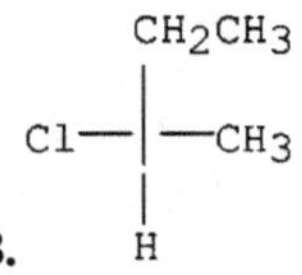 Cl—C(CH₂CH₃)(CH₃)(H)

C. CH₃—C(CH₂CH₃)(OTs)(H)

D. 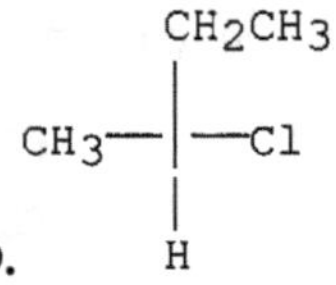 CH₃—C(CH₂CH₃)(Cl)(H)

E. 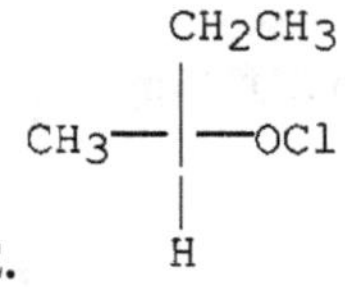 CH₃—C(CH₂CH₃)(OCl)(H)

19. Which of the following alcohols has the lowest boiling point?

A. hexanol
B. 2-methyl-1-propanol
C. propanol
D. ethanol
E. benzoic acid

20. The reaction of $(CH_3)_2CHCH_2OH$ with concentrated HBr using controlled heating yields:

A. $(CH_3)_2CHCH_2OBr$
B. $(CH_3)_2CHCH_4^+Br^-$
C. $CH_3CH_2CH_2Br$
D. $(CH_3)_2CHCH_2Br$
E. $(CH_3)_2CHCH_3$

21. When propanol is subjected to PBr₃, the compounds undergo:

A. an S_N1 elimination reaction to form propene
B. oxidation to form an aldehyde
C. an S_N1 reaction to produce an alkyl halide
D. addition, elimination, and then substitution to form bromopropane
E. an E_1 reaction to produce propene

22. Which is a product of the oxidation of $CH_3–CH_2–CH_2–O–H$?

   A. $CH_3–CH_2–CH_3$

   B. 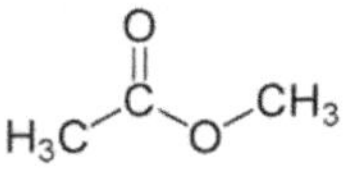 H₃C–C(=O)–O–CH₃

   C. (acetone structure)

   D. 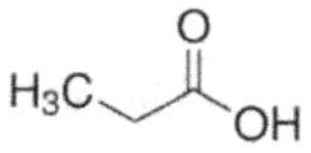 H₃C–C(=O)–OH

   E. $CH_3–CH_2–O–CH_3$

23. Which of the following has the highest boiling point?

   A. 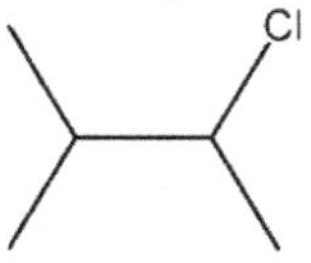 3-chloro-2-methylbutane structure

   B. 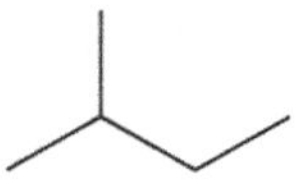 isobutane structure

   C. 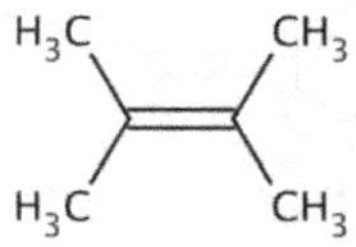 2,3-dimethyl-2-butene structure

   D. 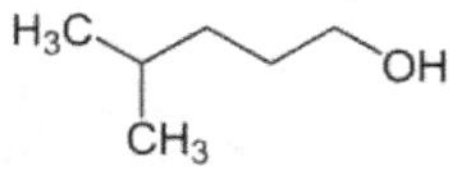 4-methyl-1-pentanol structure (H₃C–CH(CH₃)–CH₂–CH₂–OH)

   E. 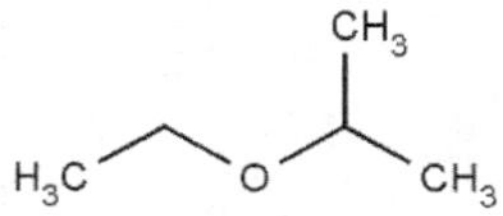 ethyl isopropyl ether structure

24. What compound is formed by the oxidation of 2-hexanol?

   A. hexanal
   B. hexanoic acid
   C. 2-hexanone
   D. 2-hexene
   E. hexyne

25. Which of the following would have the highest boiling point?

   A. 1-hexyne
   B. 1-hexene
   C. hexane
   D. 1-hexanol
   E. 1-pentanol

26. The functional group C=O is in all the species below, EXCEPT:

   A. amides
   B. ethers
   C. aldehydes
   D. ketones
   E. esters

27. Which of the following is an allylic alcohol?

A. $CH_3CH{=}CHCH_2OH$

B. $HOCH{=}CHCH_2CH_3$

C. $CH_2{=}CHCH_2CH_3$

D. $CH_2{=}CHCH_2OCH_3$

E. $CH_2{=}CHCH_2CH_2OH$

28. Which of the following molecules has the most acidic proton?

A. 2-pentanol

B. 3-pentyne

C. 2-pentene

D. pentane

E. *tert*-butanol

29. What is the major product of this reaction?

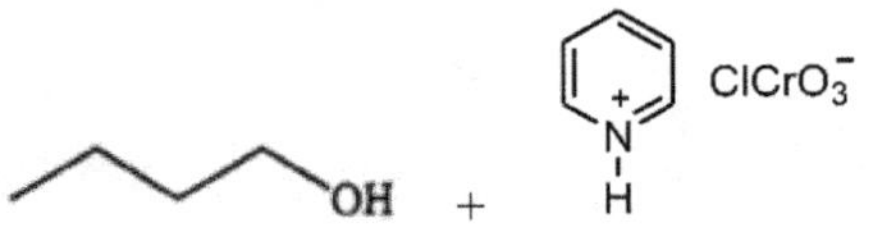

→ ?

A. 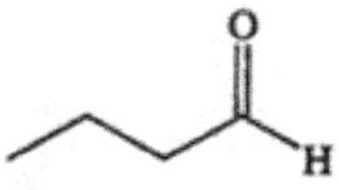

B. 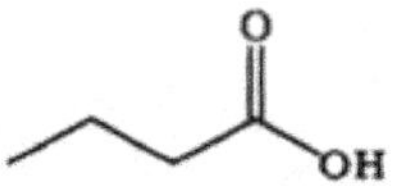

C.

D. 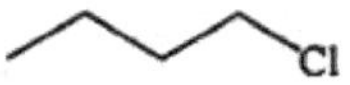

E. 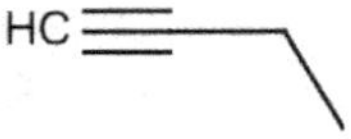

30. Which of the following alcohols dehydrates with the fastest rate?

A. 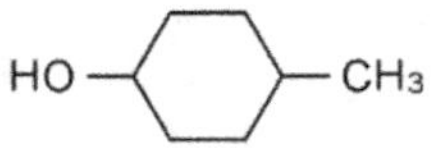

B. 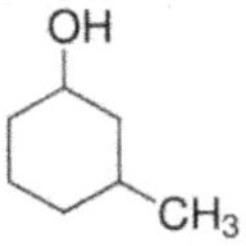

C. 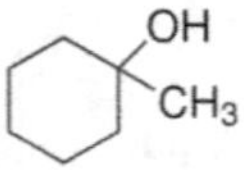

D. 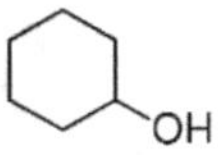

E. 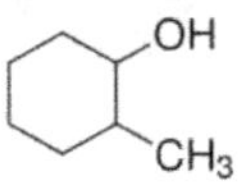

31. When (R)-2-heptanol is subjected to a two-step mechanism of tosyl chloride followed by Cl⁻, the product is:

    A. (R)-2-chloroheptane
    B. heptene
    C. (R/S)-2-chloroheptane
    D. (S)-2-chloroheptane
    E. (S)-3-chloroheptane

32. Phenol exists predominantly:

    A. in the keto form because its keto tautomer is antiaromatic
    B. in the keto form because its keto tautomer is nonaromatic
    C. in the enol form because its keto tautomer is antiaromatic
    D. in the enol form because its keto tautomer is nonaromatic
    E. in the keto form because its keto tautomer is aromatic

33. Treatment of salicylic acid with methanol and nonaqueous acid yields an:

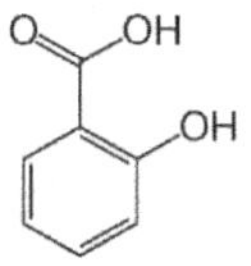

    Salicylic acid + CH_3OH + H_2SO_4 → ?

    A. acetal
    B. ether
    C. ester
    D. hemiacetal
    E. amide

34. When (R)-2-hexanol is subjected to PBr_3, the compound it produces is:

    A. (R)-2-bromohexane
    B. (S)-2-bromohexane
    C. (R/S)-2-bromohexane
    D. (S)-2-bromopentane
    E. ketone

35. Which of the following reactions yields an ester?

    A. $C_6H_5OH + CH_3CH_2Br$
    B. $CH_3COOH + C_2H_5OH + H_2SO_4$
    C. $CH_3COOH + SOCl_2$
    D. $2CH_3OH + H_2SO_4$
    E. $CH_3CH_2Br + CH_3CH_2O^-Na^+$

36. What is the major product of the reaction of 2,2-dimethylcyclohexanol with HBr?

A. 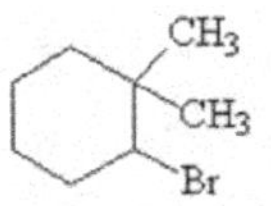 [cyclohexane with CH₃, CH₃, Br]

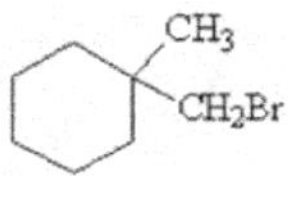

B. 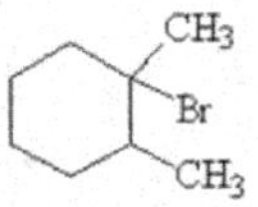 [cyclohexane with CH₃, Br, CH₃]

C. [cyclohexane with CH₃, CH₂Br]

D. 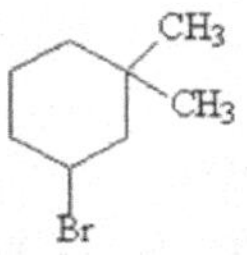 [cyclohexane with CH₃, CH₃ and Br on separate carbon]

E. 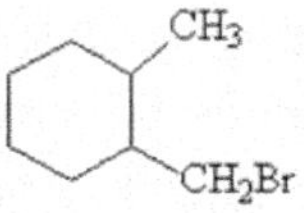 [cyclohexane with CH₃, CH₂Br]

Notes for active learning

Notes for active learning

Aldehydes and Ketones

1. Oxidation of a ketone produces:

    A. a secondary alcohol
    B. an aldehyde
    C. a carboxylic acid
    D. a primary alcohol
    E. no reaction

2. (S)-2-methylbutanal [ ] upon sitting in an acidic or basic aqueous solution.

    A. racemizes
    B. esterifies
    C. inverts completely to the R configuration
    D. hydrolyzes
    E. irreversibly forms the hydrate

3. What is an ester reduced to with diisobutylaluminum hydride (DIBAL)?

    A. 1° alcohol
    B. alkane
    C. ketone
    D. aldehyde
    E. acetal

4. Which of the following reagents quantitatively converts an enolizable ketone to its enolate salt?

    A. lithium hydroxide
    B. lithium diisopropylamide
    C. methyllithium
    D. diethylamine
    E. pyridine

5. Which type of compound is shown below?

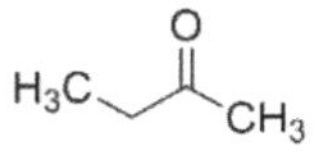

    $H_3C-C(=O)-CH_3$

    A. ester
    B. carboxylic acid
    C. ketone
    D. aldehyde
    E. amine

6. Which of the following carbonyl compounds may be synthesized from 1,3-dithiane?

    I. methyl vinyl ketone   III. 3,3-dimethyl-2-butanone
    II. 2-pentanone   IV. 2-phenylethanal

- A. I and IV only
- B. II only
- C. II and III only
- D. II and IV only
- E. I, II and IV only

7. Which of the following organic compounds is most likely NOT a ketone?

- A. estradiol (female hormone)
- B. progesterone (female hormone)
- C. androsterone (male hormone)
- D. cortisone (adrenal hormone)
- E. testosterone (male hormone)

8. The reagent(s) that convert(s) a carbonyl group of a ketone into a methylene group is:

- A. Na, NH_3, CH_3CH_2OH
- B. $LiAlH_4$
- C. $NaBH_4$, CH_3CH_2OH
- D. Zn(Hg), conc. HCl
- E. $LiAlH[OC(CH_3)_3]_3$

9. The reaction of ethylmagnesium bromide with which of the following compounds yields secondary alcohol after quenching with aqueous acid?

- A. $(CH_3)_2CO$
- B. ethylene oxide
- C. H_2CO
- D. CH_3CHO
- E. n-butyllithium

10. Which observation denotes a positive Benedict's test?

- A. A mirror-like deposit forms from a colorless solution
- B. A purple solution yields a brown precipitate
- C. A red precipitate forms from a blue solution
- D. A red-brown solution becomes clear and colorless
- E. A pale-yellow solution with an odor of chlorine changes to a purple color

11. A ylide is a molecule that can be described as a:

    A. carbanion bound to a negatively charged heteroatom
    B. carbocation bound to a positively charged heteroatom
    C. carbocation bound to a carbon radical
    D. carbocation bound to a diazonium ion
    E. carbanion bound to a positively charged heteroatom

12. Which of the following represents the correct ranking in terms of increasing the boiling point?

    A. *n*-butane < 1-butanol < diethyl ether < 2-butanone
    B. *n*-butane < 2-butanone < diethyl ether < 1-butanol
    C. 2-butanone < *n*-butane < diethyl ether < 1-butanol
    D. *n*-butane < diethyl ether < 1-butanol < 2-butanone
    E. *n*-butane < diethyl ether < 2-butanone < 1-butanol

13. Which of the following reactions does NOT yield a ketone product?

    A. 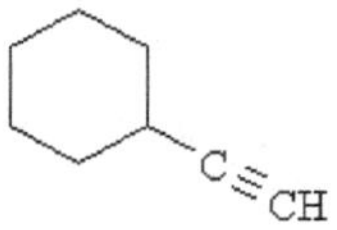 cyclohexyl-C≡CH + 1) Sia$_2$BH / THF; 2) H$_2$O$_2$ / $^-$OH →

    B. 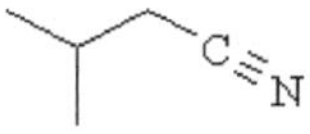 isobutyl-C≡N + 1) CH$_3$CH$_2$MgBr; 2) H$_3$O$^+$ →

    C. 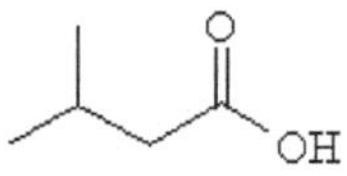 isobutyl-C(=O)OH + 1) 2 CH$_3$CH$_2$Li; 2) H$_3$O$^+$ →

    D. 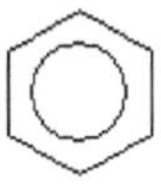 benzene + H$_3$C-C(=O)-Cl + AlCl$_3$ →

    E. 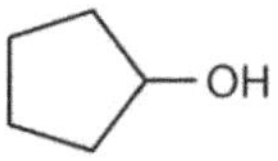 cyclopentyl-OH + PCC →

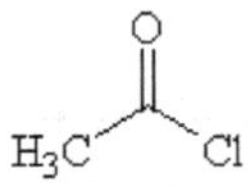

14. What reagents are needed to complete the following synthesis?

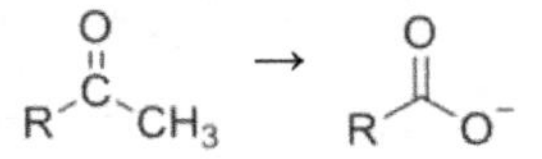

R-CO-CH₃ → R-CO-O⁻

A. 1) NaOH / heat; 2) HCl (*aq*)
B. 1) NaOH / Br₂
C. 1) warm conc. KMnO₄ / NaOH; 2) HCl(*aq*)
D. 1) Ag(NH₃)₂OH; 2) HCl (*aq*)
E. None of the above

15. A compound with an ~OH group and an ether-like ~OR group bonded to the same carbon atom is:

A. hemiacetal
B. diol
C. aldol
D. acetal
E. ether

16. Which compound gives a positive indicator with the Tollens' reagent?

A. 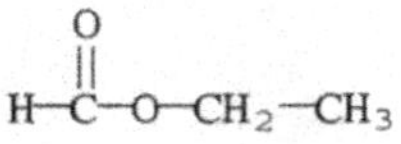 H−CO−O−CH₂−CH₃

B. 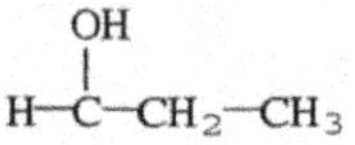 H−C(OH)−CH₂−CH₃

C. CH₃−CO−CH₂−CH₃

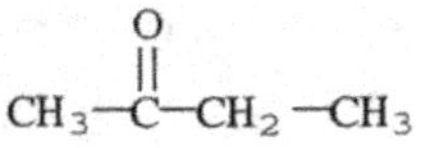

D. 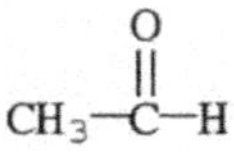 CH₃−CO−H

E. 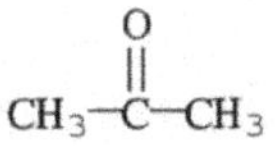 CH₃−CO−CH₃

17. Reduction of aldehydes and ketones is a [ ] reaction involving the [ ] ion(s).

A. two-step; H⁻ and H⁺
B. two-step; OH⁻ and H⁺
C. one-step; H⁻
D. one-step; H⁺
E. two-step; H⁻ and OH⁻

18. Which of the following is the best Michael acceptor?

A.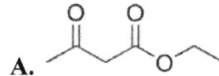

B.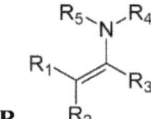

C. 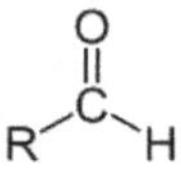 R–CHO

D.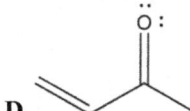

E. None of the above

19. Which of the following pairs have the most similar chemical properties?

   A. alkanes and carboxylic acids
   B. alkenes and aromatics
   C. amines and esters
   D. ketones and aldehydes
   E. ethers and alcohols

20. Which compound gives a positive Tollens test?

   A. pentane
   B. pentanoic acid
   C. 3-pentanone
   D. 2-pentanone
   E. pentanal

21. Consider the equilibrium of each of the carbonyl compounds with HCN to produce cyanohydrins. Which is the correct ranking of compounds in order of increasing K_{eq} for this equilibrium?

   A. H_2CO < cyclohexanone < CH_3CHO < 2-methylcyclohexanone
   B. CH_3CHO < 2-methylcyclohexanone < cyclohexanone < H_2CO
   C. cyclohexanone < 2-methylcyclohexanone < H_2CO < CH_3CHO
   D. cyclohexanone < 2-methylcyclohexanone < CH_3CHO < H_2CO
   E. 2-methylcyclohexanone < cyclohexanone < CH_3CHO < H_2CO

22. Which of the following functional groups represents a ketone?

A. 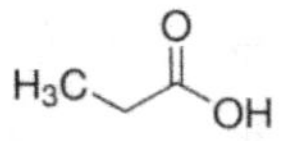 H₃C–C(=O)–OH

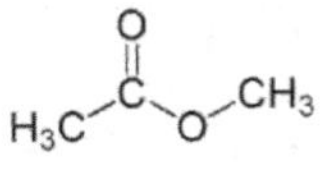

B. H₃C–C(=O)–O–CH₃

C. H₃C–C(=O)–H

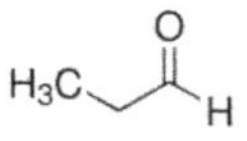

D. 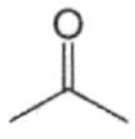 (CH₃)₂C=O

E. 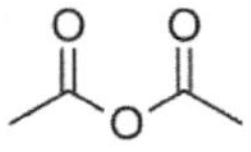 CH₃–C(=O)–O–C(=O)–CH₃

23. The following statements about oxidation of carbonyls are true, EXCEPT:

A. oxidation of aldehydes produces carboxylic acids
B. ketones do not react with mild oxidizing agents
C. Tollens' test involves the oxidation of Ag^+
D. Benedict's test involves the reduction of Cu^{2+}
E. all the statements are true

24. Which of the following classes of compounds has a carbonyl group?

A. phenol
B. ether
C. amine
D. alcohol
E. none of the above

25. The statements concerning the carbonyl group in aldehydes and ketones are true, EXCEPT:

A. in condensed form, the aldehyde group can be written as ~CHO
B. the carbonyl group is planar
C. the bond angles about the central carbon atom are 120°
D. the bond is polar with a slight negative charge on the oxygen atom
E. since the bond is polar, carbonyl groups readily form hydrogen bonds with each other

26. Which of the following compound is the most soluble in water?

A. acetone
B. cyclohexanone
C. 2-butanone
D. 3-butanone
E. benzophenone

27. Treatment of a nitrile with a Grignard reagent, followed by hydrolysis, results in:

    A. ester
    B. ketone
    C. aldehyde
    D. ether
    E. alcohol

28. Oxidation of an aldehyde produces a:

    A. tertiary alcohol
    B. secondary alcohol
    C. primary alcohol
    D. carboxylic acid
    E. ketone

29. Which of the following will alkylate a lithium enolate most rapidly?

    A. methyl bromide
    B. isopropyl bromide
    C. neopentyl bromide
    D. bromobenzene
    E. 2-methylbromobenzene

30. In Benedict's test:

    I. an aldehyde is oxidized
    II. a red/brown precipitate is formed
    III. copper (II) ion is reduced

    A. I only
    B. II only
    C. III only
    D. I and II only
    E. I, II and III

31. Of the following, which is the best solvent for an aldol addition?

    A. acetone
    B. methyl acetate
    C. propanol
    D. dimethyl ether
    E. methanol

32. The following statements about oxidation of carbonyls are true, EXCEPT:

    A. Benedict's test involves the reduction of Cu^{2+}
    B. Tollens' test involves the reduction of Ag^+
    C. oxidation of primary alcohols produces secondary alcohols
    D. oxidation of secondary alcohols produces ketones
    E. oxidation of aldehydes produces carboxylic acids

33. The reaction of ethylmagnesium bromide with which of the following compounds yields tertiary alcohol after quenching with aqueous acid?

   A. ethylene oxide
   B. (CH$_3$)$_2$CO
   C. CH$_3$CHO
   D. H$_2$CO
   E. *n*-butyllithium

34. Which of the following are enol forms of 2-butanone (CH$_3$COCH$_2$CH$_3$)?

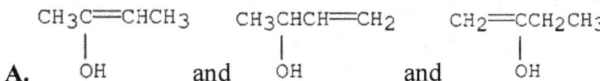

   A. CH$_3$C=CHCH$_3$ with OH, and CH$_3$CHCH=CH$_2$ with OH, and CH$_2$=CCH$_2$CH$_3$ with OH

   B. CH$_3$CHCH$_2$CH$_3$ with OH, and CH$_3$C=CHCH$_3$ with OH

   C. CH$_3$C=CHCH$_3$ with OH, and CH$_3$CHCH=CH$_2$ with OH

   D. CH$_3$CHCH=CH$_2$ with OH, and CH$_2$=CCH$_2$CH$_3$ with OH

   E. CH$_3$C=CHCH$_3$ with OH, and CH$_2$=CCH$_2$CH$_3$ with OH

35. Which statement is true regarding the major differences between aldehydes and ketones compared to other carbonyl compounds?

   A. The carbonyl group carbon atom in aldehydes and ketones is bonded to atoms that do not attract electrons strongly
   B. The polar carbon-oxygen bond in aldehydes and ketones is less reactive than the hydrocarbon portion of the molecule
   C. The carbonyl carbon in aldehydes and ketones has bond angles of 120°, unlike the comparable bond angles in other carbonyl compounds
   D. The molar masses in aldehydes and ketones are much less than in the other types of compounds
   E. The carbonyl carbon in aldehydes and ketones has bond angles of 109.5°, similar to the comparable bond angles in other carbonyl compounds

36. Which series of reactions best facilitates the following conversion?

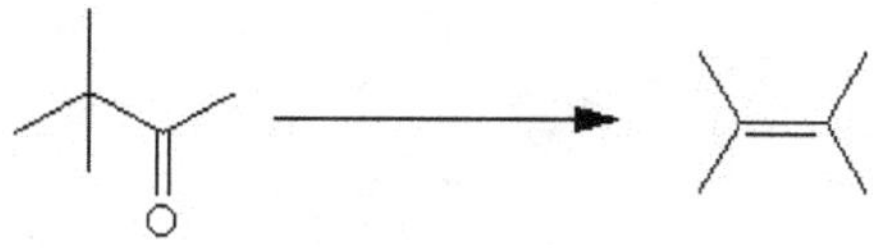

A. 1) KMnO$_4$ (*aq*); 2) Hg(OAc)$_2$ (*aq*); 3) NaBH$_4$/$^-$OH
B. 1) NaBH$_4$; 2) H$_3$PO$_4$/Δ
C. 1) H$_3$C-MgBr; 2) H$_2$O/H$_3$O$^+$
D. 1) NaBH$_4$; 2) HBr (*g*); 3) Mg/ether; 4) H$_2$O/H$_3$O$^+$
E. 1) Raney nickel; 2) H$_3$C-MgBr; 3) H$_2$O/H$_3$O$^+$

Notes for active learning

Carboxylic Acids

1. Which of the following functional groups does this organic compound contain?

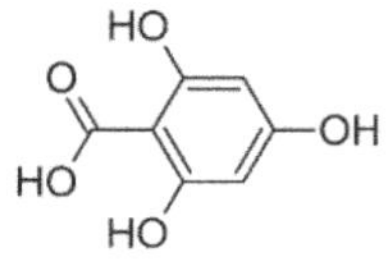

   A. amide
   B. amine
   C. carboxylic acid
   D. ester
   E. acyl halide

2. What is the major organic product of the reaction shown?

   $CH_3(CH_2)_3COOH + CH_3C(OH)HCH_3 \rightarrow$?

   A. $CH_3CH_2CH_2COOCH(CH_3)_2$
   B. $CH_3CH_2CH_2CH_2C(OH)OCH(CH_3)_2$
   C. $CH_3CH_2CH_2CH_2COOCHCH_2CH_3$
   D. $CH_3CH_2CH_2CH_2COOCH(CH_3)_2$
   E. $CH_3CH_2CH_2CH_2COCH(CH_3)_2$

3. What products are formed upon the reaction of benzoic acid with sodium hydroxide, NaOH?

   A. Sodium bicarbonate and sodium benzoate
   B. Sodium bicarbonate and benzaldehyde
   C. Sodium benzoate and water
   D. Benzaldehyde and water
   E. Toluene and water

4. Which type of compound is shown below?

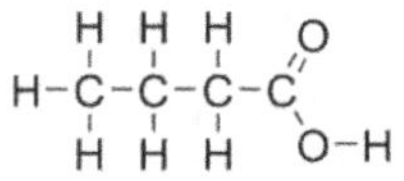

   A. ester
   B. carboxylic acid
   C. ketone
   D. aldehyde
   E. amide

5. Which carboxylic acid is used to prepare the ester shown?

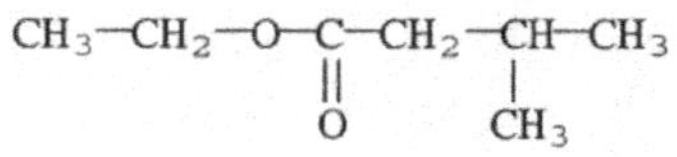

   $CH_3-CH_2-O-C-CH_2-CH-CH_3$
   with $=O$ on C and CH_3 branch

   A. $(CH_3)_2CHCH_2COOH$
   B. $(CH_3)_2CHCOOH$
   C. CH_3COOH
   D. CH_3CH_2COOH
   E. $CH_3(CH_2)_3COOH$

6. When an amine reacts with a carboxylic acid at high temperature, the major product is a(n):

A. ether
B. thiol
C. amide
D. ester
E. alcohol

7. Which acid is expected to have the highest boiling point?

A. formic
B. oxalic
C. acetic
D. benzoic
E. stearic

8. The reaction of a carboxylic acid with a base-like sodium hydroxide, NaOH, gives:

A. alcohol
B. ester
C. alkoxide salt
D. carboxylate salt
E. none of the above

9. What are the products from the reaction of ethanoic acid and methanol with sulfuric acid catalyst?

$$CH_3COOH + CH_3OH + H_2SO_4 \rightarrow \;?$$

A. $CH_3CH_2COOH + H_2O$
B. $CH_3COOCH_3 + H_2O$
C. $CH_3COCH_3 + H_2O$
D. $CH_3CH_2CHO + H_2O$
E. No reaction

10. Which fatty acid is expected to have the highest boiling point?

A. oxalic, $(CO_2H)_2$
B. formic, HCO_2H
C. benzoic, $C_6H_5CO_2H$
D. acetic, CH_3CO_2H
E. stearic, $CH_3(CH_2)_{16}CO_2H$

11. Which formula correctly illustrates the form taken by the acetic acid in a basic solution?

A. 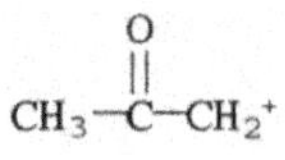 $CH_3-\overset{\overset{O}{\|}}{C}-CH_2{}^+$

B. $H_3C-\overset{\overset{:\ddot{O}:}{\|}}{C}-\ddot{\underset{..}{O}}:^-$

C. (structure with C=O and OH on ring/chain)

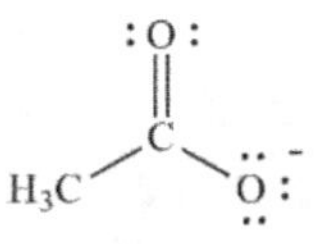

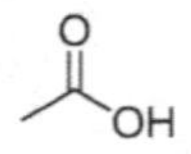

D. 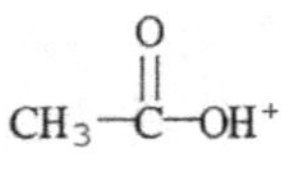 $CH_3-\overset{\overset{O}{\|}}{C}-OH^+$

E. 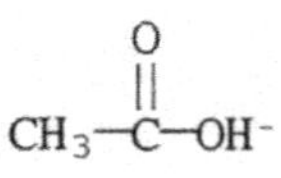 $CH_3-\overset{\overset{O}{\|}}{C}-OH^-$

12. When alcohol reacts with a carboxylic acid, the major product is a(n):

A. salt
B. ester
C. amine
D. amide
E. soap

13. Identify the carboxylic acid and alcohol from which the following ester was made.

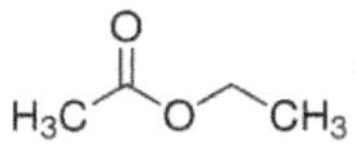

$$H_3C-C(=O)-O-CH_3$$

A. $CH_3CH_2CO_2H$ and CH_3CH_2OH
B. CH_3CO_2H and CH_3CO_2H
C. CH_3CO_2H and CH_3CH_2OH
D. $CH_3CH_2CO_2H$ and $CH_3CH_2CH_2OH$
E. None of the above

14. Which of the following has the highest boiling point?

A. ethyl alcohol, CH_3CH_2OH
B. acetic acid, CH_3COOH
C. ethane, CH_3CH_3
D. dimethyl ketone, CH_3COCH_3
E. formaldehyde, $HCHO$

15. The ion formed from a carboxylic acid is called:

A. ester cation
B. ester anion
C. carboxylate cation
D. carboxylate anion
E. amide cation

16. Which functional groups below contain(s) a hydroxyl group as a part of its structure?

    I. anhydride   II. carboxylic acid   III. ester

A. I only
B. II only
C. III only
D. I and II only
E. I, II and III

17. Which acid is expected to have the lowest boiling point?

A. formic, HCO_2H
B. oxalic, $(CO_2H)_2$
C. acetic, CH_3CO_2H
D. benzoic, $C_6H_5CO_2H$
E. stearic, $CH_3(CH_2)_{16}CO_2H$

18. Which of the following molecules is acidic?

I. (phenol, Ph–OH) II. (benzaldehyde, Ph–CHO) III. (benzoic acid, Ph–COOH)

- **A.** I only
- **B.** II only
- **C.** I and II only
- **D.** I and III only
- **E.** I, II and III

19. Which of the following compounds is the strongest acid?

- **A.** HOOCCH$_2$F
- **B.** HOOCCH$_3$
- **C.** HOOCCH$_2$Br
- **D.** HOOCCH$_2$OCH$_3$
- **E.** PhOH

20. Explain why caprylic acid CH$_3$(CH$_2$)$_6$COOH dissolves in a 5% aqueous solution of sodium hydroxide, but caprylaldehyde, CH$_3$(CH$_2$)$_6$CHO, does not.

- **A.** Caprylic acid reacts to form the water-soluble salt
- **B.** Caprylaldehyde behaves as a reducing agent, which neutralizes the sodium hydroxide
- **C.** Caprylaldehyde can form more hydrogen bonds to water than caprylic acid
- **D.** With two oxygens, caprylic acid is about twice as polar as caprylaldehyde
- **E.** Caprylaldehyde is a gas at room temperature

21. The statements about carboxylic acids are true, EXCEPT:

- **A.** they react with bases to form salts which are often more soluble than the original acid
- **B.** they form hydrogen bonds, with higher-than-expected boiling points based on molecular weight
- **C.** at low molecular weights, they are liquids with pungent odors
- **D.** they undergo substitution reactions involving the ⁻OH group
- **E.** when they behave as acids, the ⁻OH group is lost, leaving the CO⁻ ion

22. When a small amount of hexanoic acid [CH$_3$(CH$_2$)$_4$CO$_2$H, p$K_a \approx$ 4.8] is added to a separatory funnel which contains the organic solvent diethyl ether and water with a pH of 11.0, it is mainly in the [ ] phase as [ ]:

- **A.** water; CH$_3$(CH$_2$)$_4$CO$_2$H
- **B.** ether; CH$_3$(CH$_2$)$_4$CO$_2$H
- **C.** water; CH$_3$(CH$_2$)$_4$CO$_2^-$
- **D.** ether; CH$_3$(CH$_2$)$_4$CO$_2^-$
- **E.** none of the above

23. Which of the following compounds acts as an acid?

    A. CH_3COCH_3
    B. $(CH_3)_2NH$
    C. C_2H_5OH
    D. C_2H_5COOH
    E. $NaNH_2$

24. The statements concerning citric acid are true, EXCEPT it:

    A. is produced only by plants
    B. is used in many consumer products
    C. is extremely soluble in water
    D. contains three carboxylic acid groups because its carbon skeleton is branched
    E. is a weak acid

25. Which of the following molecules would be expected to be the most soluble in water?

    A. $CH_3(CH_2)_6CO_2H$
    B. $CH_3(CH_2)_{12}CO_2H$
    C. CH_3CO_2H
    D. $CH_3CH_2CH_2CO_2H$
    E. C_6H_{12}

26. Which of the following molecules is the most polar?

    A. butane
    B. butanoic acid
    C. cyclohexane
    D. ethanol
    E. 1-chlorobutane

27. The most common reactions of carboxylic acid or its derivative involve:

    A. replacement of the group bonded to the carbonyl atom
    B. oxidation of the R group
    C. addition across the double bond between carbon and oxygen
    D. replacement of the oxygen atom in the carbonyl group
    E. reduction of the R group

28. Which of the following reactions will NOT result in the formation of a carboxylic acid?

    A. Oxidation of a secondary alcohol
    B. Oxidation of a primary alcohol
    C. Hydrolysis of nitriles
    D. Grignard reagents reacting with CO_2
    E. Oxidation of an aldehyde

29. Which acid would be expected to have the lowest boiling point?

A. oxalic
B. formic
C. benzoic
D. acetic
E. oleic

30. The reaction of butanoic acid with ethanol produces:

A. butyl ethanamide
B. ethyl butanamide
C. butyl ethanoate
D. ethyl butanoate
E. butyl ethyl ester

31. Which of the following are the preferred reagents used in the following synthesis?

A. 1) NaBH$_4$/THF; 2) H$_3$O$^+$
B. 1) Mg/ether; 2) dry CO$_2$; 3) H$_3$O$^+$
C. 1) LiAlH$_4$/THF; 2) H$_3$O$^+$
D. 1) Hot KMnO$_4$; 2) H$_3$O$^+$
E. 1) LiAlH(OCH$_3$)$_3$/THF; 2) H$_3$O$^+$

32. Which of the following molecules is the strongest acid?

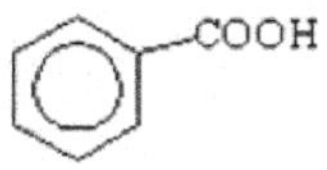

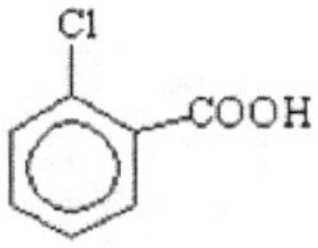

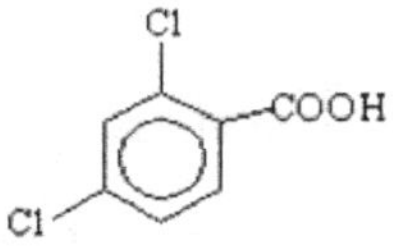

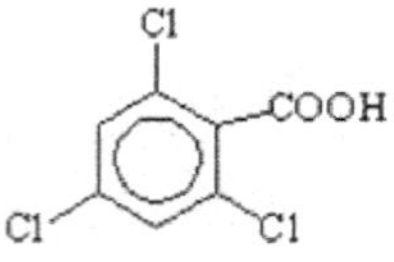

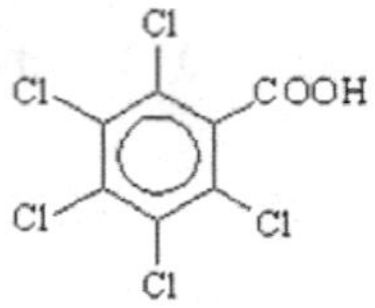

33. The water solubility of compounds containing the carboxylic acid group can be increased by reaction with:

A. water
B. sodium hydroxide
C. nitric acid
D. sulfuric acid
E. benzoic acid

34. Which of the following acids is likely to have the weakest conjugate base?

    A. $CH_3Cl_2CCO_2H$
    B. $CH_3CH_2CH_2CO_2H$
    C. $(CH_3CH_2)_3CCO_2H$
    D. $CH_3HNCH_2CH_2CH_2CO_2H$
    E. $CH_3Cl_2CCH_2CO_2H$

35. The boiling point of acetic acid is 119 °C, and methyl acetate is 57 °C. What is the reason that acetic acid boils at a much higher temperature than methyl acetate?

    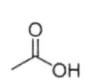
    acetic acid 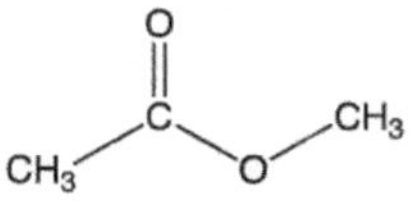   methyl acetate

    A. molecular mass
    B. presence of an ester linkage
    C. hydrophobic interactions
    D. hydrogen bonding
    E. London forces

36. The pK_a of acetic acid (CH_3COOH) is 4.8. If the pH of an aqueous solution of CH_3COOH and CH_3COO^- is 4.8, then:

    A. $[CH_3COOH] = [CH_3COO^-]$
    B. $[CH_3COOH] < [CH_3COO^-]$
    C. CH_3COOH is completely ionized
    D. $[CH_3COOH] > [CH_3COO^-]$
    E. CH_3COOH is completely nonionized

Notes for active learning

COOH Derivatives

1. The compound below is which type of compound?

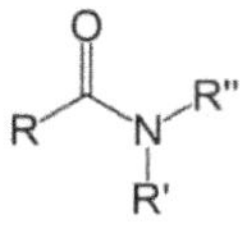

   A. amide
   B. amino acid
   C. amine
   D. aldehyde
   E. ester

2. The products of acid hydrolysis of an ester are:

   A. alcohol + water
   B. acid + water
   C. another ester + water
   D. alcohol + acid
   E. salt + water

3. Which of the following reactions is favorable, in the direction indicated, under common laboratory conditions?

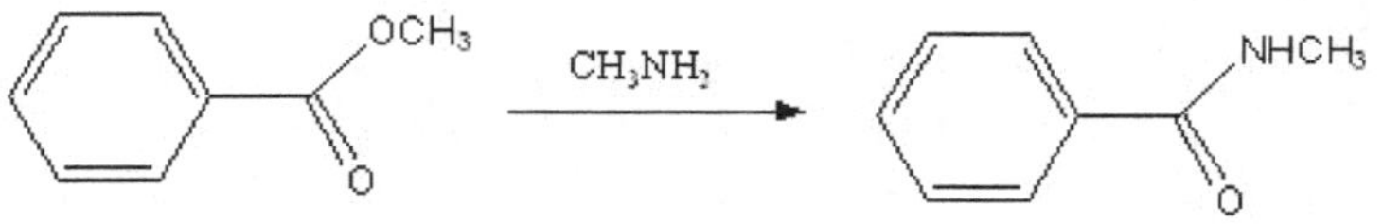

   A. methyl benzoate + CH$_3$NH$_2$ → N-methylbenzamide

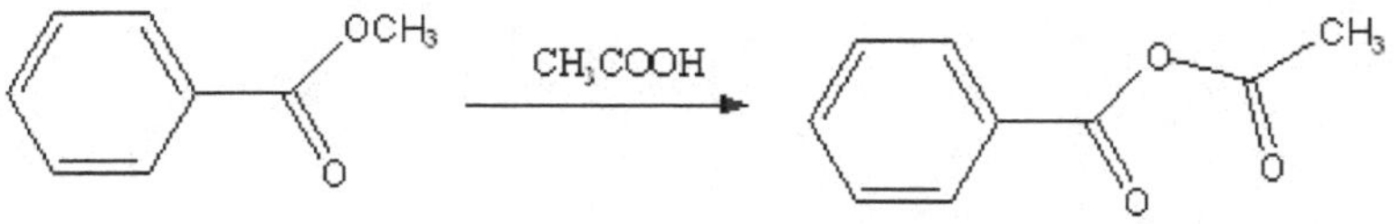

   B. methyl benzoate + CH$_3$COOH → mixed anhydride

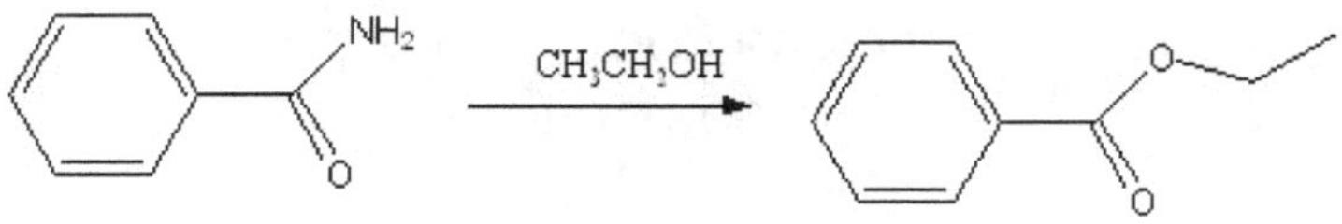

   C. benzamide + CH$_3$CH$_2$OH → ethyl benzoate

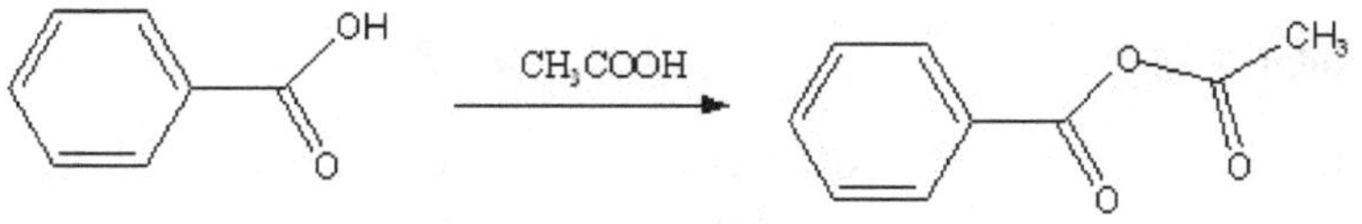

   D. benzoic acid + CH$_3$COOH → mixed anhydride

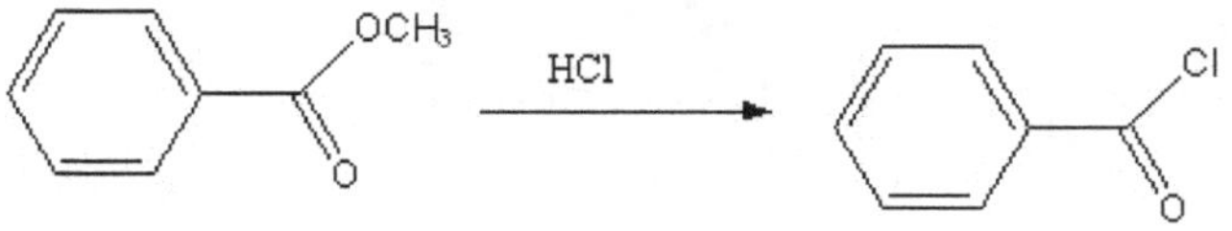

   E. methyl benzoate + HCl → benzoyl chloride

4. Hydrolysis of the ester ethyl acetate produces:

A. butanal and ethanol
B. ethanal and acetic acid
C. ethanol and acetic acid
D. butanoic acid
E. butanol

5. What class of compound has the following general structure?

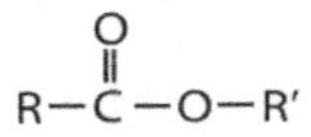

$$R-\overset{\overset{O}{\|}}{C}-O-R'$$

A. ester
B. ketone
C. aldehyde
D. anhydride
E. carboxylic acid

6. What functional group is NOT present in the following structure for thyroxine (i.e., thyroid hormone)?

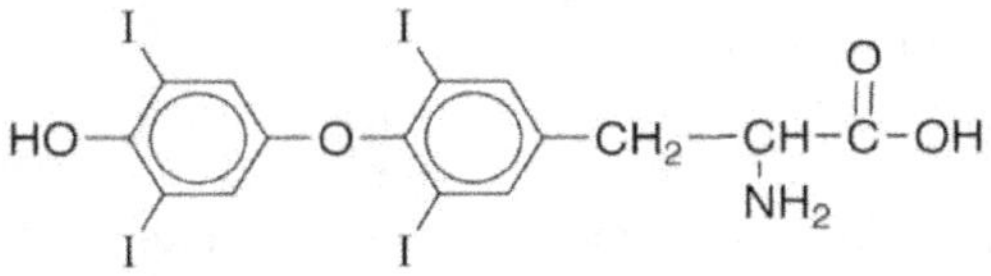

A. organic halide
B. ether
C. carboxylic acid
D. anhydride
E. phenol

7. What are the major organic products of the reaction shown?

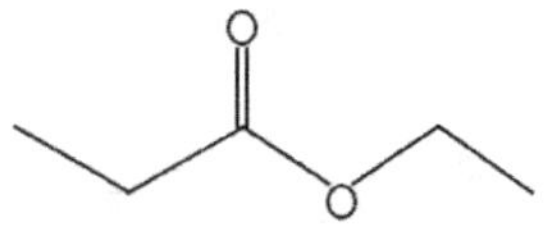

+ $H_2O / H_2SO_4 \rightarrow$?

A. $CH_3COOH + HOCH_2CH_2CH_3$
B. $CH_3CH_2CH_2OH + CH_3CH_2OH$
C. $CH_3CH_2COOH + HOCH_2CH_3$
D. $CH_3CH_2CH_2COO^- + {}^+H_2OCH_2CH_3$
E. $CH_3CH_2COOH + CH_3COOH$

8. What are the products of this reaction?

$CH_3CH_2CONHCH_2CH_3 + NaOH \rightarrow$?

A. $CH_3CH_2COO^-Na^+ + CH_3CH_2NH_2$
B. $CH_3CH_2CH_2OH + CH_3CH_2NH_3^-Na^+$
C. $CH_3CH_2COO^-Na^+ + CH_3CH_2NH_3^+Cl^-$
D. $CH_3CH_2COOH + CH_3CH_2NH_2$
E. None of the above

9. Which product is formed during a reaction of acetic acid (CH_3COOH) with methylamine (CH_3NH_2)?

   A. ethylammonium hydroxide
   B. methanol
   C. methylacetamine
   D. methylacetamide
   E. methylammonium acetate

10. Esters and amides are most easily made by nucleophilic acyl substitution reactions on:

    A. alcohols
    B. acid anhydrides
    C. carboxylates
    D. carboxylic acids
    E. acid chlorides

11. The reaction of an amine and a carboxylic acid produces what kind of compound?

    A. amine
    B. amide
    C. anhydride
    D. ketone
    E. ester

12. Penicillin can be taken orally if the compound can survive degradation by the low pH of the stomach. The amide bond between the *R* group and 6-APA of dicloxacillin is stable at low pH because:

    A. high hydroxide concentration promotes amide stability
    B. amide bonds do not easily undergo hydrolysis
    C. hydrogen bonding has stabilization effects
    D. aromatic compounds are unaffected by changes in pH
    E. none of the above

13. Identify the functional group:

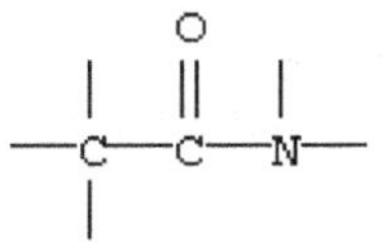

    $$-\overset{|}{\underset{|}{C}}-\overset{O}{\overset{\|}{C}}-\overset{|}{N}-$$

    A. anhydride
    B. amine
    C. amide
    D. ester
    E. nitrile

14. Which of the following is a product of this reaction?

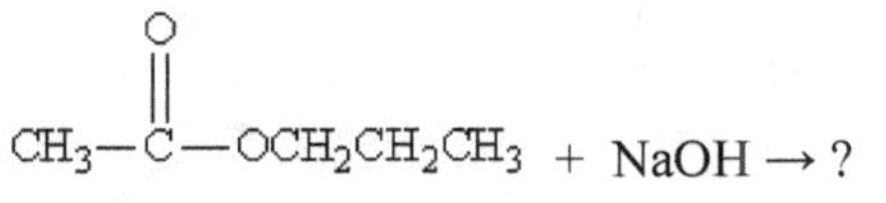

    $$CH_3-\overset{O}{\overset{\|}{C}}-OCH_2CH_2CH_3 + NaOH \rightarrow ?$$

    A. CH_3COOH
    B. $CH_3CH_2COO^-\ Na^+$
    C. $CH_3CH_2–OH$
    D. $CH_3COO^-\ Na^+$
    E. none of the above

15. Which sequence correctly ranks each carbonyl group in order of increasing reactivity toward nucleophilic addition?

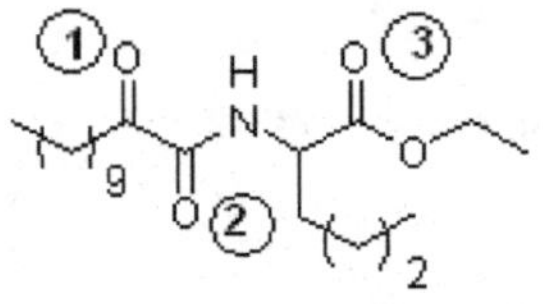

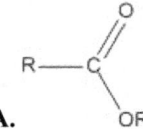

A. 1 < 2 < 3
B. 2 < 3 < 1
C. 3 < 1 < 2
D. 1 < 3 < 2
E. 3 < 2 < 1

16. Which functional group(s) below indicate(s) two atoms connected by a triple bond?

    I. ester   II. nitrile   III. alkyne

A. I only
B. II only
C. III only
D. II and III only
E. I, II and III

17. Which of these molecules is an ester?

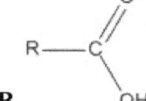

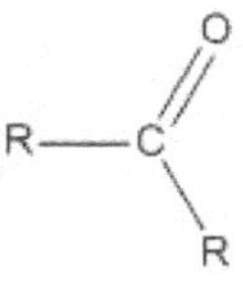

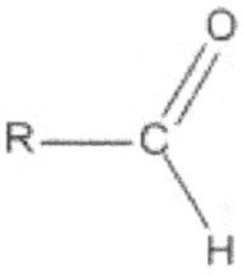

E. None of the above

18. The products of basic hydrolysis of an ester are:

A. acid + water
B. alcohol + water
C. alcohol + acid
D. another ester + water
E. carboxylate salt + alcohol

19. Which of the following products might be formed if benzoyl chloride was treated with excess CH_3CH_2MgBr?

A. $C_6H_5CH_2CHO$
B. $C_6H_5C(CH_2CH_3)OHCH_2CH_3$
C. C_6H_5COOH
D. $C_6H_5COOCH_2CH_3$
E. none of the above

20. The compound below has which functional groups?

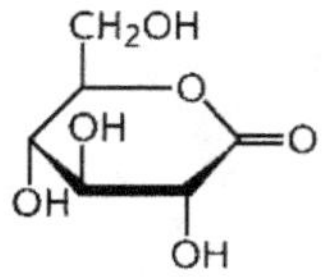

    A. lactone and alcohol
    B. ether and alcohol
    C. ester and hemiacetal
    D. ester and acetal
    E. ketone and aldehyde

21. Benzoyl chloride, PhC(O)Cl, reacts with water to form benzoic acid, $PhCO_2H$. In this addition-elimination:

    A. water acts as a nucleophile, benzoyl chloride acts as an electrophile, and chloride acts as a leaving group
    B. water acts as an electrophile, benzoyl chloride acts as a nucleophile, and chloride acts as a leaving group
    C. water and benzoyl chloride function as electrophiles
    D. there are no nucleophiles or electrophiles
    E. water acts as the electrophile, and benzoyl chloride acts as a nucleophile

22. The reaction of benzoic acid with thionyl chloride followed, by treatment with ammonia, yields which of the following compounds?

    A. *p*-chlorobenzamide
    B. *m*-chlorobenzamide
    C. benzamide
    D. *p*-aminobenzaldehyde
    E. 3-chloro-4-aminobenzaldehyde

23. The compound capsaicin below has which functional groups?

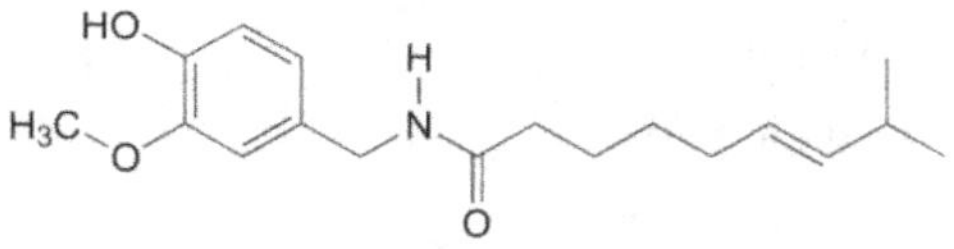

    A. aldehyde and amine
    B. carboxylic acid and amine
    C. nitro and amine
    D. amide and ether
    E. ester and amine

24. What is the name of the product formed by the reaction of propanoic acid with ethanol?

    A. ethyl propanoate
    B. pentanal
    C. pentyl ester
    D. ethylpropylketone
    E. pentanone

25. Which of the following type of bond is depicted below?

A. amide bond
B. glycosidic bond
C. ester bond
D. ether bond
E. hydrogen bond

26. The compound below has which functional groups?

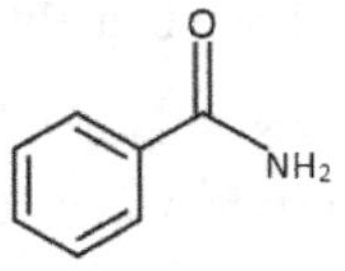

A. aromatic and alcohol
B. aromatic and amine
C. aromatic and amide
D. aromatic and carboxylic acid
E. aromatic and aldehyde

27. What are the products of an acid-catalyzed hydrolysis reaction of amides and water?

A. an alcohol and an alkane
B. an amine and a ketone
C. a carboxylic acid and an amine salt
D. an ester and an ether
E. a carboxylic acid and an alcohol

28. Which of the following compounds is the most susceptible to nucleophilic attack by ⁻OH?

A. propionyl bromide
B. benzyl bromide
C. propanal
D. butanoic acid
E. none of the above

29. What is the major organic product of this reaction?

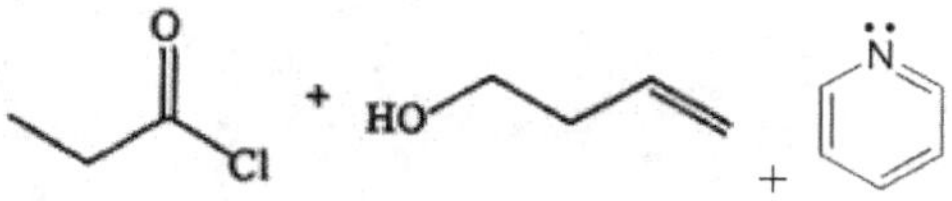

A. 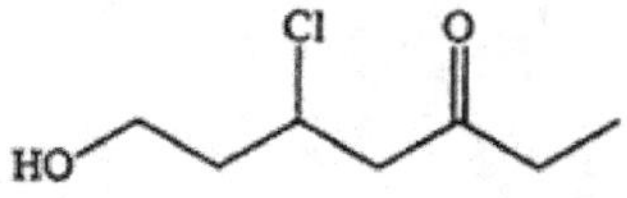 (structure with HO, Cl, and ketone)

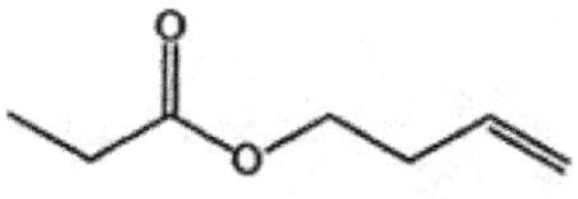

B. 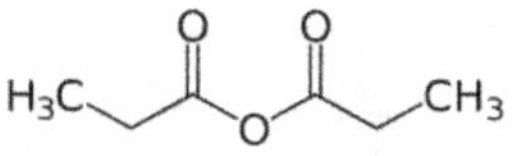 H₃C-C(O)-O-C(O)-CH₃ type anhydride

C. (propanoate ester with allyl group)

D. 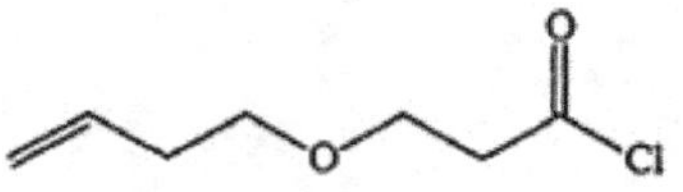 (allyl ether with acid chloride)

E. 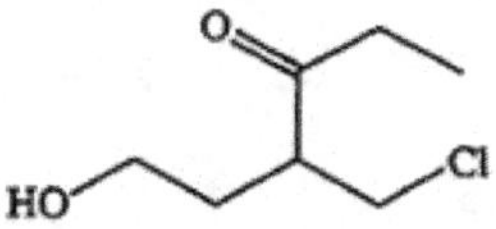 (structure with HO and Cl on ketone)

30. What reagent(s) are needed to complete the following reaction?

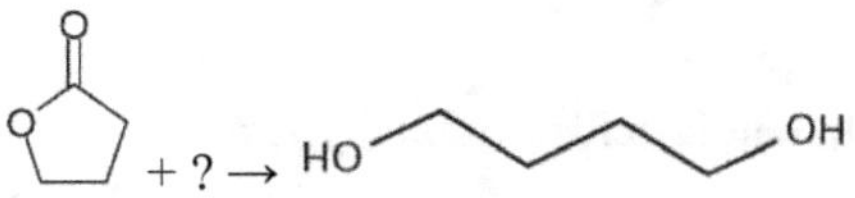

γ-butyrolactone + ? → HO-(CH₂)₄-OH

A. H₂, Pd
B. 1) DIBAL–H; 2) H₃O⁺
C. 1) LiAlH₄; 2) H₃O⁺
D. 1) NaBH₄; 2) H₃O⁺
E. 1) H₂, Pd; 2) H₃O⁺

31. What is the major ring-containing product of this reaction?

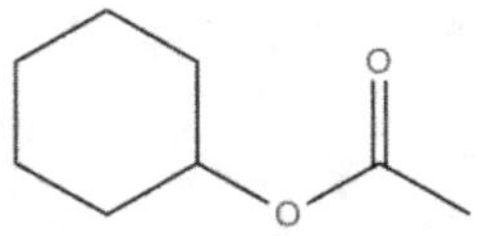

cyclohexyl ester + KOH, H₂O/Δ → H⁺ workup → ?

A.  cyclohexene

B. 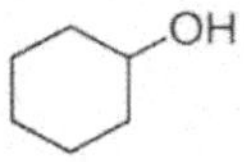 cyclohexanol (with OH)

C. 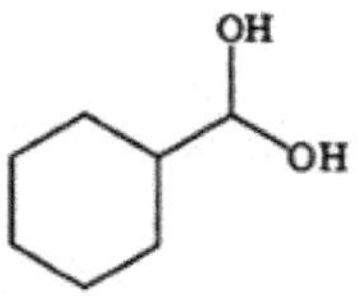 cyclohexyl CH(OH)₂

D. 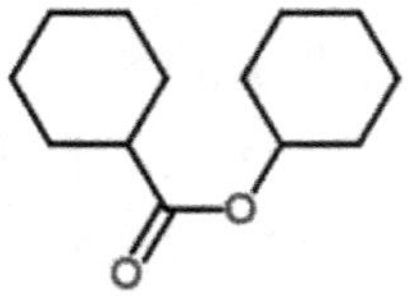 cyclohexyl ester of cyclohexanecarboxylic acid

E. 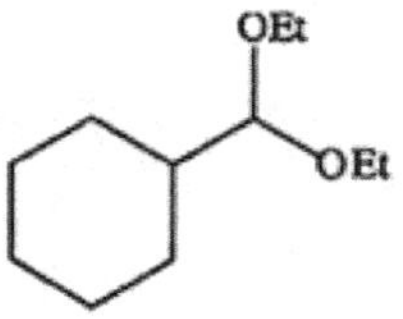 cyclohexyl CH(OEt)₂

32. What is the name of the product formed by the reaction of butanoic acid with methylamine?

A. N-methylbutamide
B. N-methylbutanamide
C. pentanone
D. pentylamine
E. N,N-methylbutamide

33. Amides are less basic than amines because the:

A. carbonyl group donates electrons by resonance
B. carbonyl group withdraws electrons by resonance
C. nitrogen does not have a lone pair of electrons
D. nitrogen has a full positive charge
E. amides do not contain nitrogen

34. Which of the following products may be formed in the reaction below?

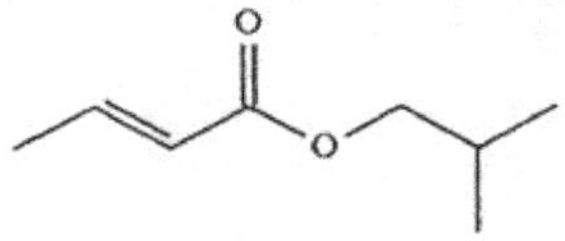

$+ H_2O / H_3O^+ \rightarrow ?$

- A. HOCH$_2$CH(CH$_3$)$_2$
- B. CH$_3$CH(CH$_3$)$_2$
- C. HOOCCH$_2$CH(CH$_3$)$_2$
- D. CH$_3$CH=CHCHO
- E. HOCH$_2$CH$_2$CH(CH$_3$)$_2$

35. Which is the most reactive of the four derivatives of a carboxylic acid?

- A. anhydride
- B. acid bromide
- C. amide
- D. ester
- E. thioester

36. What are the products of a hydrolysis reaction of an ester?

- A. Alcohol and alkane
- B. Carboxylic acid and alcohol
- C. Ether and alkene
- D. Ketone and aldehyde
- E. Alcohol and alkene

Notes for active learning

Amines

1. Which compound is a primary amine?

   A. N, N-dimethylethylamine
   B. isopropylamine
   C. diethylamine
   D. trimethylamine
   E. N-ethyl-N-methylpropylamine

2. Which organic functional group is important for its basic properties?

   A. carbonyl
   B. hydroxyl
   C. amine
   D. aromatic
   E. phenol

3. The following is an example of a:

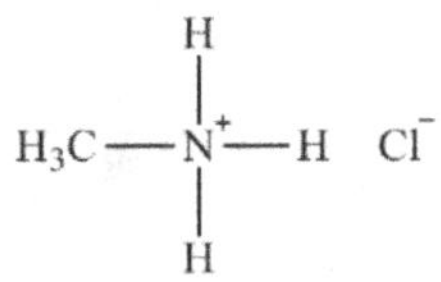

   $$H_3C-\overset{H}{\underset{H}{N^+}}-H \quad Cl^-$$

   A. primary ammonium salt
   B. tertiary ammonium
   C. quaternary amide salt
   D. tertiary amine
   E. secondary ammonium

4. Which of the following can be synthesized from an arenediazonium salt?

   I. C_6H_5Br II. C_6H_5CN III. C_6H_5OH

   A. I only
   B. I and II only
   C. I and III only
   D. II and III only
   E. I, II and III

5. Methyl bromide can generate the corresponding methylamine through alkyl halide ammonolysis. The nitrile reduction pathway generates the corresponding:

   A. ethylmethylamide
   B. ethylamine
   C. reduced methylene group
   D. dehalogenated methyl group
   E. none of the above

6. Amines are classified by:

A. number of carbons present in the molecule
B. number of carbons attached to the carbon bonded to the nitrogen
C. number of hydrogens attached to the nitrogen
D. number of alkyl groups attached to the nitrogen
E. none of the above

7. *p*-Toluidine is somewhat soluble in water due to the polarity of:

A. its aromatic ring structure
B. its amine group
C. *p*-toludinoic acid
D. benzene
E. its methyl group

8. Which of the following does NOT contain a polar carbonyl group?

A. an amine
B. a carboxylic acid
C. an ester
D. a ketone
E. none of the above

9. When comparing amine compounds of different classes but similar molar masses, which type most likely has the highest boiling point?

A. primary amines
B. secondary amines
C. tertiary amines
D. quaternary ammonium salts
E. amine oxides

10. Which of the compounds listed is the strongest organic base that functions as a proton acceptor?

A. 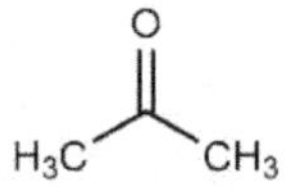 $H_3C-CO-CH_3$
B. 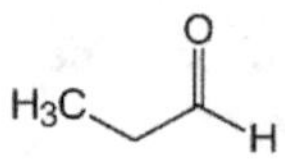 H_3C-CHO
C. 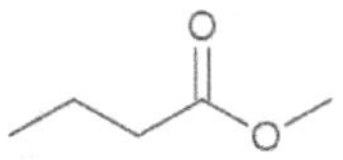 $CH_3CH_2-COO-CH_3$
D. 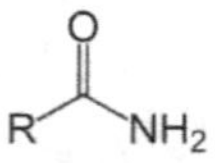 $R-CO-NH_2$
E. $CH_3-CH_2-NH_2$

11. Which of the following behaves like a base?

 I. $CH_3CONHCH_3$ II. $(CH_3)_2NH$ III. $C_2H_5CONHCH_3$

A. I only
B. II only
C. III only
D. I and II only
E. I, II and III

12. The reaction of an amine with water is best represented by:

    A. R–NH$_2$ + H$_2$O ↔ R–NH$_3^+$ + OH$^-$
    B. R–NH$_2$ + 2 H$_2$O ↔ R–NH$_4^{2+}$ + 2 OH$^-$
    C. R–NH$_2$ + 2 H$_2$O ↔ R–N^{2-} + 2 H$_3$O$^+$
    D. R–NH$_2$ + H$_2$O ↔ R–NH$^-$ + H$_3$O$^+$
    E. R–NH$_2$ + H$_2$O ↔ R–N^{2-} + M$^+$ + H$_3$O$^+$

13. Which of the following compounds has the lowest boiling point?

    A. dimethylamine
    B. *sec*-butylamine
    C. diethylamine
    D. *n*-butylamine
    E. ethanolamine

14. Which of these molecules is a tertiary amine?

    A. RNH$_2$
    B. R$_2$NH
    C. R$_3$N
    D. R$_3$NH$^+$
    E. R$_4$N$^+$

15. Which formula best represents the form an amine takes in acidic solution?

    A. RNH$_2^-$
    B. RNH$_2^+$
    C. RNH$_2$
    D. RNH$^-$
    E. RNH$_3^+$

16. The functional group C=O is in all the species below, EXCEPT:

    A. amines
    B. amides
    C. carboxylic acids
    D. aldehydes
    E. esters

17. A quaternary ammonium salt does NOT carry out substitution reactions with alkyl halides because the nitrogen atom is not:

    A. an electrophile
    B. a nucleophile
    C. negatively charged
    D. saturated
    E. a radical

18. Amines are most similar in chemical structure and behavior to:

    A. sodium hydroxide
    B. a primary alcohol
    C. the hydronium ion
    D. ammonia
    E. water

19. Why is an amine salt more soluble in water than the corresponding free amine?

A. The negative charge on the nitrogen atom increases water solubility
B. It has a higher molecular weight than the corresponding amine
C. It is ionic and therefore more soluble than covalent compounds with the same structure
D. All amines are insoluble in water
E. None of the above

20. Based on the properties of the attached functional group, which compound below interacts most strongly with water, thus making it the most soluble compound?

A. $CH_3–CH_2–I$
B. $CH_3–O–CH_3$
C. $CH_3–CH_2–NH_2$
D. $CH_3–CH_2–H$
E. $CH_3–CH_2–S–H$

21. Assuming roughly equivalent molecular weights, which of the following has the highest boiling point?

A. alcohol
B. ether
C. tertiary amine
D. quaternary ammonium salt
E. alkyl chloride

22. The compound trimethylamine is a(n) [ ] and has the formula [ ]:

A. base, $(CH_3)_3N$
B. acid, $(CH_3)_2NH$
C. base, $(CH_3)_2NH$
D. acid, $(CH_3)_3N$
E. base, $(CH_3)NH$

23. Which compound is an example of an amine salt?

A. sulfanilamide
B. thioacetamide
C. dimethylammonium bromide
D. histamine
E. pyridoxine

24. Which statement about the differences between an amine and an amide is NOT correct?

A. Amides act as proton acceptors, while amines do not
B. Amines are basic, and amides are neutral
C. Amines form ammonium salts when treated with acid; amides do not
D. The lone pair of electrons on amides is held more tightly than the lone pair on amines
E. Amine nitrogen atoms are sp^3 hybridized, while amides have sp^2 hybridization

25. What is the conjugate base of CH_3NH_2?

   A. NH_2^-
   B. NH_4^+
   C. CH_3NH^-
   D. $CH_3NH_3^+$
   E. none of the above

26. A functional group containing nitrogen is found in:

   A. carboxylic acids
   B. amines
   C. alcohols
   D. alkenes
   E. ethers

27. Which of the following is NOT correct about amines?

   A. Amines are bases (proton acceptors)
   B. Amines are converted to ammonium salts by reaction with HCl
   C. Amines are organic derivatives of ammonia
   D. Amines are very water soluble
   E. Amines are often odorous compounds

28. What is the conjugate acid of CH_3NH_2?

   A. NH_4^+
   B. NH_2^-
   C. $CH_3NH_3^+$
   D. CH_3NH^-
   E. CH_3^-

29. Which amine has the lowest boiling point?

   A. 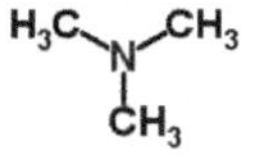 trimethylamine (H₃C-N(CH₃)-CH₃)
   B. 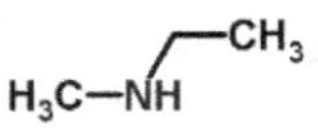 H₃C—NH—CH₃ (N-methyl... methylethylamine)
   C. 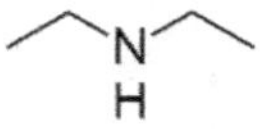 diethylamine (CH₃CH₂-NH-CH₂CH₃)
   D. 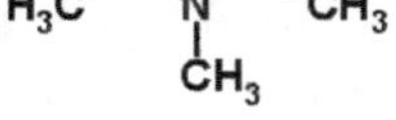 N,N-dimethylethylamine (H₃C-N(CH₃)-CH₂CH₃... with CH₃)
   E. 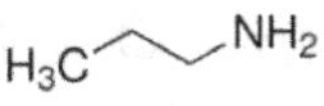 H₃C—CH₂—NH₂

30. In water, does the molecule lysergic acid diethylamide act as:

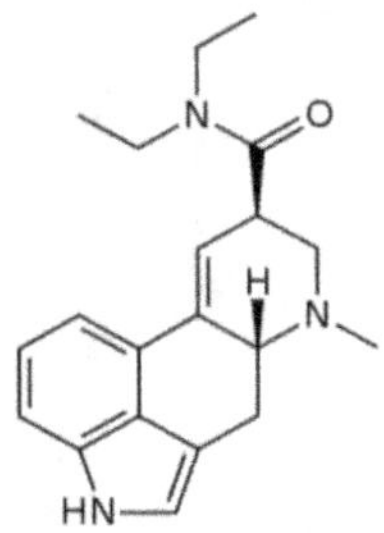

I. acid   II. base   III. neither acid nor base

A. I only
B. II only
C. III only
D. I and II only
E. requires more information

31. Which of the following amines is the most basic in the gas phase?

A. NH_3
B. $(CF_3)_3N$
C. $(CH_3)_3N$
D. H_2NCH_3
E. $HN(CH_3)_2$

32. All of the following are properties of amines, EXCEPT:

A. amines react with acids to form amides at low temperatures
B. amines with low molecular weights are soluble in water
C. amines that form hydrogen bonds have higher boiling points relative to their molecular mass
D. amines frequently have offensive odors
E. amines function as bases in many reactions

33. Which of the compounds shown form hydrogen bonds between a mixture of the molecules?

A. $(CH_3)_3N$
B. $CH_3CH_2OCH_3$
C. $CH_3CH_2CH_2F$
D. $CH_3CH_2CH_2CH_3$
E. $CH_3NHCH_2CH_3$

34. Dodecylamine, $CH_3(CH_2)_{10}CH_2NH_2$, is insoluble in water but can be converted to a water-soluble form. Which of the species below represents a water-soluble form of this compound?

A. $CH_3(CH_2)_{10}CH_2NHCH_3$
B. $CH_3(CH_2)_{10}CH_2NH_3^+Cl^-$
C. $CH_3(CH_2)_{10}CH_2NH_2$–OH
D. $CH_3(CH_2)_{10}CH_2NH_2$–Cl
E. $CH_3(CH_2)_{10}CH_2NH_2$–O–CH_3

35. Which of the following molecules represents a tertiary amine?

A. 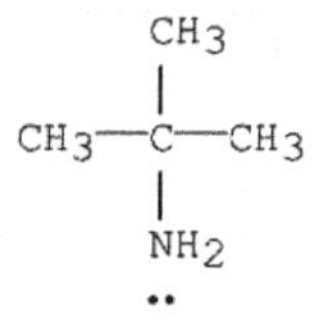 CH3—C(CH3)(CH3)—NH2 (central C with CH3 above, NH2 below)

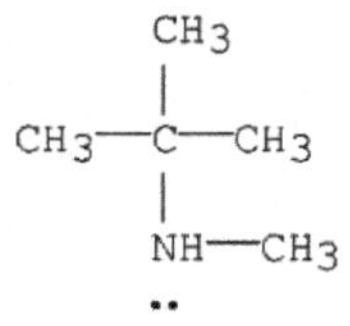

B. CH3—C(CH3)(CH3)—NH—CH3

C. CH3—CH(CH3)—CH2—NH2

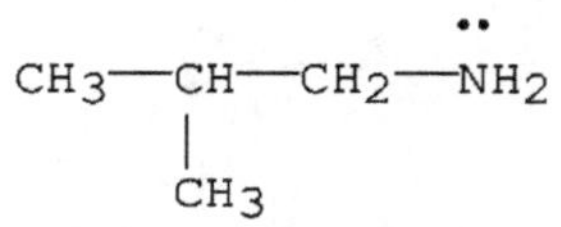

D. 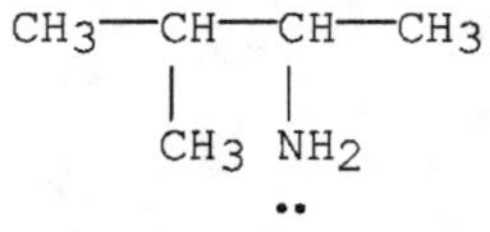 CH3—CH(CH3)—CH(NH2)—CH3

E. 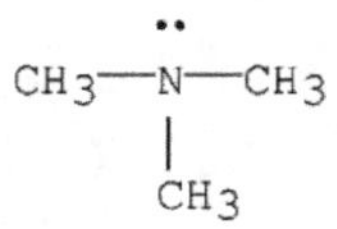 CH3—N(CH3)—CH3

36. Which of the following sequences ranks the following isomers in order of increasing boiling points?

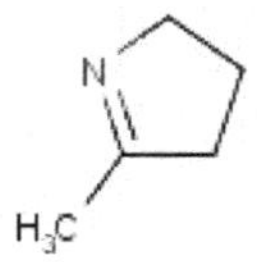

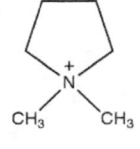

1 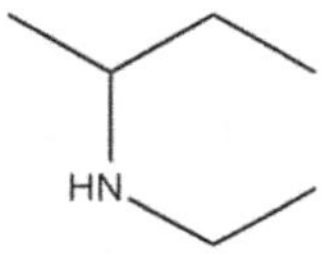   2   3

A. 1 < 3 < 2
B. 3 < 2 < 1
C. 2 < 1 < 3
D. 2 < 3 < 1
E. 1 < 2 < 3

Notes for active learning

Amino Acids, Peptides, Proteins

1. Name the amino acid produced when propanoic acid is subjected to the following sequence of reagents:

   1) PBr₃, Br₂ 2) H₂O 3) NH₃, Δ

   A. alanine
   B. aspartic acid
   C. glutamic acid
   D. valine
   E. asparagine

2. What type of molecular attraction is expected to dominate between two threonine amino acids within a folding polypeptide chain?

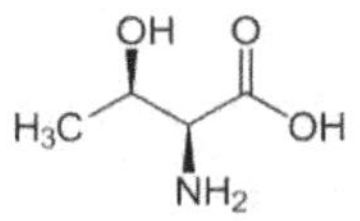

   A. dipole–induced dipole
   B. induced dipole–induced dipole
   C. ion–dipole
   D. dipole–dipole
   E. London forces

3. Which of the standard amino acids is achiral?

   A. lysine
   B. proline
   C. valine
   D. alanine
   E. glycine

4. Which of the following amino acids does NOT contain an aromatic R-group?

   A. tyrosine
   B. phenylalanine
   C. tryptophan
   D. histidine
   E. none of the above

5. The linear sequence of amino acids along a peptide chain determines its:

   A. primary structure
   B. secondary structure
   C. tertiary structure
   D. quaternary structure
   E. none of the above

6. Polar *R* groups, along with acidic and basic *R* groups, are [ ] because they are attracted to water molecules.

   A. unreactive
   B. ionized
   C. hydrophobic
   D. hydrophilic
   E. none of the above

7. An amino acid whose *R* group is predominantly hydrocarbon is classified as:

   A. acidic
   B. basic
   C. nonpolar
   D. polar
   E. isoelectric

8. There are [ ] different types of major biomolecules used by humans.

   A. a few dozen
   B. four
   C. several thousand
   D. several million
   E. a few hundred

9. Amino acids are linked to one another in a protein by which of the following bonds?

   A. amide bonds
   B. carboxylate bonds
   C. ester bonds
   D. amine bonds
   E. glycosidic bonds

10. Which of the following amino acids is most likely present in the hydrophobic binding region of a protein?

   A. tyrosine
   B. glutamine
   C. valine
   D. serine
   E. histidine

11. Identify the functional groups in the following compound.

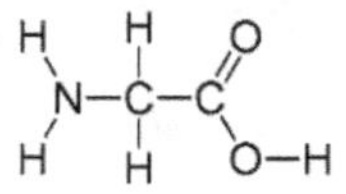

   A. alcohol and ketone
   B. alcohol, amine and ketone
   C. amine and carboxylic acid
   D. amine, hydroxide and ketone
   E. amine and aldehyde

12. The pH at which the positive and negative charges of an amino acid balance is:

   A. isotonic point
   B. isobestic point
   C. isobaric point
   D. isoelectric point
   E. isomer point

13. The proton on the nitrogen of saccharin is much more acidic than the proton on the amide nitrogen of aspartame. This is best explained by the fact that:

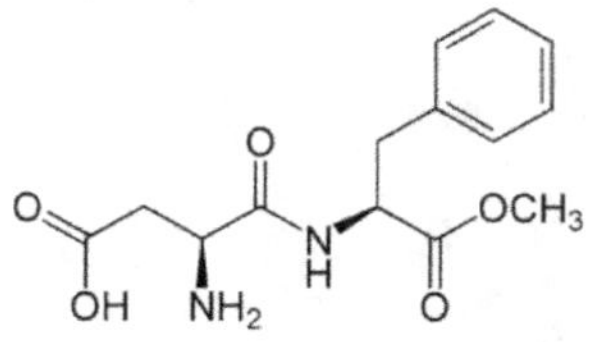

Aspartame (Asp-Phe) 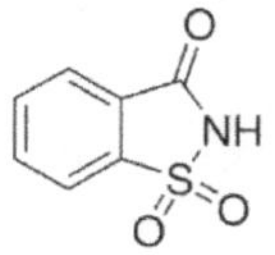   *Saccharin*

   A. extra resonance stabilization of the resulting anion is provided by the sulfone group
   B. the saccharin nitrogen is *sp* hybridized
   C. the adjacent sulfone group stabilizes the anion because it is an electron-donating group
   D. cyclic amides are more acidic than acyclic amides
   E. none of the above

14. 2,4-dinitrofluorobenzene is used in protein analysis to determine the:

   A. most reactive amino acid in the protein
   B. amino acid at the *C*-terminus
   C. amino acid at the *N*-terminus
   D. most frequent amino acid in the protein
   E. none of the above

15. Insulin is an example of a(n):

   A. transport protein
   B. storage protein
   C. structural protein
   D. enzyme
   E. hormone

16. What type of protein structure corresponds to a spiral alpha-helix of amino acids?

   A. primary
   B. secondary
   C. tertiary
   D. quaternary
   E. none of the above

17. At pH 8, which of the following is true for aspartame shown below?

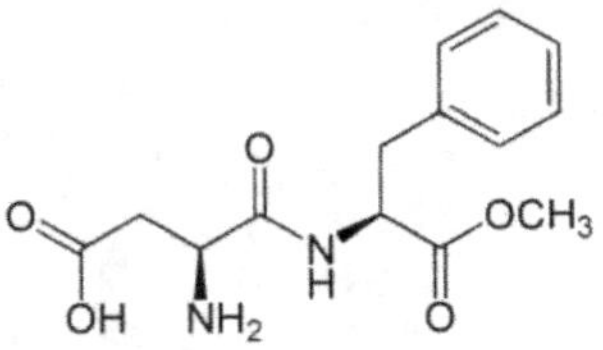

A. The acid and amino groups are protonated
B. The acid and ester groups are deprotonated
C. The acid group is deprotonated, and the amino group is protonated
D. The amino group is deprotonated, and the acid group is protonated
E. None of the above

18. Non-polar *R* groups on amino acids are said to be [ ] because they are not attracted to water molecules.

A. ionized
B. unreactive
C. hydrophilic
D. hydrophobic
E. none of the above

19. Which of the following amino acids has its isoelectric point at the lowest pH?

A. arginine
B. aspartic acid
C. valine
D. glycine
E. methionine

20. In biosynthesis, which amino acid serves as the source of the amino group for other amino acids?

A. D-phenylalanine
B. L-phenylalanine
C. racemic phenylalanine
D. L-glutamic acid
E. D-glutamic acid

21. Proteins are characterized by the fact that they:

A. always have quaternary structures
B. retain their conformation above 35-40 °C
C. have a primary structure formed by covalent linkages
D. are composed of a single peptide chain
E. none of the above

22. The reaction mechanism by which 2,4-dinitrofluorobenzene reacts with a protein is:

    A. addition
    B. S$_N$2
    C. nucleophilic aromatic substitution
    D. electrophilic aromatic substitution
    E. none of the above

23. Members of which class of biomolecules is the building blocks of proteins?

    A. fatty acids
    B. amino acids
    C. glycerols
    D. monosaccharides
    E. nucleic acids

24. If a hair stylist is about to apply a reducing agent to a client with fine hair who wants to have his hair curly, should the reducing agent be regular strength, concentrated, or diluted?

    A. diluted, so as not to cause the hair to fall apart completely
    B. concentrated in order to add more disulfide cross-linking
    C. concentrated in order to add more hydrogen bonding
    D. concentrated because, in thin hair, each strand is made of fewer cysteine amino acids
    E. regular strength

25. All of the bonds below are apparent in the secondary and tertiary structure of a protein, EXCEPT:

    A. electrostatic interactions
    B. peptide bonds
    C. hydrogen bonding
    D. hydrophobic interactions
    E. none of the above

26. Proteins are polymers consisting of which monomer units?

    A. keto acids
    B. amide
    C. amino acids
    D. ketones
    E. polyaldehydes

27. Proteins migrate the smallest distance during gel electrophoresis when the protein is at:

    A. high pH
    B. neutral pH
    C. low pH
    D. their isoelectric point
    E. high and low pH

28. Why might a change in pH cause a protein to denature?

A. The hydrogen bonds between the hydrophobic portions of the protein collapse due to extra protons
B. The disulfide bridges open
C. The functional groups that give the protein its shape becomes protonated or deprotonated
D. The water hardens and causes the protein's shape to change
E. All the above

29. Which of the following compounds is an amino acid?

A. 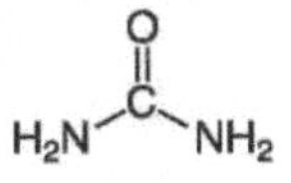 $H_2N-C(=O)-NH_2$

B. $H_2N-CH_2-C(=O)-OH$

C. $H_3N^+-C(=O)-O^-$

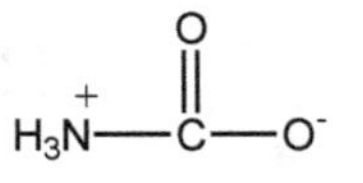

D. 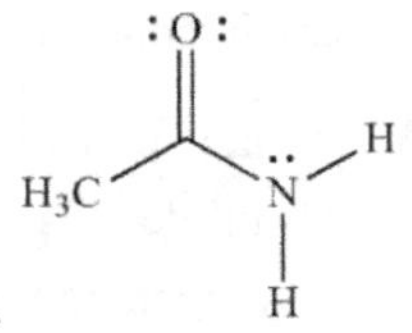 $H_3C-C(=O)-NH-H$ (with lone pairs on O and N)

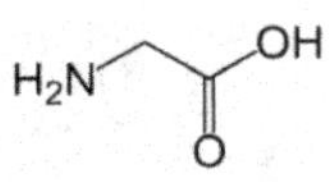

E. 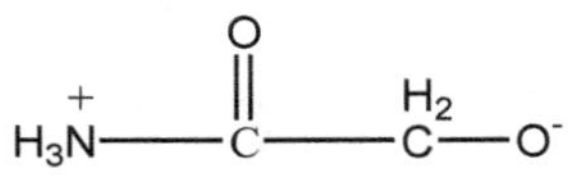 $H_3N^+-C(=O)-CH_2-O^-$

30. The most basic functional group of aspartame is the:

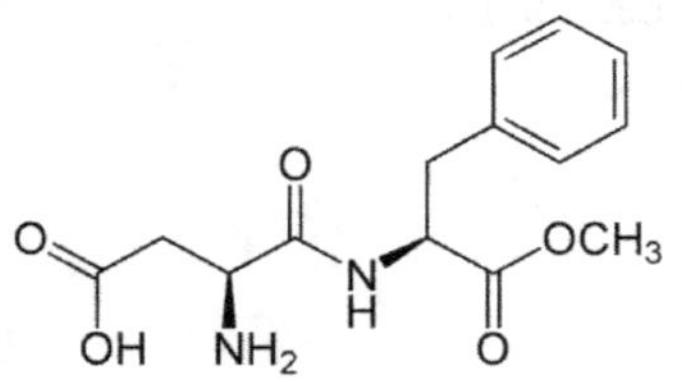

Aspartame (Asp-Phe)

A. amide nitrogen
B. amino group
C. ester carbonyl oxygen
D. aromatic ring
E. amide oxygen

31. What is the major product of the following reaction series?

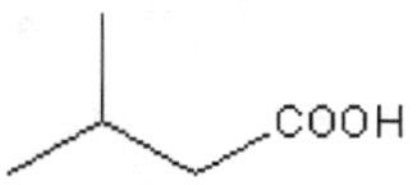

(CH₃)₂CHCH₂COOH + PBr₃ → ? + excess NH₃ → ?

A. Ala   B. Gly   C. Leu   D. Ile   E. Val

32. The protein conformation determined mostly from interactions between R groups is:

A. tertiary structure
B. quaternary structure
C. primary structure
D. secondary structure
E. none of the above

33. What is the best description of the linkage shown in the following polypeptide?

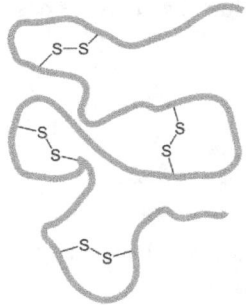

A. disulfide linkage
B. glucosidic linkage
C. hemiacetal
D. acetal
E. amide

34. A particular polypeptide can be represented as Gly-Ala-Ala-Phe-Cys-Gly-Ala-Cys-Phe-Cys. How many peptide bonds are there in this polypeptide?

A. 9
B. 10
C. 8
D. 6
E. 11

35. Hydrophobic interactions help to stabilize the [ ] structure(s) of a protein.

A. primary
B. secondary
C. secondary and tertiary
D. tertiary and quaternary
E. secondary and quaternary

36. The side chains, or *R* groups, of amino acids can be classified into each of the following categories EXCEPT:

A. acidic
B. basic
C. non-polar
D. polar
E. isoelectric

37. Considering the acid/base character of many amino acid side chains, why do changes in pH interfere with the function of proteins?

    I. At high pH, the acidic side chains of amino acids such as glutamic acid and tyrosine become negatively charged as they lose a hydrogen ion

    II. A change in pH can result in a change in the type of molecular attractions that occur among amino acids within a polypeptide

    III. At low pH, the alkaline side chains of amino acids such as lysine and histidine become positively charged as they gain a hydrogen ion

- **A.** I only
- **B.** II only
- **C.** III only
- **D.** I and II only
- **E.** I, II and III

38. A particular polypeptide can be represented as Gly-Ala-Phe-Cys-Gly-Ala-Phe-Cys. How many sites with positive or negative charges are in this polypeptide?

- **A.** 8
- **B.** 7
- **C.** 4
- **D.** 2
- **E.** 9

39. Which of the following amino acids does NOT contain a basic side chain?

    I. arginine   II. lysine   III. threonine

- **A.** I only
- **B.** II only
- **C.** III only
- **D.** I and III only
- **E.** II and III only

40. The isoelectric point of an amino acid is the:

- **A.** pH at which it exists in the zwitterion form
- **B.** pH equal to its pK_a
- **C.** pH at which it exists in the basic form
- **D.** pH at which it exists in the acid form
- **E.** pH equal to its pK_b

41. Collagen is an example of a(n):

- **A.** storage protein
- **B.** transport protein
- **C.** enzyme
- **D.** structural protein
- **E.** hormone

42. The coiling of a chain of amino acids describes a protein's:

    A. primary structure
    B. secondary structure
    C. tertiary structure
    D. quaternary structure
    E. none of the above

43. The laboratory conditions typically used to hydrolyze a protein are:

    A. dilute acid and room temperature
    B. dilute base and room temperature
    C. concentrated acid and heat
    D. concentrated base and heat
    E. exposure to 340 nm light

44. Which of the following is an essential amino acid?

    A. aniline
    B. valine
    C. glycine
    D. serine
    E. proline

45. What is the name given to the localized bending and folding of a polypeptide backbone of a protein molecule?

    A. primary structure
    B. secondary structure
    C. tertiary structure
    D. quaternary structure
    E. zymogen structure

46. Which of the following macromolecules are composed of polypeptides?

    A. amino acids
    B. proteins
    C. carbohydrates
    D. fats
    E. steroids

47. Which amino acid can form covalent sulfur-sulfur bonds?

    A. proline
    B. methionine
    C. cysteine
    D. glycine
    E. phenylalanine

48. Which amino acid is a secondary amine with its nitrogen and alpha carbon joined as part of a ring?

    A. lysine
    B. histidine
    C. alanine
    D. aspartic acid
    E. proline

Notes for active learning

Lipids

1. The molecule shown can be classified as a(n):

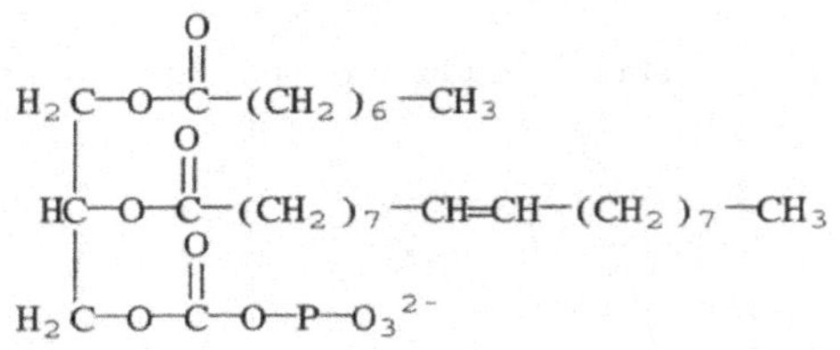

$$H_2C-O-C(=O)-(CH_2)_6-CH_3$$
$$HC-O-C(=O)-(CH_2)_7-CH=CH-(CH_2)_7-CH_3$$
$$H_2C-O-C(=O)-O-P-O_3^{2-}$$

- **A.** sphingolipid
- **B.** eicosanoid
- **C.** wax
- **D.** glycerophospholipid
- **E.** steroid

2. Which molecule is a fatty acid?

- **A.** CH_3COOH
- **B.** $CH_2=CHCOOH$
- **C.** $(CH_3)_2CH(CH_2)_3COOH$
- **D.** $CH_3(CH_2)_7CH=CH(CH_2)_7COOH$
- **E.** None of the above

3. What is the purpose of the plasma membrane?

- **A.** Storing of the genetic material of the cell
- **B.** Retaining water in the cell to prevent it from dehydrating
- **C.** Acting as a cell wall to give the cell structure and support
- **D.** Acting as a boundary, but letting molecules in and out
- **E.** None of the above

4. Given a single triglyceride containing three identical and fully saturated fatty acid residues, which of the following terms accurately describes it?

- **A.** optically inactive
- **B.** *meso*
- **C.** optically active
- **D.** chiral
- **E.** racemic

5. The fatty acids below contain between sixteen and eighteen carbons and range from saturated to three double bonds. Which has the lowest melting point?

- **A.** palmitic acid (saturated)
- **B.** oleic acid (one double bond)
- **C.** linoleic acid (two double bonds)
- **D.** linolenic acid (three double bonds)
- **E.** all have approximately the same melting point

6. Which of the following statements correctly describe(s) the relationship between fatty acid structure and its melting point?

   I. Saturated fatty acids melting points increase gradually with the molecular weights
   II. As the number of double bonds in a fatty acid increase, its melting point decreases
   III. The *trans* double bond in the fatty acid has a greater effect on its melting point than does the *cis* double bond

   A. I only
   B. II only
   C. III only
   D. I and II only
   E. I, II and III

7. The hydrocarbon end of a soap molecule is:

   A. hydrophilic and attracted to grease
   B. hydrophobic and attracted to grease
   C. hydrophilic and attracted to water
   D. hydrophobic and attracted to water
   E. neither hydrophobic nor hydrophilic

8. It is important to have cholesterol in one's body because:

   A. it breaks down extra fat lipids
   B. it serves as the starting material for the biosynthesis of most other steroids
   C. it is the starting material for the building of glycogen
   D. the brain is made almost entirely of cholesterol
   E. it mainly serves as an energy reserve

9. Cholesterol belongs to the [ ] group of lipids.

   A. prostaglandin
   B. triacylglycerol
   C. saccharides
   D. steroid
   E. wax

10. In chemical terms, soaps are best described as:

   A. simple esters of fatty acids
   B. mixed esters of fatty acids
   C. salts of carboxylic acids
   D. long chain acids
   E. bases formed from glycerol

11. The function of cholesterol in a cell membrane is to:

   A. act as a precursor to steroid hormones
   B. take part in the reactions that produce bile acids
   C. maintain structure due to its flat rigid characteristics
   D. attract hydrophobic molecules to form solid deposits
   E. none of the above

12. Which of the following terms best describes the compound below?

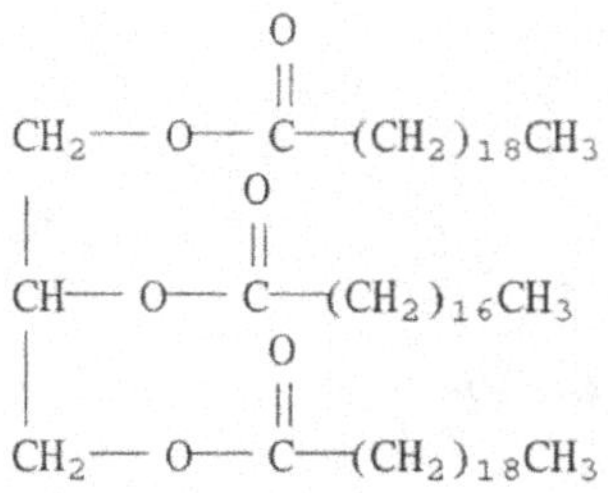

$$\begin{array}{l} CH_2-O-\overset{O}{\overset{\|}{C}}-(CH_2)_{18}CH_3 \\ CH-O-\overset{O}{\overset{\|}{C}}-(CH_2)_{16}CH_3 \\ CH_2-O-\overset{O}{\overset{\|}{C}}-(CH_2)_{18}CH_3 \end{array}$$

A. unsaturated triglyceride
B. saturated triglyceride
C. terpene
D. prostaglandin
E. lecithin

13. Which fatty acid is a saturated fatty acid?

A. oleic acid
B. linoleic acid
C. arachidonic acid
D. myristic acid
E. none of the above

14. Commercially, liquid vegetable oils are converted to solid fats such as margarine by:

A. oxidation
B. hydration
C. hydrolysis
D. hydrogenation
E. saponification

15. The biochemical roles of lipids are:

A. short-term energy storage, transport of molecules, and structural support
B. storage of excess energy, component of cell membranes, and chemical messengers
C. catalysis, protection against outside invaders, motion
D. component of cell membranes, catalysis, and structural support
E. neurotransmitters, hormones, transport of molecules

16. Which fatty acid composition yields a triglyceride that is most likely an oil?

A. 3 palmitic acid units
B. 2 palmitic and 1 oleic acid units
C. 2 linoleic and 1 stearic acid units
D. 3 stearic acid units
E. 2 palmitic and 1 stearic acid units

17. Which of the following molecules is an omega-3 fatty acid?

A. oleic acid
B. linolenic acid
C. linoleic acid
D. palmitic acid
E. none of the above

18. Which of the following is a lipid?

A. lactose
B. aniline
C. nicotine
D. estradiol
E. collagen

19. Which of the following terms best describes the interior of a soap micelle in water?

A. hard
B. saponified
C. hydrophobic
D. hydrophilic
E. hydrogenated

20. Which best describes the lipid shown below?

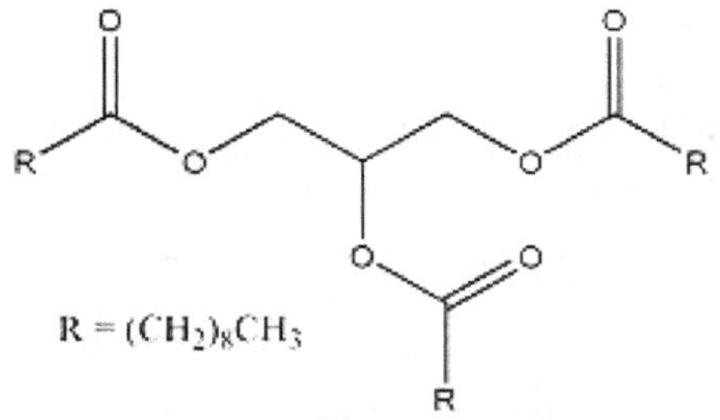

$R = (CH_2)_8CH_3$

A. saturated fatty acid
B. unsaturated fatty acid
C. wax
D. triglyceride
E. lecithin

21. Lipids are compounds soluble in:

A. glucose solution
B. organic solvents
C. distilled water
D. normal saline solution
E. oxygen

22. Which statement regarding fatty acids is NOT correct? Fatty acids:

A. are always liquids
B. are long-chain carboxylic acids
C. are usually unbranched chains
D. usually have an even number of carbon atoms
E. none of the above

23. Unsaturated triacylglycerols are usually [ ] because [ ]?

A. liquids ... they have relatively short fatty acid chains
B. liquids ... the kinks in their fatty acid chains prevent their fitting closely
C. liquids ... they contain impurities from their natural sources
D. solids ... they have relatively long fatty acid chains
E. solids ... the similar zig-zag shape of their fatty acid chains allows them to fit closely

24. Two families of fatty acids with significance in nutrition are Ω-6 and Ω-3 fatty acids. Which of the following structures would be classified as a Ω-3 fatty acid?

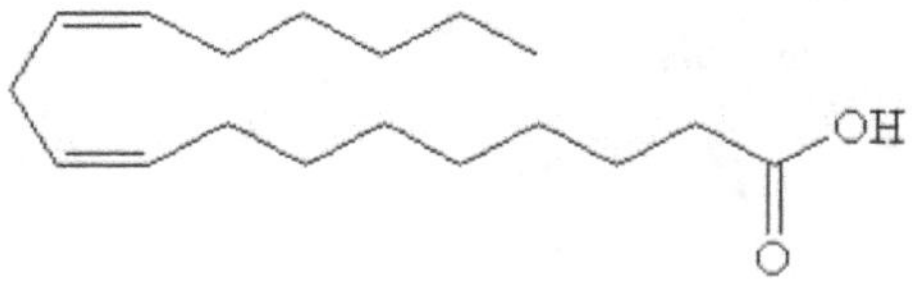

A.

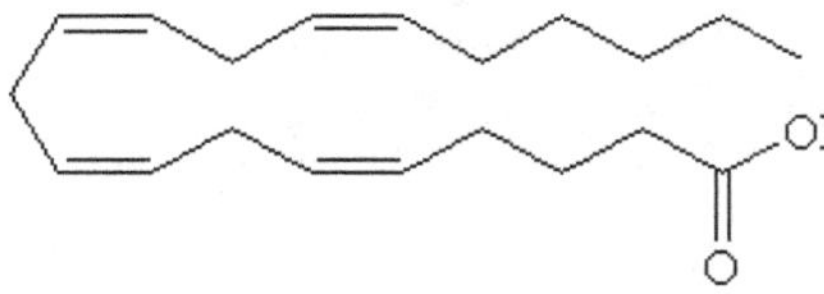

B.

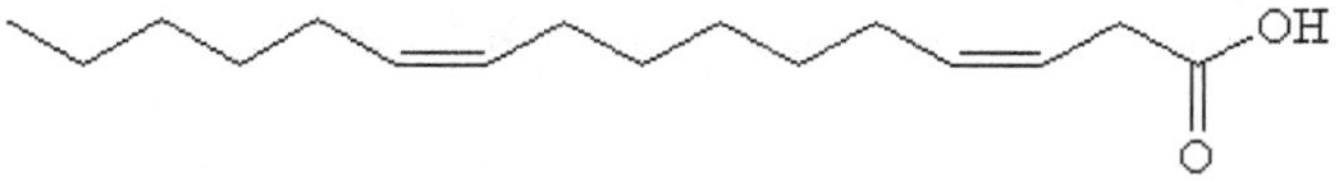

C.

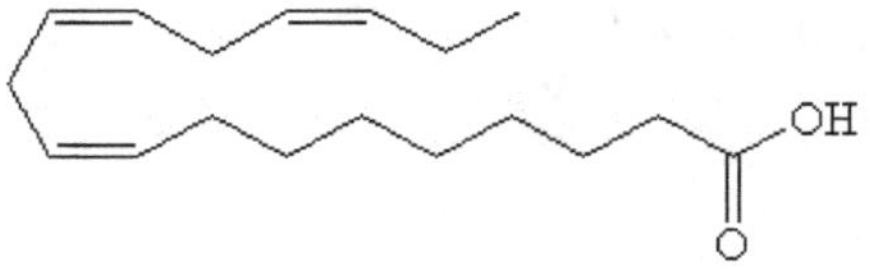

D.

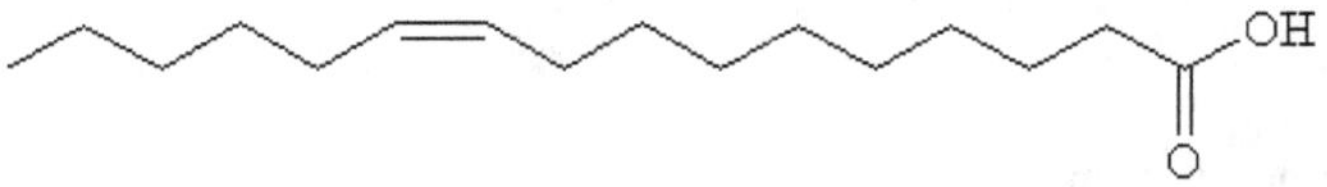

E.

25. Triacylglycerols are compounds that contain combined:

A. cholesterol and other steroids
B. fatty acids and phospholipids
C. fatty acids and glycerol
D. fatty acids and choline
E. lecithin and choline

26. The name of the reaction that occurs when a fat reacts with sodium hydroxide and water is:

A. oxidation
B. hydration
C. reduction
D. hydrogenation
E. saponification

27. Which of the following molecules produce fat from an esterification reaction?

   I. $CH_3(CH_2)_{14}COOH$
   II. $CH_3(CH_2)_7CH=CH(CH_2)_7COCH_2CH_3$
   III. $HOCH_2CHO$
   IV. $HOCH_2CH(OH)CH_2OH$

A. I and IV only
B. II and III only
C. II and IV only
D. I and III only
E. I and II only

28. Oils are generally [ ] at room temperature and are obtained from [ ]:

A. liquids ... plants
B. liquids ... animals
C. solids ... plants
D. solids ... animals
E. none of the above

29. Lipids are naturally occurring compounds which all:

A. contain fatty acids as structural units
B. are water-insoluble but soluble in nonpolar solvents
C. contain ester groups
D. contains cholesterol
E. are unsaturated

30. Hydrogenation of vegetable oils converts them into what type of molecule?

A. esters
B. ethers
C. hemiacetals
D. saturated fats
E. polymers

31. A polyunsaturated fatty acid contains more than one:

A. long carbon chain
B. carbonyl group
C. hydroxyl group
D. carboxyl group
E. double bond

32. The molecule shown can be classified as a(an):

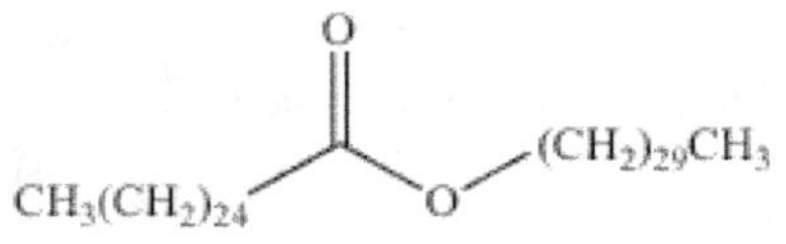

$CH_3(CH_2)_{24}$—C(=O)—O—$(CH_2)_{29}CH_3$

A. steroid
B. eicosanoid
C. wax
D. glycerophospholipid
E. sphingolipid

33. The chemical bond that links glycerol to fatty acid is an example of what type of linkage?

A. ester
B. ionic
C. ether
D. peptide
E. hydrogen bond

34. A molecule that has hydrophobic and hydrophilic portions is:

A. amphoteric
B. enantiomeric
C. amphiprotic
D. amphipathic
E. allotropic

35. The products of the base-catalyzed breakdown of fat are:

A. salts of fatty acids
B. salts of fatty acids and glycerol
C. esters of fatty acids
D. terpenes
E. steroids

36. The potassium or sodium salt of a long chain carboxylic acid is a(n):

A. emollient
B. ester
C. triglyceride
D. soap
E. none of the above

37. A molecule that has polar and nonpolar parts is:

A. an isomer
B. amphipathic
C. an enantiomer
D. hydrophobic
E. hydrophilic

38. Saturated fats are [ ] at room temperature and are obtained from [ ]?

A. liquids; plants
B. liquids; animals
C. solids; plants
D. solids; animals
E. solids; plants and animals

39. Which of the following is NOT found in a lipid wax?

A. saturated fatty acid
B. long-chain alcohol
C. glycerol
D. ester linkage
E. unsaturated fatty acid

40. How many molecules of fatty acid are needed to produce one molecule of fat or oil?

A. 1
B. 1.5
C. 2
D. 3
E. 6

41. When dietary triglycerides are hydrolyzed, the products are:

A. glycerol and lipids
B. carbohydrates
C. amino acids
D. alcohols and lipids
E. glycerol and fatty acids

42. Unsaturated fatty acids have lower melting points than saturated fatty acids because:

A. their molecules fit closely
B. *cis* double bonds give them an irregular shape
C. they have fewer hydrogen atoms
D. they have more hydrogen atoms
E. *trans* double bonds give them an irregular shape

43. Which of the following is commonly known as glycerol?

A. 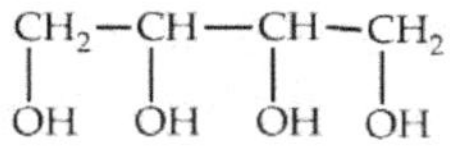 CH$_2$—CH—CH—CH$_2$ with OH on each carbon
B. 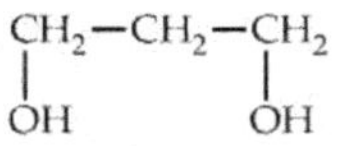 CH$_2$—CH$_2$—CH$_2$ with OH on first and third carbon
C. 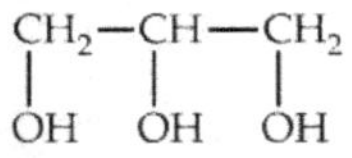 CH$_2$—CH—CH$_2$ with OH on each carbon
D. 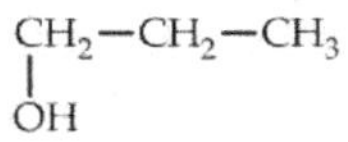 CH$_2$—CH$_2$—CH$_3$ with OH on first carbon
E. 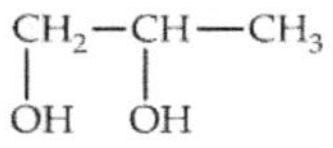 CH$_2$—CH—CH$_3$ with OH on first two carbons

44. How many fatty acids are in a phospholipid molecule?

A. 0
B. 1
C. 2
D. 3
E. variable

45. Which molecule is NOT a fatty acid?

A. CH$_3$(CH$_2$)$_{14}$COOH
B. CH$_3$CH$_2$(CH=CHCH$_2$)$_3$(CH$_2$)$_6$COOH
C. (CH$_3$)$_2$CH(CH$_2$)$_3$COOH
D. CH$_3$(CH$_2$)$_7$CH=CH(CH$_2$)$_7$COOH
E. None of the above

46. The chemical composition of lipids are esters of glycerol with three:

A. long-chain alcohols and fatty acids
B. identical saturated fatty acids
C. identical unsaturated fatty acids
D. predominantly saturated fatty acids
E. predominantly unsaturated fatty acids

47. Which of the following is NOT a function of lipids within the body?

- **A.** cushioning to prevent injury
- **B.** insulation
- **C.** energy reserve
- **D.** precursor for glucose catabolism
- **E.** precursor for the synthesis of androgens

48. Which of the following lipids is an example of a simple lipid?

- **A.** oil
- **B.** wax
- **C.** fat
- **D.** terpene
- **E.** none of the above

Notes for active learning

Carbohydrates

1. The conversion of cyclic glucose between the alpha and beta form is:

A. dimerization
B. cyclization
C. mutarotation
D. polymerization
E. hydrolysis

2. Common reducing reactions of monosaccharides are due to:

A. their cyclic structures
B. the presence of at least one hydroxyl group
C. the presence of more than one hydroxyl group
D. the presence of a carbonyl group, usually on carbon #1
E. the presence of at least one chiral carbon atom

3. What type of biological compound is characterized by an aldehyde or ketone and alcohol functional groups?

A. nucleic acid
B. fatty acid
C. sugar
D. amino acid
E. none of the above

4. Identify the C_3 epimer of the sugar below drawn in its open-chain (acyclic) Fischer projection.

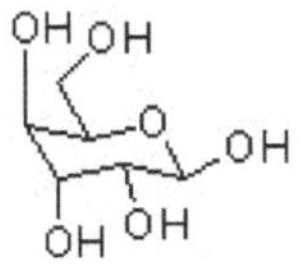
OH
OH
O
OH
OH
OH

A. 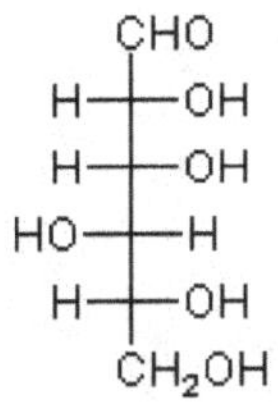
CHO
H OH
H OH
HO H
H OH
CH₂OH

B. 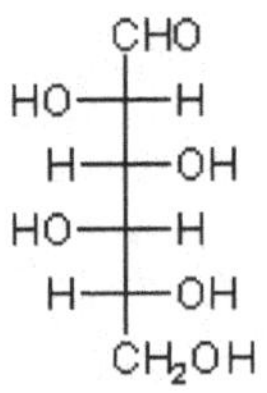
CHO
HO H
H OH
HO H
H OH
CH₂OH

C. 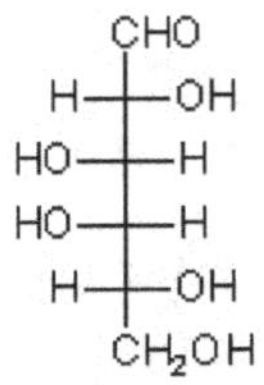
CHO
H OH
HO H
HO H
H OH
CH₂OH

D. 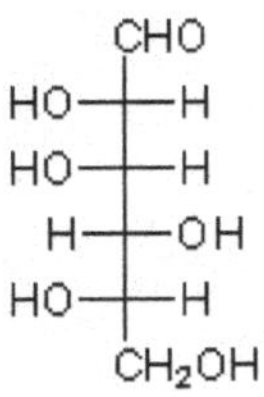
CHO
HO H
HO H
H OH
HO H
CH₂OH

E. None of the above

5. An amylose is a form of starch that has:

A. α(1→4) and β(1→4) glycosidic linkages between glucose units
B. glycosidic linkages joining glucose units
C. only β(1→4) glycosidic linkages between glucose units
D. only α(1→4) glycosidic linkages between glucose units
E. carbon-carbon glycosidic linkages joining glucose units

6. Fructose can be classified as a(n):

A. aldoketose
B. aldopentose
C. aldohexose
D. ketopentose
E. ketohexose

7. All of the statements concerning monosaccharides are correct, EXCEPT:

A. the number of possible stereoisomers is 2^n, where n is the number of chiral carbon atoms in the molecule
B. monosaccharides with 5 or 6 carbon atoms exist in solution in the cyclic form
C. the two different cyclic forms of a monosaccharide are tautomers
D. a molecule is classified as a D or L isomer by the position of the hydroxyl group on the chiral center farthest from the carbonyl group
E. monosaccharides have the general formula $C_n(H_2O)_n$, but this only describes the number and kinds of atoms, not their structure

8. Which of the following statements describes most monosaccharides?

A. They are unsaturated compounds
B. They are rarely monomers in nature
C. They are composed of carbon, hydrogen, and oxygen, with each carbon bound to at least one oxygen
D. They are insoluble
E. None of the above

9. Galactose has the structure shown below. It can be classified as a(n):

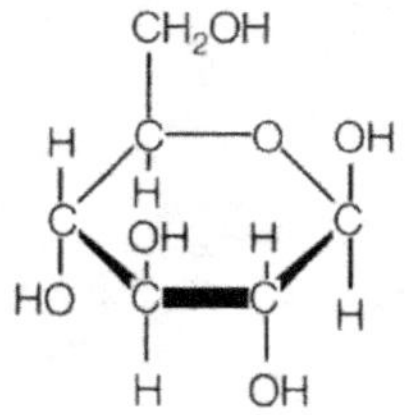

A. monosaccharide
B. disaccharide
C. ketose
D. ribose
E. ketone

10. The three elements in carbohydrates are [ ], [ ] and [ ]:

   A. nitrogen, oxygen, hydrogen
   B. carbon, hydrogen, oxygen
   C. carbon, hydrogen, water
   D. nitrogen, oxygen, carbon
   E. carbon, nitrogen, hydrogen

11. When a monosaccharide forms a cyclic hemiacetal, the carbon atom that contained the carbonyl group is identified as the [ ] carbon atom because [ ]:

   A. D ... the carbonyl group is drawn to the right
   B. L ... the carbonyl group is drawn to the left
   C. anomeric ... its substituents can assume an α or β position
   D. acetal ... it forms bonds to an ~OR and an ~OR'
   E. enantiomeric ... depending on its position, the resulting ring can have a mirror image

12. Which of the following is a non-reducing sugar?

   A. mannose
   B. sucrose
   C. lactose
   D. glucose
   E. galactose

13. Which of the following is correct regarding the classification of carbohydrate isomers?

   A. Each D-aldohexose has exactly two anomers
   B. There are 16 D-aldohexose stereoisomers
   C. There are 8 aldohexose stereoisomers
   D. Glucose has the same number of stereoisomers as fructose
   E. None of the above

14. Which of the following contains α(1→6) branches?

   A. cellulose
   B. sucrose
   C. amylose
   D. glycogen
   E. maltose

15. Humans cannot digest cellulose because they:

   A. have intestinal flora which uses up β(1→4) glycosidic bonds
   B. are poisoned by β(1→4) glycosidic bonds
   C. are allergic to β(1→4) glycosidic bonds
   D. lack the necessary enzymes to break β(1→4) glycosidic bonds
   E. cannot digest chlorophyll

16. Mutarotation is a process where:

A. glucose undergoes reaction to form an equilibrium mixture of anomers
B. glucose reacts with alcohol forming a cyclic acetal
C. the aldehyde group present in sugar is converted to a hemiacetal
D. two glucose molecules react to form a disaccharide
E. glucose isomerizes to fructose

17. Fructose does not break apart into smaller units because it is a(n):

A. monosaccharide
B. polysaccharide
C. hexose
D. aldose
E. disaccharide

18. Two cyclic isomeric sugars that only differ in the position of the ~OH group attached to the hemiacetal carbon are:

A. enantiomers
B. mutarotation
C. anomers
D. epimers
E. diastereomers

19. Which molecule is a reducing sugar?

A. sucrose
B. maltose
C. starch
D. glycogen
E. amylopectin

20. The glycosidic bond that connects the two monosaccharides in lactose is:

A. α(1→6)
B. α,β(1→2)
C. α(1→4)
D. β(1→4)
E. none of the above

21. If one of the carboxylic acids of tartaric acid is reduced to an aldehyde and the other is replaced with a CH₂OH group, which of the following results?

A. aldotetrose
B. aldotriose
C. ketotriose
D. aldopentose
E. none of the above

22. The cyclic structure shown below is classified as a(n):

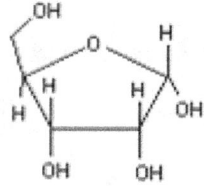

A. ketose
B. aldehyde
C. pentose
D. hexose
E. acetal

23. An acyclic sugar shown by a Fischer projection is classified as a D-isomer if the hydroxyl group on the chiral carbon:

A. nearest to the carbonyl group points to the left
B. nearest to the carbonyl group points to the right
C. farthest from the carbonyl group points to the left
D. farthest from the carbonyl group points to the right
E. α on the carbonyl group points to the left

24. What is the structural difference between deoxyribose and ribose?

A. methyl group
B. hydroxyl group
C. carboxyl group
D. carbonyl group
E. none of the above

25. Each of the following yields a positive Benedict's test for reducing sugars, EXCEPT:

A. α-1,1-glucose-glucose
B. β-1,4-glucose-glucose
C. glucose
D. fructose
E. maltose

26. How many stereoisomers do carbohydrates have?

A. 2^n, where n is the number of chiral centers
B. $2^n - 1$, where n is the number of chiral centers
C. $2n$, where n is the number of chiral centers
D. $2n - 1$, where n is the number of chiral centers
E. $2n - 2$, where n is the number of chiral centers

27. Which of the following molecules is a disaccharide?

A. fructose
B. cellulose
C. amylose
D. glucose
E. lactose

28. Ribose can be classified as a(n):

A. aldoketose
B. aldopentose
C. aldohexose
D. ketopentose
E. ketohexose

29. Which functional group is not usually found in carbohydrates?

A. hydroxyl
B. ether
C. amide
D. aldehyde
E. ketone

30. The linkage between Subunits 2 and 3 in acarbose is best described as the following?

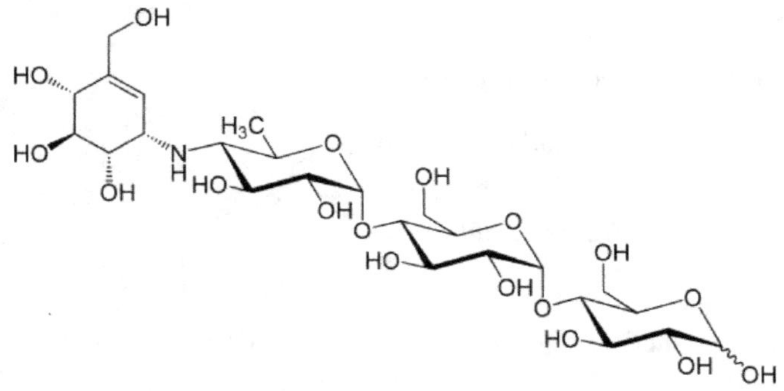

A. β-(1→4)
B. β-(1→2)
C. α-(1→6)
D. α-(1→4)
E. α-(1→2)

31. A monosaccharide consisting of 5 carbon atoms, one of which is a ketone, is classified as a(n):

A. aldohexose
B. ketotetrose
C. aldotetrose
D. aldopentose
E. ketopentose

32. The monosaccharide shown below is a(n)

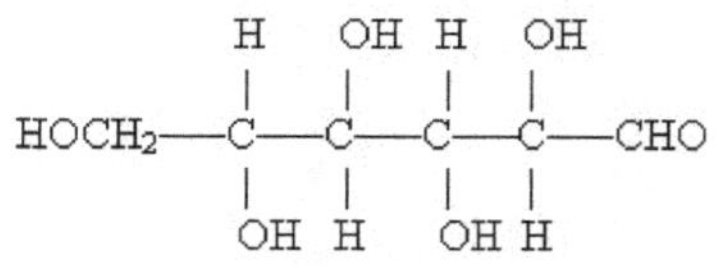

HOCH₂—C(H)(OH)—C(OH)(H)—C(H)(OH)—C(OH)(H)—CHO

A. aldohexose
B. aldopentose
C. ketohexose
D. ketopentose
E. aldoheptose

33. In the sucrose molecule shown below, which bond joins the disaccharide?

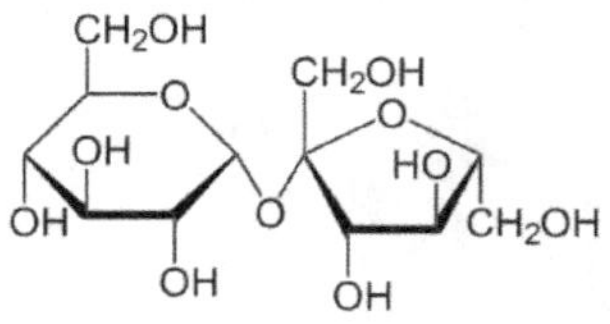

A. α-glucosidic linkage
B. β-glucosidic linkage
C. acetal
D. hemiacetal
E. ester

34. What is the major biological function of the glycogen biomolecule?

A. It is used to synthesize disaccharides
B. It is the building block of proteins
C. It stores glucose in animal cells
D. It is a storage form of sucrose
E. It stores glucose in plant cells

35. What is the minimal number of chiral centers necessary for a *meso* carbohydrate?

A. 0
B. 1
C. 2
D. 3
E. 4

36. A carbohydrate that gives two molecules when it is completely hydrolyzed is a:

A. polysaccharide
B. starch
C. monosaccharide
D. disaccharide
E. trisaccharide

37. How many degrees of unsaturation are present in acarbose?

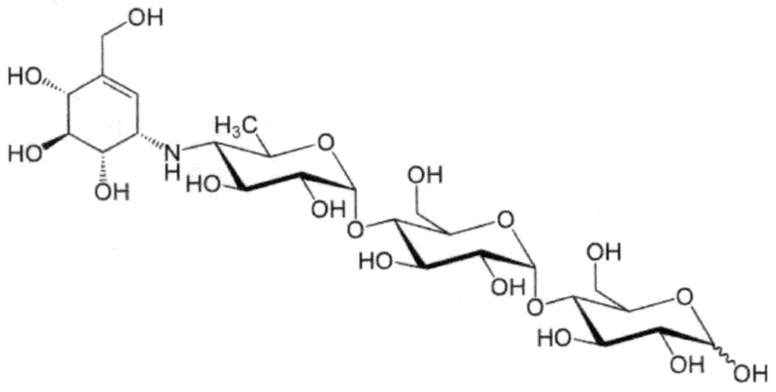

- A. 1
- B. 3
- C. 4
- D. 5
- E. 6

38. Which group of carbohydrates CANNOT be hydrolyzed to give smaller molecules?

- A. oligosaccharides
- B. trisaccharides
- C. disaccharides
- D. monosaccharides
- E. polysaccharides

39. D-ribulose has the following structural formula. To what carbohydrate class does ribulose belong?

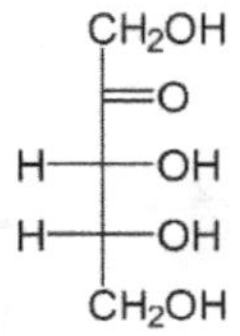

- A. ketotetrose
- B. ketopentose
- C. aldotetrose
- D. aldopentose
- E. ketohexose

40. Carbohydrate can be defined as a molecule:

- A. composed of carbon atoms bonded to water molecules
- B. composed of amine groups and carboxylic acid groups bonded to a carbon skeleton
- C. composed mostly of hydrocarbons and soluble in non-polar solvents
- D. that is an aldehyde or ketone and has more than one hydroxyl group
- E. ending in ~ase

41. The diagram below shows a step in which of the following processes?

α-D-glucopyranose ⇌ β-D-glucopyranose

A. anomerization
B. mutarotation
C. hemiacetal formation
D. aldehyde formation
E. oxidation

42. The compound shown is best described as:

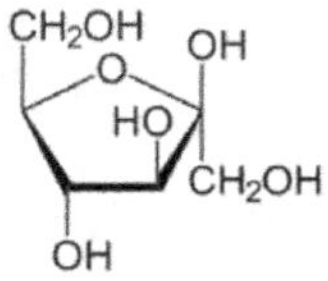

A. furanose form of an aldopentose
B. pyranose form of an aldopentose
C. pyranose form of a ketopentose
D. furanose form of a ketopentose
E. pyranose form of an aldohexose

43. To what class of compounds does glucose belong?

A. cyclic acetal
B. cyclic hemiacetal
C. cyclic ketone
D. cyclic aldehyde
E. cyclic acid

44. Glucose undergoes an isomerization to yield fructose 1,6-bisphosphate. Which of the following is an isomerization reaction?

A. $CH_3CH_2COCl + H_2O \rightarrow CH_3CH_2COOH + HCl$
B. $CH_2CH_2OHCOOCH_2 + CH_3OH \rightarrow CHCH_2OHCOOCH_3 + CH_3CH_2OH$
C. $CH_2OHCOCH_2CH_2CH_3 \rightarrow CHOCHOHCH_2CH_2CH_3$
D. $CH_3CH_2CH_2CHOHCH_3 \rightarrow CH_3CH_2CHCHCH_3$
E. None of the above

45. Disaccharides are best characterized as:

A. two monosaccharides linked by a nitrogen bond
B. two peptides linked by a hydrogen bond
C. two monosaccharides linked by an oxygen bond
D. two amino acids linked by a peptide bond
E. two glycogens linked by a fatty acid

46. If Benedict's reagent is used to test for reducing sugars, tartaric acid yields:

A. positive result only in the open-chain configuration
B. ambiguous result
C. negative result
D. positive result
E. positive result in basic solution

47. Maltose is a:

A. trisaccharide
B. polysaccharide
C. monosaccharide
D. disaccharide
E. phosphosaccharide

48. Which molecule shown is a D-isomer?

A.
CHO
H—C—OH
HO—C—H
H—C—OH
H—C—OH
CH$_2$OH

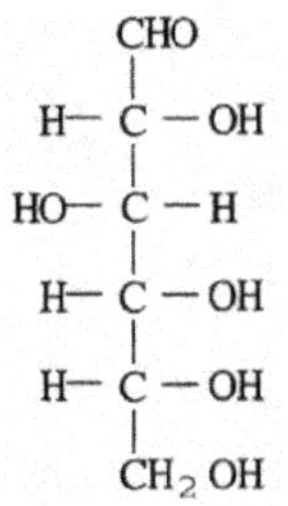

B.
CHO
HO—C—H
HO—C—H
HO—C—H
HO—C—H
CH$_2$OH

C.
CHO
HO—C—H
HO—C—H
H—C—OH
HO—C—H
CH$_2$OH

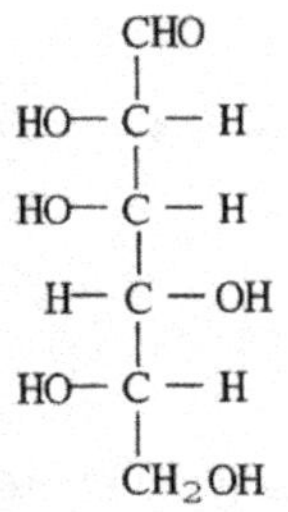

D.
CHO
HO—C—H
H—C—H
HO—C—OH
HO—C—H
CH$_2$OH

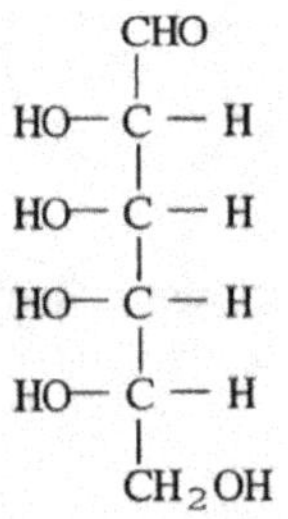

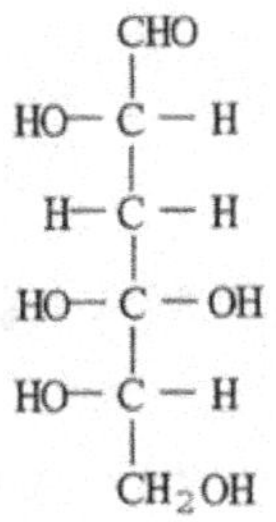

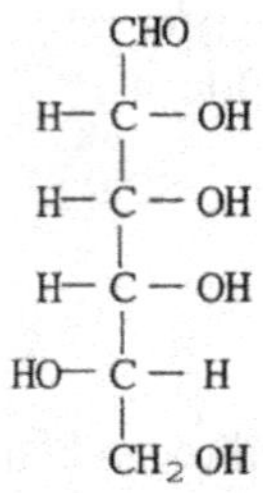

E.
CHO
H—C—OH
H—C—OH
H—C—OH
HO—C—H
CH$_2$OH

Notes for active learning

Notes for active learning

Nucleic Acids

1. Which of the following statements describes a cellular activity during transcription and translation?

   A. Protein is formed
   B. Transfer RNA is used to link peptides
   C. The DNA helix uncoils
   D. Nucleotides bind to complementary bases
   E. None of the above

2. The two strands of DNA in the double helix are held by:

   A. dipole–dipole attractions
   B. metallic bonds
   C. ionic bonds
   D. covalent bonds
   E. none of the above

3. What is the major difference between nucleotides of deoxyribonucleic acid and ribonucleic acid?

   A. The ribose nucleic acid is missing a hydroxyl group on the sugar
   B. The ribose nucleic acid is missing a hydroxyl group on the nitrogenous base
   C. The deoxyribose nucleic acid is missing a hydroxyl group on the sugar
   D. The deoxyribose nucleic acid is missing a hydroxyl group on the nitrogenous base
   E. Only the nitrogenous bases are different

4. In biochemical reactions, the reduction of carbonyl groups is carried out by:

   A. pyruvic acid
   B. NADH
   C. $LiAlH_4$
   D. $NaBH_4$
   E. lactic acid

5. What is the complementary DNA sequence to ATATGGTC?

   A. CGCGTTGA
   B. GCGCAACT
   C. TATACCAG
   D. TUTUCCAG
   E. UAUACCAG

6. What happens to DNA when placed into an aqueous solution at physiological pH?

   A. Individual DNA molecules repel each other due to the presence of positive charges
   B. DNA molecules bind to negatively charged proteins
   C. Individual DNA molecules attract each other due to the presence of positive and negative charges
   D. Individual DNA molecules repel each other due to the presence of negative charges
   E. DNA molecules bind to neutral proteins

7. What is the process by which a DNA molecule synthesizes a complementary single strand of RNA?

A. translation
B. transcription
C. replication
D. duplication
E. none of the above

8. Consider the following types of compounds:

I. amino acid
II. nitrogen-containing base
III. phosphate group
IV. five-carbon sugar

From which of the above compounds are the monomers (i.e., nucleotides) of nucleic acids formed?

A. I only
B. I and II only
C. II and IV only
D. II, III and IV only
E. I, II, III and IV

9. The nucleotide sequence, T-A-G, stands for

A. threonine-alanine-glutamine
B. thymine-adenine-guanine
C. tyrosine-asparagine-glutamic acid
D. thymine-adenine-glutamine
E. none of these

10. The two new DNA molecules formed in replication:

A. contain one parent and one daughter strand
B. both contain only the parent DNA strands
C. both contain only two new daughter DNA strands
D. are complementary to the original DNA
E. are identical, with one containing parental strands and the other containing daughter strands

11. Which of the following is found in an RNA nucleotide?

I. phosphoric acid II. nitrogenous base III. ribose sugar

A. I only
B. II only
C. III only
D. I and II only
E. I, II and III

12. Which of the following descriptions of the nucleoside uridine does NOT apply to the molecule's structure?

    A. The uracil base is directly bonded to the 1' position of ribofuranose in the α position
    B. The ribofuranose moiety is only in the D configuration
    C. Nitrogen, at position 1 in the uracil base, is directly bonded to the ribofuranose moiety
    D. The 5'~OH group is replaced with phosphate(s) in the nucleotide structure
    E. None of the above

13. Which of these nitrogenous base pairs is in DNA?

    A. adenine-guanine
    B. adenine-uracil
    C. adenine-cytosine
    D. guanine-cytosine
    E. guanine-thymine

14. If one strand of a DNA double helix has the sequence AGTACTG, what is the sequence of the other strand?

    A. GACGTCA
    B. AGTACTG
    C. GTCATGA
    D. TCATGAC
    E. AGUACUG

15. The main role of DNA is to provide instructions on how to build:

    I. lipids   II. carbohydrates   III. proteins

    A. I only
    B. II only
    C. III only
    D. I and II only
    E. I, II and III

16. Nucleic acids are polymers of [ ] monomers.

    A. monosaccharide
    B. fatty acid
    C. DNA
    D. nucleotide
    E. none of the above

17. During DNA transcription, a guanine base on the template strand codes for which base on the growing RNA strand?

    A. guanine
    B. thymine
    C. adenine
    D. cytosine
    E. uracil

College Organic Chemistry Practice Questions with Detailed Explanations

18. Cellular respiration produces the same products as:

A. do nucleic acids
B. do campfires
C. does catabolism
D. does anabolism
E. none of the above

19. What intermolecular force connects strands of DNA in the double helix?

A. hydrogen bonds
B. ionic bonds
C. amide bonds
D. ester bonds
E. none of the above

20. How are codons and anticodons related?

A. Codons are the base pairs on a tRNA that bind to complementary strands of DNA and produce proteins
B. Anticodons are the codons on the mRNA used to bind to DNA
C. Codons start the process of transcription; anticodons end the process
D. Codons and anticodons are complementary base pairs that encode for an amino acid
E. None of the above

21. What type of biological compound is a polymer composed of sugar, a base, and phosphoric acid?

A. nucleic acid
B. lipid
C. carbohydrate
D. protein
E. none of the above

22. What is the term for how a DNA molecule synthesizes an identical molecule of DNA?

A. transcription
B. translation
C. duplication
D. replication
E. none of the above

23. The bonds that link the base pairs in the DNA double helix are [ ] bonds?

A. hydrophobic
B. hydrogen
C. peptide
D. ionic
E. ester

24. Translation is the process whereby:

A. protein is synthesized from DNA
B. protein is synthesized from mRNA
C. DNA is synthesized from DNA
D. DNA is synthesized from mRNA
E. mRNA is synthesized from DNA

25. What is the main function of the structural difference of the sugar that makes up RNA compared to the sugar of DNA?

   A. It stabilizes the RNA outside the nucleus
   B. It acts as an energy source to produce proteins
   C. It allows the RNA to be easily digested by enzymes
   D. It keeps the RNA from binding tightly to DNA
   E. All the above

26. The double helix of DNA is stabilized mainly by:

   A. ionic bonds
   B. covalent bonds
   C. ion–dipole bonds
   D. ester bonds
   E. hydrogen bonds

27. Which of the following is the correct listing of DNA constituents in the order of increasing size?

   A. Nucleotide, codon, gene, nucleic acid
   B. Nucleic acid, nucleotide, codon, gene
   C. Nucleotide, codon, nucleic acid, gene
   D. Gene, nucleic acid, nucleotide, codon
   E. Gene, codon, nucleotide, nucleic acid

28. The number of adenines in a DNA molecule is equal to the number of thymines because:

   A. adenines are paired opposite of guanine in a DNA molecule
   B. of the strong attraction between the nucleotides of adenine and thymine
   C. the structure of adenine is similar to uracil
   D. adenine is paired to cytosine in a DNA molecule
   E. none of the above

29. What is the sugar component in RNA called?

   A. fructose
   B. galactose
   C. glucose
   D. ribose
   E. deoxyribose

30. What is the process in which the DNA double helix unfolds, and each strand serves as a template for synthesizing a new strand?

   A. translation
   B. replication
   C. transcription
   D. complementation
   E. restriction digestion

31. If NADH is the reduced form of the high-energy intermediate dinucleotide, which of the following is the oxidized form of this vital biomolecule?

   A. NAD^{+2}
   B. NAD
   C. NAD^+
   D. $NADH_2$
   E. $NADH^+$

32. Which of the following illustrates the direction of flow for protein synthesis?

   A. RNA → protein → DNA
   B. DNA → protein → RNA
   C. RNA → DNA → protein
   D. DNA → RNA → protein
   E. protein → RNA → DNA

33. The attractive force between the cyclic amine bases in DNA is/are:

   A. disulfide bridges
   B. hydrogen bonding
   C. hydrophobic stacking
   D. ionic interactions of salt bridges
   E. ion-diploe interaction

34. The three-base sequence in mRNA specifying the amino acid is a(n):

   A. rRNA
   B. anticodon
   C. codon
   D. tRNA
   E. nucleotide

35. Which of the following is an RNA codon for protein synthesis?

   I. GUA   II. CGU   III. ACG

   A. I only
   B. II only
   C. III only
   D. I and II only
   E. I, II and III

36. During DNA replication, an adenine base on the template strand codes for which base on the complementary strand?

   A. thymine
   B. guanine
   C. cytosine
   D. adenine
   E. uracil

37. Which of the following is NOT part of a nucleotide?

   A. cyclic nitrogenous base
   B. fatty acid
   C. phosphate group
   D. cyclic sugar
   E. oxygen

38. Which of the following types of linkage is in a nucleic acid?

   A. phosphate linkage
   B. ester linkage
   C. glycoside linkage
   D. peptide linkage
   E. amide linkage

39. Which of the following codes for an amino acid during protein synthesis?

   A. RNA nucleotide
   B. RNA trinucleotide
   C. DNA nucleotide
   D. DNA trinucleotide
   E. none of the above

40. How does RNA differ from DNA?

   A. RNA is double-stranded, while DNA is single-stranded
   B. RNA is a polymer of amino acids, while DNA is a polymer of nucleotides
   C. RNA contains uracil, while DNA contains thymine
   D. In RNA G pairs with T, while in DNA G pairs with C
   E. DNA has an additional alcohol group compared to RNA

41. The one cyclic amine base that occurs in DNA but not in RNA is:

   A. cystine
   B. guanine
   C. thymine
   D. uracil
   E. adenine

42. In the synthesis of mRNA, an adenine in the DNA pairs with:

   A. guanine
   B. thymine
   C. uracil
   D. adenine
   E. cytosine

43. Nucleic acids determine the:

   A. quantity and type of prions
   B. number of mitochondria in a cell
   C. sequence of amino acids
   D. pH of the cell nucleus
   E. catabolism rate for food

44. What type of biological compound is characterized by alcohol and amine functional groups?

   A. nucleic acid
   B. lipid
   C. carbohydrate
   D. protein
   E. amino acid

45. DNA is a(n):

A. peptide

B. protein

C. nucleic acid

D. enzyme

E. steroid

46. Which of the following amine bases is NOT present in DNA?

A. adenine

B. cytosine

C. guanine

D. uracil

E. thymine

47. In transcription:

A. both strands of the DNA are copied

B. uracil pairs with thymine

C. a double helix containing one parent strand and one daughter strand is produced

D. the mRNA produced is identical to the parent DNA

E. the mRNA contains the genetic information from DNA

48. The two strands of the double helix of DNA are held by:

A. disulfide bridges

B. ionic bonds

C. hydrogen bonds

D. covalent bonds

E. sugar-phosphate bonds

Notes for active learning

Notes for active learning

Answer Keys and Detailed Explanations

If you benefited from this book, we would appreciate if you left a review on Amazon, so others can learn from your input. Reviews help us understand our customers' needs and experiences while keeping our commitment to quality.

Answer Keys

Organic Chemistry Nomenclature

1: D	11: D	21: A	31: A	41: A
2: C	12: A	22: A	32: B	42: A
3: A	13: D	23: B	33: B	43: C
4: A	14: A	24: A	34: A	44: C
5: B	15: D	25: D	35: C	45: D
6: C	16: B	26: B	36: B	46: B
7: B	17: A	27: D	37: D	
8: D	18: E	28: B	38: A	
9: E	19: B	29: B	39: E	
10: B	20: B	30: C	40: C	

Covalent Bond

1: D	11: B	21: C	31: C	41: B
2: B	12: E	22: E	32: C	42: E
3: A	13: E	23: C	33: A	43: C
4: D	14: B	24: D	34: D	44: A
5: B	15: B	25: B	35: B	45: B
6: C	16: E	26: D	36: A	46: B
7: D	17: D	27: C	37: C	47: A
8: D	18: C	28: B	38: D	48: C
9: B	19: E	29: B	39: D	
10: B	20: E	30: E	40: B	

Stereochemistry

1: C	11: B	21: D	31: A	41: D
2: D	12: D	22: B	32: A	42: A
3: D	13: A	23: D	33: C	43: B
4: A	14: D	24: C	34: A	44: B
5: C	15: A	25: A	35: B	45: A
6: C	16: D	26: B	36: A	46: B
7: E	17: D	27: B	37: A	47: D
8: E	18: C	28: E	38: E	
9: B	19: A	29: D	39: C	
10: B	20: C	30: C	40: C	

Molecular Structure and Spectra

1: D	11: C	21: B	31: B
2: D	12: A	22: C	32: A
3: D	13: C	23: A	33: D
4: A	14: B	24: C	34: B
5: A	15: C	25: A	35: B
6: E	16: B	26: A	36: B
7: D	17: B	27: E	
8: E	18: C	28: A	
9: B	19: A	29: D	
10: C	20: A	30: E	

Alkanes and Alkyl Halides

1: C	11: D	21: C	31: D
2: D	12: D	22: D	32: B
3: B	13: D	23: B	33: A
4: D	14: C	24: C	34: B
5: E	15: B	25: C	35: D
6: A	16: D	26: A	36: E
7: B	17: D	27: D	
8: D	18: B	28: C	
9: C	19: B	29: A	
10: B	20: D	30: A	

Alkenes

1: C	11: A	21: E	31: B
2: B	12: C	22: A	32: A
3: D	13: A	23: C	33: A
4: D	14: A	24: C	34: A
5: D	15: C	25: A	35: C
6: A	16: D	26: A	36: A
7: A	17: A	27: E	
8: D	18: B	28: D	
9: A	19: D	29: A	
10: C	20: C	30: B	

Alkynes

1: E	11: C	21: B	31: D
2: C	12: D	22: C	32: A
3: C	13: B	23: D	33: C
4: A	14: A	24: C	34: C
5: A	15: C	25: D	35: C
6: D	16: B	26: B	36: B
7: B	17: D	27: D	
8: A	18: B	28: A	
9: C	19: C	29: C	
10: C	20: C	30: B	

Aromatic Compounds

1: C	11: C	21: B	31: A
2: B	12: A	22: D	32: A
3: E	13: D	23: C	33: B
4: D	14: C	24: D	34: C
5: A	15: D	25: C	35: A
6: D	16: E	26: C	36: D
7: C	17: A	27: A	
8: D	18: D	28: D	
9: A	19: C	29: D	
10: D	20: E	30: D	

Alcohols

1: B	11: B	21: D	31: D
2: E	12: C	22: D	32: D
3: C	13: C	23: D	33: C
4: E	14: B	24: C	34: B
5: A	15: A	25: D	35: B
6: C	16: B	26: B	36: B
7: B	17: E	27: A	
8: D	18: B	28: A	
9: B	19: D	29: A	
10: A	20: D	30: C	

Aldehydes and Ketones

1: E	11: E	21: E	31: D
2: A	12: E	22: D	32: C
3: D	13: A	23: C	33: B
4: B	14: B	24: E	34: E
5: C	15: A	25: E	35: A
6: D	16: D	26: A	36: B
7: A	17: A	27: B	
8: D	18: D	28: D	
9: D	19: D	29: A	
10: C	20: E	30: E	

Carboxylic Acids

1: C	11: B	21: E	31: C
2: D	12: B	22: C	32: E
3: C	13: C	23: D	33: B
4: B	14: B	24: A	34: A
5: A	15: D	25: C	35: D
6: C	16: B	26: B	36: A
7: E	17: A	27: A	
8: D	18: D	28: A	
9: B	19: A	29: B	
10: E	20: A	30: D	

COOH Derivatives

1: A	11: B	21: A	31: B
2: D	12: B	22: C	32: B
3: A	13: C	23: D	33: B
4: C	14: D	24: A	34: A
5: A	15: B	25: A	35: B
6: D	16: D	26: C	36: B
7: C	17: A	27: C	
8: A	18: E	28: A	
9: E	19: B	29: C	
10: E	20: A	30: C	

Answer Keys

Amines

1: B	11: B	21: D	31: C
2: C	12: A	22: A	32: A
3: A	13: A	23: C	33: E
4: E	14: C	24: A	34: B
5: B	15: E	25: C	35: E
6: D	16: A	26: B	36: A
7: B	17: B	27: D	
8: A	18: D	28: C	
9: A	19: C	29: A	
10: E	20: C	30: B	

Amino Acids, Peptides, Proteins

1: A	11: C	21: C	31: E	41: D
2: D	12: D	22: C	32: A	42: B
3: E	13: A	23: B	33: A	43: C
4: E	14: C	24: A	34: A	44: B
5: A	15: E	25: B	35: D	45: B
6: D	16: B	26: C	36: E	46: B
7: C	17: C	27: D	37: E	47: C
8: B	18: D	28: C	38: D	48: E
9: A	19: B	29: B	39: C	
10: C	20: D	30: B	40: A	

Lipids

1: D	11: C	21: B	31: E	41: E
2: D	12: B	22: A	32: C	42: B
3: D	13: D	23: B	33: A	43: C
4: A	14: D	24: D	34: D	44: C
5: D	15: B	25: C	35: B	45: C
6: D	16: C	26: E	36: D	46: D
7: B	17: B	27: A	37: B	47: D
8: B	18: D	28: A	38: D	48: D
9: D	19: C	29: B	39: C	
10: C	20: D	30: D	40: D	

Carbohydrates

1: C	11: C	21: A	31: E	41: B
2: D	12: B	22: C	32: A	42: D
3: C	13: A	23: D	33: A	43: B
4: C	14: D	24: B	34: C	44: C
5: D	15: D	25: A	35: C	45: C
6: E	16: A	26: A	36: D	46: C
7: C	17: A	27: E	37: D	47: D
8: C	18: C	28: B	38: D	48: A
9: A	19: B	29: C	39: B	
10: B	20: D	30: D	40: D	

Nucleic Acids

1: D	11: E	21: A	31: C	41: C
2: A	12: A	22: D	32: D	42: C
3: C	13: D	23: B	33: B	43: C
4: B	14: D	24: B	34: C	44: A
5: C	15: C	25: C	35: E	45: C
6: D	16: D	26: E	36: A	46: D
7: B	17: D	27: A	37: B	47: E
8: D	18: B	28: B	38: A	48: C
9: B	19: A	29: D	39: B	
10: A	20: D	30: B	40: C	

Organic Chemistry Nomenclature – Detailed Explanations

1. D is correct.

The longest carbon chain is composed of seven carbon atoms.

There are chlorine, ethyl, and methyl substituents.

2. C is correct.

The longest carbon chain is composed of six carbon atoms and is the cyclohexene substructure.

The molecule contains a methyl group (~CH$_3$) in the fourth position.

3. A is correct.

One option is to draw the atoms to determine the atom count.

Alternatively, using a subscript of n for the number of carbons, the degrees of unsaturation can be determined from the following formulae:

Alkane: C_nH_{2n+2} = 0 degrees of unsaturation

Alkene: C_nH_{2n} = 1 degree of unsaturation

Alkyne: C_nH_{2n-2} = 2 degrees of unsaturation

Rings = 1 degree of unsaturation

The reference molecule has 2 rings and therefore 2 degrees of unsaturation:

C_nH_{2n-2}

$C_8H_{16-2} = C_8H_{14}$

4. A is correct.

The *para* notation means two substituents are on opposite sides of the aromatic rings in a C_1–C_4 relationship.

When the bromine atoms are adjacent (C_1–C_2), the isomer is *ortho*.

When the bromine atoms are in a C_1–C_3 relationship, the isomer is *meta*.

5. B is correct.

The longest carbon chain in the molecule has 4 carbon atoms, and therefore the root is *but–*.

An alkene is positioned at the second carbon with a suffix of *–ene*.

6. C is correct.

The longest carbon chain is composed of five carbon atoms.

The highest priority group is an aldehyde.

The substituent groups are the hydroxymethyl group, the ethyl group, the alkene, and the alkyne.

7. B is correct.

The cyclohexane is the longest carbon chain, and the alcohol is the highest priority group, making the root name *cyclohexanol*.

The chlorine and methyl groups are the substituents.

8. D is correct.

The alcohol of this aromatic compound has the highest priority, and the carbon count starts at this position. The root (and suffix) name of the compound is "phenol."

The ethyl group is numbered as three instead of four because the numbering favors the lower position values.

9. E is correct.

The longest carbon chain of the molecule is the 3-carbon propane.

There are three substituents in this molecule: two methyl groups are on the nitrogen atom, and one methyl group is at the second position.

10. B is correct.

The highest priority groups are *trans* to one another.

The longest carbon chain has seven carbon atoms, and the alkene is in the fourth position (fourth carbon from the alcohol, which is the highest priority group).

11. D is correct.

The longest carbon chain contains five carbons.

The highest priority group is the alcohol, so the name has *–ol* as the suffix and *pent–* as the root name.

The methyl substituent is in the second position.

12. A is correct.

Propyl substituents contain a three-carbon chain. If the point of attachment (indicated by the squiggle line) is at the second carbon, the group is an *iso*propyl group.

B: *tert-* butyl

C: *sec*-butyl

D: isobutyl

E: neopentyl

isopropyl   sec-butyl   isobutyl

tert-butyl   Isopentyl or isoamyl   neopentyl   tert-pentyl or tert-amyl

Common names of alkyl substituents (recognized by IUPAC)

13. D is correct.

A: the molecule is *ortho*-fluorobenzoic acid.

B: the substituents should be given the lowest numbering and alphabetized to give 2-chloro-1,3-dinitrobenzene.

C: the substituents should be alphabetized to give 1-bromo-2-iodobenzene.

14. A is correct.

The longest chain is the cyclohexane; the chlorines are on the same side (*cis*) and are three carbons (1,3) apart.

15. D is correct.

The longest carbon chain has five carbon atoms.

The alcohol is attached to the carbon in the second position and the methyl to the carbon in the fourth position.

16. B is correct.

The longest carbon chain is a four-carbon cyclo derivative (i.e., cyclopentane).

There are four substituents located at the first and third positions in the molecule.

17. A is correct.

Draw the line formula of the carbon chain from the proper interpretation of the subscripts.

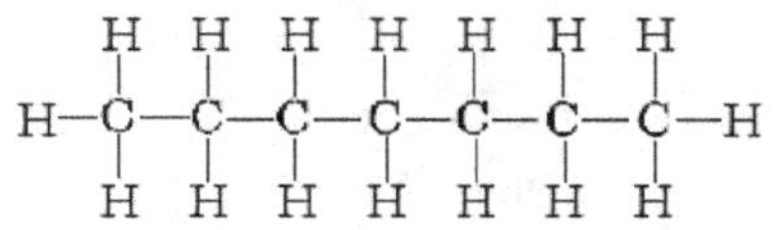

$$H-\underset{H}{\overset{H}{C}}-\underset{H}{\overset{H}{C}}-\underset{H}{\overset{H}{C}}-\underset{H}{\overset{H}{C}}-\underset{H}{\overset{H}{C}}-\underset{H}{\overset{H}{C}}-\underset{H}{\overset{H}{C}}-H$$

There are two methyl groups and five methylene (~CH_2~) in the molecule for seven carbons.

Use the following formula to calculate the degrees of unsaturation:

C_nH_{2n+2}: for an alkane (0 degrees of unsaturation)

18. E is correct.

The longest carbon chain is composed of five carbon atoms.

There are two alkenes, so the root name is "pentadiene."

There is a methyl group in the second position along the carbon chain.

19. B is correct.

Because the group has three carbon atoms, the group is a *propyl group*.

Propyl substituents exist as the *n*-propyl (i.e., *normal* or straight chain) or *isopropyl group*.

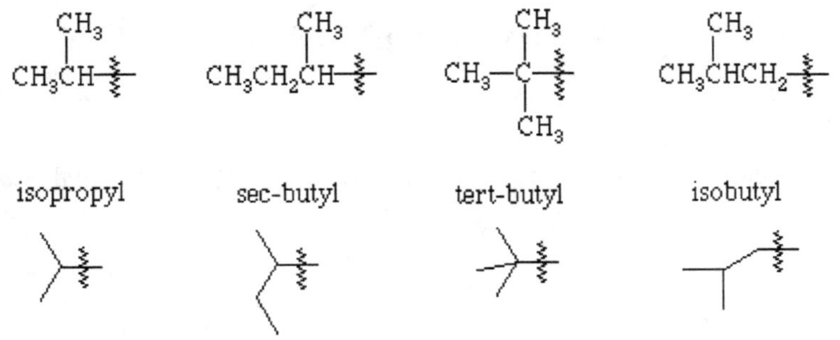

CH₃CH– with CH₃	CH₃CH₂CH– with CH₃	CH₃–C– with CH₃ and CH₃	CH₃CHCH₂– with CH₃
isopropyl	sec-butyl	tert-butyl	isobutyl

Sample common names for organic substituents used in the nomenclature

20. B is correct.

The longest carbon chain has six carbon atoms.

There are two methyl substituents in the molecule, and there is an isopropoxide substituent at the C4 position.

The stereochemistry of the alkene is *E*.

21. A is correct.

The molecule has four carbon atoms and two alkenes, hence the root name butadiene.

The double bonds of alkenes are in the first and third positions of the carbon chain.

22. A is correct.

The carboxylic acid (–*oic* acid) is considered the 1 position.

The hydroxyl group is in the *ortho* or the two positions.

23. B is correct.

The word *acetone* has two important components.

The *ace* is similar to *acetyl* and indicates that the structure includes a methyl group bonded to a carbonyl group.

The suffix –*one* indicates that the carbonyl group is a ketone.

24. A is correct.

The name above is the common name for the compound (i.e., toluene is the suffix for a benzene ring with a methyl substituent).

For IUPAC, the longest carbon chain in the compound is the six-membered benzene ring.

There are two substituent groups present in the molecule: the ethyl group and the methyl group.

The IUPAC name is 1-ethyl-3-methylbenzene

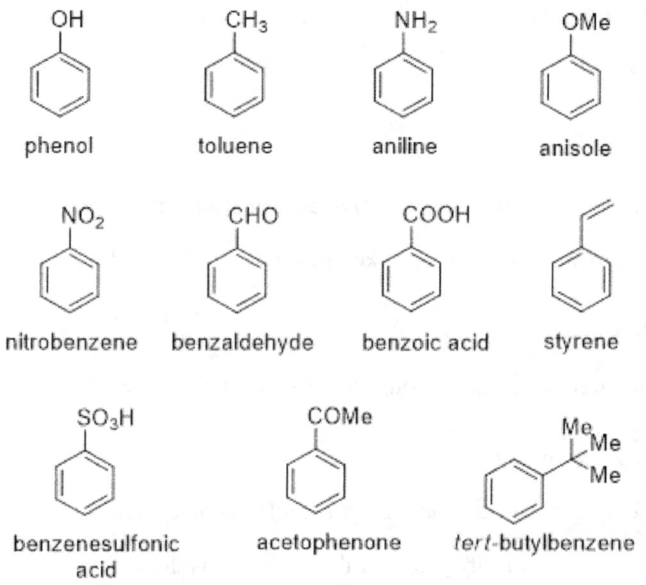

Common names for benzene derivatives: phenol, toluene, aniline, anisole, nitrobenzene, benzaldehyde, benzoic acid, styrene, benzenesulfonic acid, acetophenone, tert-butylbenzene

25. D is correct.

Cyclopropane is the only compound listed that contains three carbons.

All other answer choices have four carbons.

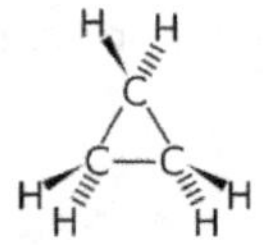

Cyclopropane with the stereochemistry of hydrogens indicated

26. B is correct.

The longest carbon chain in the molecule is six carbon atoms long.

The two methyl substituents are at the second and fourth positions in the chain.

27. D is correct.

The longest carbon chain that includes the double bond (i.e., alkene functional group) is a five-carbon molecule.

The chain is numbered with the alkene given the lowest number.

Ethyl (i.e., 2 carbon) substituent is at the second position in the carbon chain.

28. B is correct.

The longest carbon chain has 5 carbons and is cyclopentane – the root name of the molecule.

It possesses two substituent groups: the chloride and the methyl group.

Stereochemistry is indicated in the structure; the *cis-* notation is necessary for the compound's name.

29. B is correct.

Pentanal:

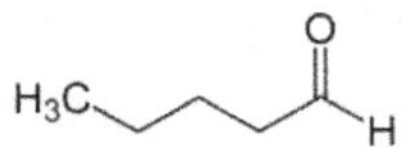

The longest carbon chain is five carbon atoms; the molecule contains an aldehyde, so the suffix is –*al*.

The suffix –*one* signifies a ketone, and ~*oic acid* is for carboxylic acid.

30. C is correct.

The three substituent groups attached to the nitrogen atom include the two methyl and the *tert*-butyl group.

31. A is correct.

The longest carbon chain for this molecule is cyclohexene.

The highest priority group of the molecule is the carboxylic acid, and the carbon atom it is bonded to should be labeled as carbon one.

Therefore, the *oxo*~ (i.e., prefix for the ketone) group is positioned at carbon two.

32. B is correct.

The longest carbon chain is composed of seven carbon atoms.

The remaining carbon groups are substituent methyl groups located at the second, fourth, and fifth positions along the carbon chain.

33. B is correct.

The longest chain of carbon atoms is the cyclohexane ring, hence the molecule's name.

There are two methyl substituents located at the first and second positions in the ring.

The groups are on the same side of the ring, so they have a *cis* orientation.

34. A is correct.

The longest carbon chain is 8 carbon atoms. The only substituent is the isopropyl group.

35. C is correct.

When numbering the longest carbon chain of a molecule, start on the end resulting in the lowest numbering for the substituent groups.

The correct molecule should have two methyl groups in the second position and one in the third position.

Detailed Explanations: Organic Chemistry Nomenclature

36. B is correct.

The longest carbon chain in the compound is composed of four carbon atoms.

The highest priority group in the molecule is the amine.

The amine is attached to carbon number 2 (i.e., *sec*-position).

The 5 common names recognized by IUPAC are isopropyl, isobutyl, *sec*-butyl, *tert*-butyl, and neopentyl.

H₃C—CH(NH₂)—CH₃ (with additional CH₃)

2-butanamine is the IUPAC name for the molecule.

37. D is correct.

Draw the four-carbon chain with the double bond at the second position in the chain.

The second and third positions have a chlorine atom and are the highest priority substituents of the alkene.

They must be oriented on the same side of the double bond because the molecule is *cis*.

38. A is correct.

The longest continuous carbon chain is six atoms long, making it a substituted hexane chain.

The molecule contains a ketone carbonyl (designated by the suffix –*one*), with the carbon atoms numbered from the end of the chain closest to the carbonyl.

The ethyl substituent is located at carbon 3, while the carbonyl is at carbon 2.

Therefore, the IUPAC name for this molecule 3-ethylhexan-2-one.

39. E is correct.

The longest carbon chain is seven carbon atoms and includes an alkene.

The alkene is the highest priority functional group and is assigned the lowest number (i.e., 1 in this example).

Therefore, the molecule has a chlorine substituent in the fourth position.

40. C is correct.

The longest carbon chain is six carbon atoms.

The two substituents are the chlorine atom and the methyl group.

The highest priority group is chlorine and therefore assumes the lowest number.

41. A is correct.

The suffix ~*oate* signifies an ester functional group as the highest priority group in the molecule.

42. A is correct.

The longest carbon chain has seven carbon atoms; it has one chlorine and one methyl substituent.

IUPAC recognizes 5 common names in the nomenclature of organic molecules:

t-butyl neopentyl isopropyl

sec-butyl isobutyl

43. C is correct.
The longest carbon chain in the molecule is composed of seven carbon atoms.

The methyl substituent is at the fifth carbon of the chain.

The alkene *pi* bond is between carbon one and carbon two.

44. C is correct.

Neutral carbon atoms maintain bonds to four other atoms.

Therefore, an acyclic hydrocarbon cannot terminate with a methylene (~CH_2~) group.

Methyl groups are at the ends of alkanes and have a formula of ~CH_3.

Using a subscript of n for the number of carbons, the *degrees of unsaturation* can be determined from the following formulae:

Alkane: C_nH_{2n+2} = 0 degrees of unsaturation

Alkene: C_nH_{2n} = 1 degree of unsaturation

Alkyne: C_nH_{2n-2} = 2 degrees of unsaturation

$CH_3CH_3CH_3$ has 3 carbons and, according to the formula C_nH_{2n+2}, should have 8 hydrogens.

This molecule has 9 hydrogens; it is impossible because it would require a carbon with 5 bonds.

A: $CH_3CHCH_3CH_2CH_3$ has 5 carbons and, according to the formula C_nH_{2n+2}, should have 12 hydrogens.

B: $CH_3CH_2CH_2CH_2CH_3$ has 5 carbons and, according to the formula C_nH_{2n+2}, should have 12 hydrogens.

D: $CH_3CH_2CH_2CH_3$ has 4 carbons and, according to the formula C_nH_{2n+2}, should have 10 hydrogens.

E: CH_3CH_3 has 2 carbons and, according to the formula C_nH_{2n+2}, should have 6 hydrogens.

45. D is correct.

Number this five-carbon chain from the highest-priority functional group – the alcohol (functional groups with higher oxidation states are higher priority). Thus, alcohol is attached to carbon 1.

The C=C double bond is between carbons 2 and 3, and the C≡C triple bond is between carbons 4 and 5.

The stereochemistry of the alkene is '*E*' (highest priority groups are on opposite sides of the alkene).

The priority groups are ranked according to the Cahn-Ingold-Prelog rules for prioritization based on the atomic number of atoms attached to the alkene.

Cis–trans relationship cannot be used to describe the molecule because the substituents across the double bond are different.

46. B is correct.

The longest carbon chain in the molecule contains six carbon atoms.

The highest (and only) functional group is the amide.

Notes for active learning

Covalent Bond – Detailed Explanations

1. D is correct.

The *benzylic* position is one carbon away from benzene or an aromatic ring. Without the aromatic ring, the cation is considered an *allylic* carbocation when one carbon from a double bond.

Vinyl means on the double bond.

2. B is correct.

The hydrogen atom bonds by overlapping its $1s$ orbital with the orbital of a bonding partner. The carbon atom is bonded to three atoms and is positively charged; indications that the carbon atom is sp^2 hybridized.

3. A is correct.

Using the formula for calculating the degrees of unsaturation reveals that 2 unsaturation elements (i.e., rings or *pi* bonds) exist.

Structures that contain atoms with satisfied octets are favored.

Molecules with charged carbon atoms tend to be less stable.

4. D is correct.

Wohler's experiment is significant because it demonstrated that organic compounds could be created from inorganic compounds. This result ran counter to the belief that the material that composed life was different from the matter of nonliving things, a theory called vitalism.

5. B is correct.

The *allylic cation* is a carbocation one *sigma* bond away from a double bond (i.e., alkene).

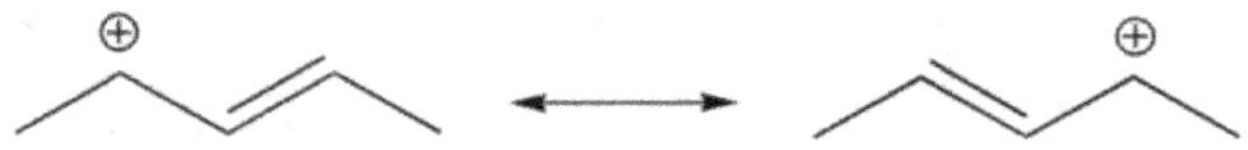

Resonance hybrids of an allylic carbocation

Methylene groups are points of saturation in the molecule that can prevent the conjugation of nearby alkenes.

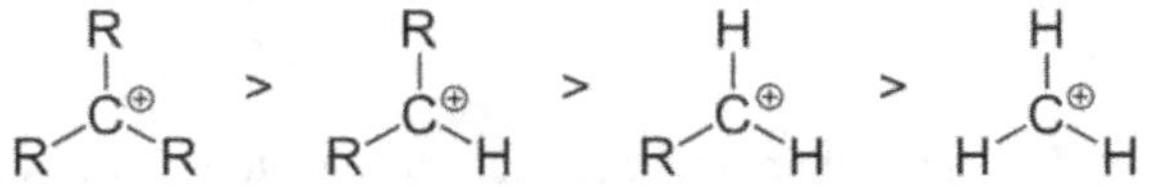

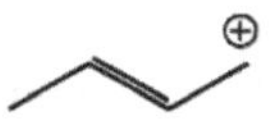

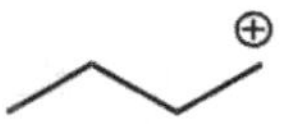

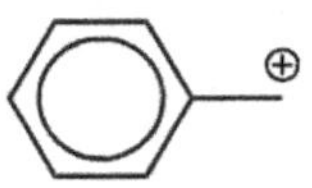

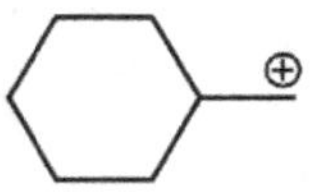

6. C is correct.

Propene is an example of an alkene, which have bonding angles of about 120 degrees.

This molecular geometry affords the substituent groups the greatest amount of spatial separation to minimize the intramolecular Van der Waals repulsions among them.

7. D is correct.

The electronegativity difference decides the distribution of electrons between bonded atoms.

This unequal distribution of electrons creates the polarity (∂^+ and ∂^-) of the bond.

8. D is correct.

The molecular formula given above is for an acyclic alkyne.

Acyclic alkynes have a linear geometry (i.e., 180 degrees), and the carbon atoms have a *sp* hybridization.

Cyclic alkynes have two fewer hydrogen atoms because those bonds are replaced with carbon-carbon bonds to form a ring.

9. B is correct.

Tertiary carbocations are more stable than *secondary* and *primary benzylic* carbocations, which are more stable than *primary* or *vinylic* carbocations.

Methyl cations are the least stable of the carbocations.

10. B is correct.

Nodes only form between the orbitals of atoms if the orbital phases have the opposite sign.

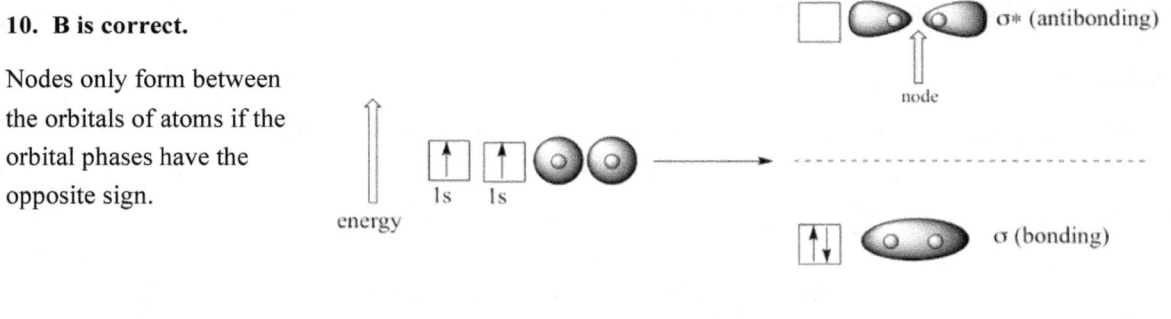

Atomic orbitals   *Molecular orbitals*

11. B is correct.

As the number of covalent bonds (i.e., triple > double > single) increases between carbon atoms, the bond order and strength increase.

Stronger bonds are shorter, so the two carbon atoms that make up the triple bond have the shortest bond.

12. E is correct.

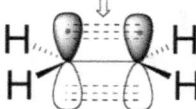

pi bond overlap between the unhybridized orbitals is indicated by the arrow

In organic molecules, the overlap of *pi* bonds is responsible for the formation of carbon-carbon *pi* bonds.

The orbitals of double and triple bonds (alkenes and alkynes, respectively) are unhybridized and are more reactive than the *sigma* bonds of single bonds.

Therefore, the reactions of alkenes typically involve breaking the alkene double bond.

13. E is correct.

Reasonable resonance forms do not allow nuclei to change positions and satisfy the octet rule.

Hydrogen can only have one bond, and carbon can only have four bonds based on the valence numbers.

Negative charges should be placed on more electronegative atoms.

14. B is correct.

The nitrogen atom does not contain a formal charge (the formal charge calculation for this atom is 0).

However, the charge is slightly negative because the nitrogen atom withdraws electron density from substituent groups *via* induction, and the lone pair represents a region of electron density.

15. B is correct.

Compounds that tend to be polar possess electronegative heteroatoms and more carbon-heteroatom bonds. If these groups can ionize to form charges, then they become even more polar.

16. E is correct.

For symmetric compounds like acetylene, no molecular dipole moment exists because the electron density is evenly distributed on either side of the triple bond.

Asymmetric molecules have molecular dipoles.

17. D is correct.

Except for structure D, the other structures are intermediates for electrophilic aromatic substitution.

These resonance intermediates typically have one cation and no negative charges in the ring.

continued…

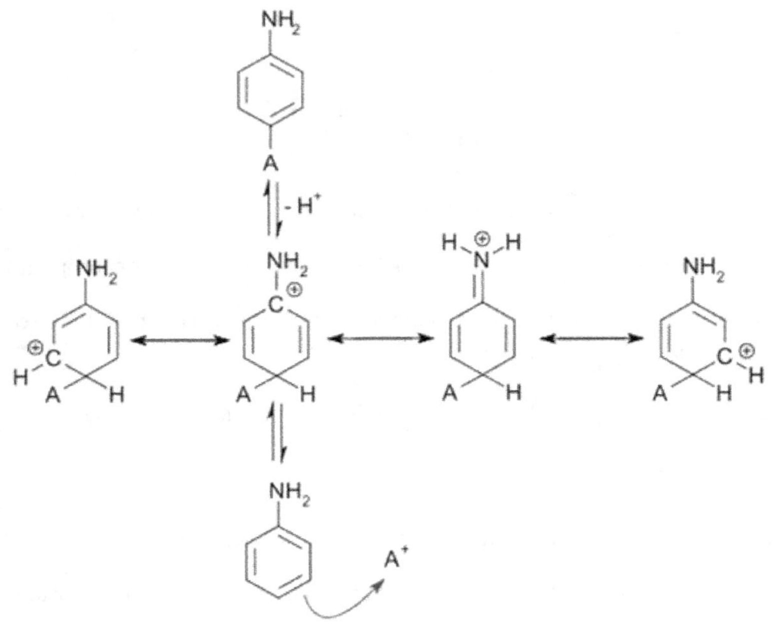

Resonance hybrids of aniline during electrophilic aromatic substitution

18. C is correct.

The bonds between atoms (i.e., intramolecular) are stronger than bonds between molecules (i.e., intermolecular).

Examples of intermolecular bonds are hydrogen, dipole-dipole, dipole-induced dipole, and van der Waals.

Sigma bonds are single bonds and involve overlap along the internuclear axis between the atoms. S*igma* bonds are stronger than the *pi* bond of a double bond.

A: *hydrogen bonding* is a common intramolecular bond in which a *hydrogen* atom of one molecule is attracted to an electronegative atom (nitrogen, oxygen, or fluorine).

B: *dipole-dipole* bonds are attractive forces between the positive end of one polar molecule and the negative end of another polar molecule.

D: *ionic bonds* result from the complete transfer of valence electron(s) between atoms; it generates two oppositely charged ions. The metal loses electrons to become a positively charged cation, whereas the nonmetal accepts those electrons to become a negatively charged anion.

Ionic bonds can be disrupted in water and much weaker in aqueous solutions than in a dry environment.

19. E is correct.

The nitrogen lone pairs of amines (e.g., pyrrole) are sp^3 hybridized unless there is a *pi* bond between the nitrogen and carbon atom (e.g., pyridine).

Pyrrole has an sp^3 hybridized nitrogen (lone pair in the ring)

The nitrogen lone pair on pyridine is in an sp^2 hybridized orbital and is not a part of the aromatic system.

Pyridine has a sp^2 hybridized nitrogen (lone pair outside the ring)

20. E is correct.

Only the hydrogen atom can bond to other atoms with its unhybridized $1s$ orbital.

The *sigma* bonding is described by the hybridization of their atomic orbitals before bonding with other atoms.

Therefore, when a carbon atom forms a *sigma* bond, it does so by overlapping one of its hybridized orbitals.

However, *pi* bonding involves the indirect overlap of unhybridized p orbitals.

21. C is correct.

Multiple bonds and rings introduce degrees of unsaturation.

Using a subscript of n for the number of carbons, the *degrees of unsaturation* is determined by:

Alkane: C_nH_{2n+2} = 0 degrees of unsaturation

Alkene: C_nH_{2n} = 1 degree of unsaturation

Alkyne: C_nH_{2n-2} = 2 degrees of unsaturation

Rings = 1 degree of unsaturation

Double bonds = 1 degree of unsaturation

There is one degree of unsaturation for this compound (the ring).

With no other degrees of unsaturation present, there are no alkenes or double bonds present in the molecule.

22. E is correct.

Hydrogen atoms only bond with other atoms using their $1s$ orbital.

The nitrogen atoms hybridize their atomic orbitals to produce sp^3 orbital hybrids for bonding.

sp^3 orbitals are used because the nitrogen atom forms bonds of equal length with 4 other atoms.

23. C is correct.

The bond dipole in H–F is the largest because of the large difference in electronegativity between hydrogen and fluorine. The electrostatic attraction pulls the atoms closer, so the bond is the shortest and the strongest.

The bond dipole in H–I is the smallest because the electronegativity between the atoms is the lowest; therefore, H–I bond is the longest and weakest.

24. D is correct.

Tertiary carbocations are more stable than secondary carbocations, which are more stable than primary.

Carbocation stability: 3° > 2° > 1° > methyl

The more substituted the cation, the more it benefits from hyperconjugation stabilizing factors.

25. B is correct.

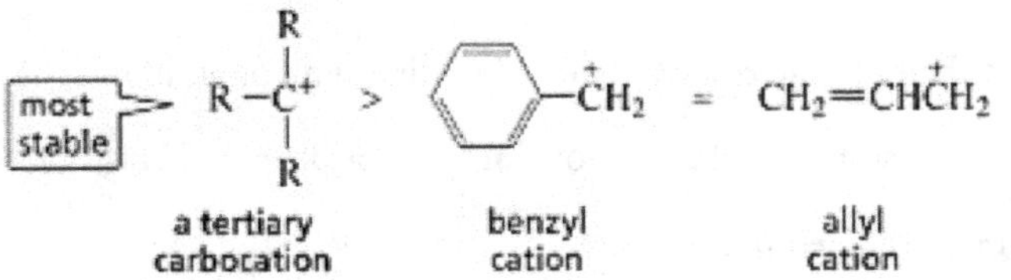

Carbocations are stabilized by resonance and by hyperconjugation.

The phenyl ring offers additional stability due to resonance structures with the delocalization of the pi electrons.

The primary non-conjugated carbocation (shown below) will be the least stable (except vinyl cation).

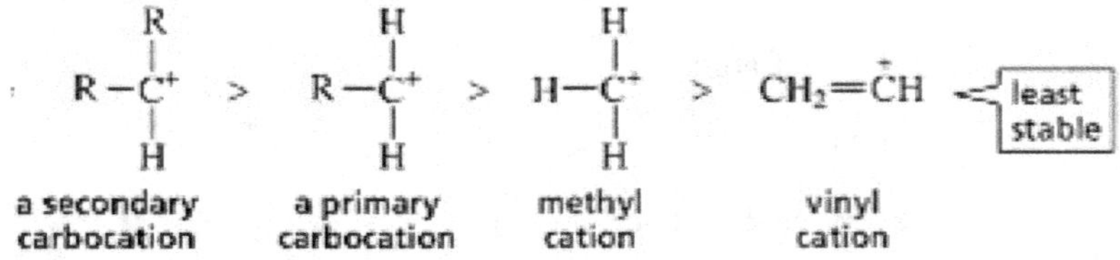

Relative stabilities of carbocations

26. D is correct.

The hybridized orbital is a combination of one *s* and three *p* orbitals.

Therefore, the energy of the hybridized orbital is between the energy of the combined orbitals.

The *s* orbital has lower energy than the *p* orbital.

27. C is correct.

The two allylic cations are resonance forms of each other.

Resonance involves the movement of conjugated *pi* systems, not of the atoms.

A and B are constitutional isomers (i.e., same molecular formula, but different connectivity).

28. B is correct.

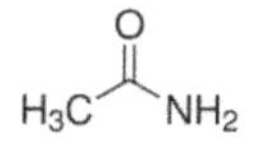

acetamide

The conjugation observed (shown below) for carboxylic acids and their derivatives (i.e., acetamide) causes the carbonyl to adopt a bond length that is between a C–O single and C=O double bond.

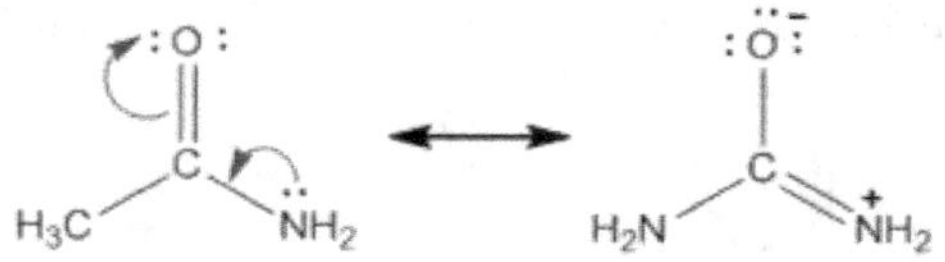

Resonance structures of acetamide

The orbitals involved in resonance are the filled, donating *p* orbital of the nitrogen atom and the adjacent electron-accepting carbonyl *pi** orbital.

29. B is correct.

Acetylene is the common name for ethyne (C_2H_2). Alkynes are linear with a bond angle of 180°.

H—C≡C—H

Acetylene has a triple bond and therefore contains sp hybridized carbons.

A carbon of a triple bond is *sp* hybridized (i.e., *sp* + 2 unhybridized *p* orbitals).

Atoms in triple bonds use *sp* hybridized orbitals from the 2*s* orbital merging (i.e., hybridizing) with a 2*p* orbital.

A: 1,3,5–heptatriene contains *sp²* hybridized orbitals (i.e., double bonds) and *sp³* orbitals for single bonds. Bond angles of 120° are trigonal planar and originate from *sp²* orbitals. The molecule has single bonds, *sp³*; a portion of the molecule is tetrahedral.

C: 2-butyne is an alkyne, but only two carbons are *sp* hybridized, while the remaining carbon is *sp³* hybridized, which results from the combination of the 2*s* and the 2*p* orbitals.

Four *sp³* carbons form and the bond angle is 109.5° with tetrahedral geometry.

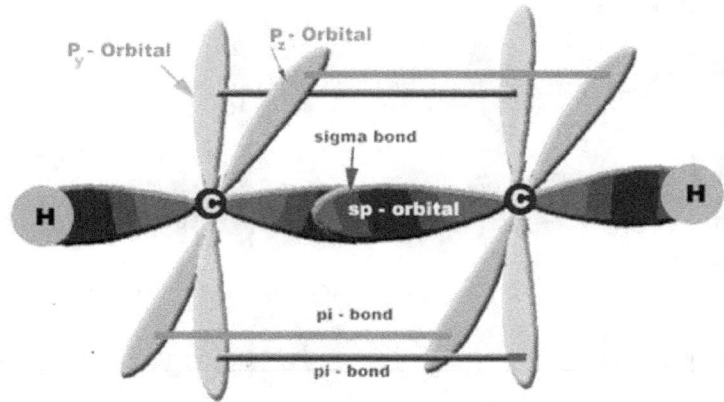

Molecular orbital structure for ethyne showing one sigma bond and two pi bonds.

D: dichloromethane has carbon *sp³* hybridized because it is attached to two hydrogens and two chlorine atoms.

The geometry is tetrahedral for the greatest separation between the substituents, and the bond angle is 109.5°.

continued…

E: 1,3-hexadiene contains *sp*² hybridized orbitals (i.e., double bonds) and *sp*³ for single bonds. Bond angles of 120° are trigonal planar and originate from *sp*² orbitals.

The molecule contains single bonds, *sp*³; therefore, a portion of the molecule is tetrahedral.

30. E is correct.

Six *sigma* bonds connect the carbon atoms in benzene.

Furthermore, the delocalized *pi* electron density in the ring; by three *pi* bonds in resonance.

Two resonance Kekule structures of benzene

31. C is correct.

Draw each bond where the electrons are distributed to satisfy the octets of the carbon and heteroatoms:

CH₃C≡N

The nitrile has a triple bond composed of a *sigma* bond and two *pi* bonds.

32. C is correct.

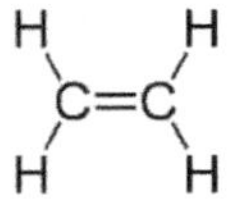

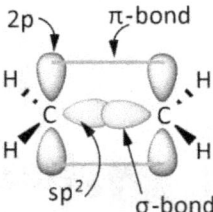

ethene

Hybridization and sigma and pi bonds indicated

Hybridization	Bond angle	Geometry
sp^3	109.5°	tetrahedral
$sp^2 + p$ (unhybridized)	120°	trigonal planar (flat)
$sp + p + p$ (two unhybridized)	180°	linear

33. A is correct.

In carbon–carbon double bonds, there is an overlap of *sp²* orbitals and a *p* orbital on the adjacent carbon atoms.

The *sp²* orbitals overlap head-to-head as a *sigma* (σ) bond, whereas the *p* orbitals overlap sideways as a *pi* (π) bond.

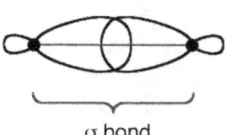

Sigma bond formation showing electron density along the internuclear axis

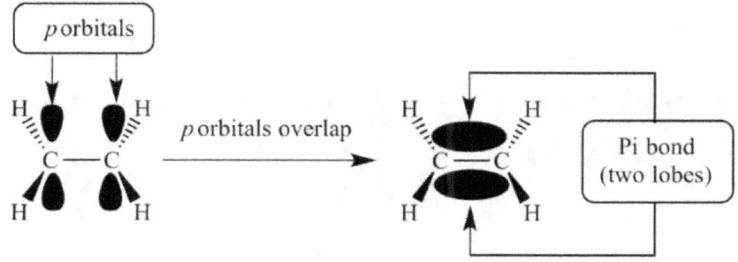

Two pi orbitals showing the pi bond formation during sideways overlap – note the absence of electron density (i.e., node) along the internuclear axis.

Bond lengths and strengths (σ or π) depend on the size and shape of the atomic orbitals and the density to overlap effectively.

The σ bonds are stronger than π bonds because head-to-head orbital overlap involves more shared electron density than sideways overlap.

The σ bonds formed from two 2*s* orbitals are shorter than those formed from two 2*p* or two 3*s* orbitals.

Carbon, oxygen, and nitrogen are in the second period ($n = 2$), while sulfur (S), phosphorus (P), and silicon (Si) are in the third period.

S, P, and Si use 3*p* orbitals to form π bonds, while C, N, and O use 2*p* orbitals.

The 3*p* orbitals are much larger than 2*p* orbitals, and a reduced probability for overlap of the 2*p* orbital of C and the 3*p* orbital of S, P, and Si.

B: S, P, and Si can hybridize, but these elements can combine *s* and *p* orbitals and (unlike C, O, and N) have *d* orbitals.

C: S, P and Si (ground-state electron configurations) have partially occupied *p* orbitals which form bonds.

D: carbon combines with elements below the second row of the periodic table.

For example, carbon commonly forms bonds with higher principal quantum number ($n > 2$) halogens (e.g., F, Cl, Br and I).

34. D is correct.

Draw the C–H bonds for this compound.

Each carbon atom of cyclohexane bonds to two hydrogen atoms with is one degree of unsaturation (the ring).

Six *sigma* bonds exist in the ring, so the number of σ bonds is 18.

35. B is correct.

There are four regions of electron density around the nitrogen atom (including the lone pair). Therefore, the nitrogen atom is sp^3 hybridized.

The bonding angles of molecules that possess nonbonding lone pairs of electrons are slightly smaller than what is predicted by the hybridization state. The nonbonding electrons exert a greater repulsive force than the bonding electrons between the central atom and the substituent groups.

However, bulky ethyl substituents increase the bond angle from approximately 107° to approximately 109.5°.

Ammonia has a bond angle of approximately 107° due to the electrostatic repulsion of the nitrogen's lone pair on the hydrogen atoms.

In amines, as substituents become larger (e.g., $(CH_3CH_2)_3N$), the bond angle between bulky groups increases, and the molecular shape approaches a tetrahedral with a bond angle of approximately 109.5°.

36. A is correct.

Three degenerate *p* orbitals exist for an atom with an electron configuration in the second (n = 2) shell or higher. The first (n = 1) shell only has an *s* orbital. The *d* orbitals become available from the third shell (n = 3).

37. C is correct.

The dipole moment is determined by the magnitudes of the individual bond dipoles and the spatial arrangement of the substituents on the molecule.

The dipole moment is greatest when there is a large difference in electronegativity of the bonded atoms. Therefore, a carbon–carbon bond (i.e., same electronegativity) has no dipole moment, while carbon–halogen bonds have moderately large dipole moments because of the electronegativity difference between carbon and halogen.

(1*R*,2*S*)-1,2-dichloro-1,2-diphenylethane is effectively *cis* due to restricted rotation.

continued…

(1R,2S)-1,2-dichloro-1,2-diphenylethane has two phenyl rings attached on one side and two chlorine groups attached on the other side.

The (R/S)–designation indicates that the two highest priority substituents (i.e., chlorine) are attached to the same side (priority according to molecular weight, so chlorine has higher priority). The highly electronegative chlorines pull electron density, creating a net dipole.

A: (1S,2S)-1,2-dichloro-1,2-diphenylethane contains only single bonds. The (R/S)–designation indicates that the two highest priority substituents (i.e., chlorine) are attached on opposite sides (with restricted rotation due to the size of the phenyl substitutes). The highly electronegative chlorines pull electron density in different spatial orientations, canceling a net dipole.

B: 1,2-dichlorobutane has carbon–chlorine highly polar bonds, but free rotation about the carbon–carbon single bond cancels any net dipole.

D: (E)-1,2-dichlorobutene differs from the Z configuration (or 1R,2S in the correct answer) because the two highest priority substituents are on opposite sides of the double bond. As the chlorine pulls electron density, dipoles cancel for a net dipole of zero.

E: (Z)-1,2-dibromobutene is the Z configuration (or 1R,2S in the correct answer) because the two highest priority substituents are on the same side of the double bond. However, bromine is less electronegative than chlorine and therefore has a smaller net dipole.

38. D is correct.

Benzene with sigma and pi bonds shown.

39. D is correct.

Formal charge = group # − nonbonding electrons − ½ bonding electrons

Nitrogen is in group V on the periodic table.

The ammonium cation has four bonds or eight bonding electrons.

The formal charge for the nitrogen atom is 5 − 0 − 8/2 = +1.

40. B is correct.

Pyrrolidine is not an aromatic compound, so the lone pair of electrons on nitrogen is available for bonding (i.e., function as a base).

The molecule is a secondary alkyl amine, and the nitrogen atom has sp^3 hybridization.

41. B is correct.

The carbonyl carbon is trigonal planar because the double-bonded carbon is sp^2 hybridized.

I: each of the two methyl carbons is sp^3 hybridized and tetrahedral.

III: none of the carbons have an unshared pair of electrons (i.e., carbanions) because carbanions are highly reactive and observed in a limited number of examples (e.g., Grignard reagent, Gillman reagent, acetylide anion) and are not present in stable molecules.

42. E is correct.

The allylic cation can delocalize the cation at the most substituted position is the most stable molecule.

The other allylic cations are not as stable because the cation is less substituted in the other resonance forms.

43. C is correct.

Resonance structures are derived from the movement of lone pairs and *pi* electrons.

Generated negative charges are placed on the more electronegative atoms and the positive charges on the less electronegative atoms.

44. A is correct.

The oxygen atom contains four regions of electron density (i.e., two lone pairs and two methyl substituents) and adopts a tetrahedral configuration.

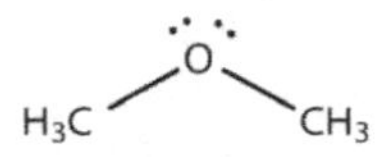

$$H_3C-O-CH_3$$

Dimethyl ether has an angle between substituent (i.e., methyl) groups of approximately 109.5 degrees.

45. B is correct.

There are four bonding patterns of carbon described by the three hybridization bonding models.

When carbon forms four single (σ) bonds, it is sp^3 hybridized.

When carbon forms one double (π) bond and two single (σ) bonds, it is sp^2 hybridized.

Carbon is sp hybridized when it forms one triple (2–π) bond and one single (σ) bond, or it forms two double (π) bonds.

O=CH–CH$_2$–CH=C=C=CH$_2$

sp^2 sp^3 sp^2 sp sp sp^2

46. B is correct.

Full arrowheads show the movement of a pair of electrons (compared to single-headed – or *fishhook* – arrows for radical reactions).

The only movement of the *pi* (π) electrons is responsible for the stable diene structures to the right. The *pi* electrons must be in conjugation.

47. A is correct.

There are four regions of electron density around the nitrogen atom (including the lone pair). Therefore, the nitrogen atom is *sp³* hybridized.

Electron geometry describes the geometry of the electron pairs, groups, and domains on the central atom, whether they are bonding or non-bonding. Molecular geometry is the name of the shape used to describe the molecule.

When atoms bond to a central atom, they do it in a way that maximizes the distance between bonding electrons. This gives the molecule its overall shape. If no lone pairs of electrons are present, the electronic geometry is the same as the molecular shape.

A lone pair occupies more space than bonding electrons, so the net effect is to bend the shape of the molecule (electron geometry conforms to predicted shape).

The shape of this molecule is trigonal pyramidal with bonds of approximately 109.5 degrees due to the bulky ethyl substituents.

The three substituents (i.e., ethyl) and the lone pair on the nitrogen result in the pyramidal shape consistent with VESPER theory.

48. C is correct.

Tertiary carbocations are more stable than secondary or primary carbocation because they experience more hyperconjugation effects from neighboring C–H bonds.

Resonance stabilization lowers the energy of the cation.

Notes for active learning

Stereochemistry – Detailed Explanations

1. C is correct.

Draw the different isomers of butene.

This compound can be drawn with the double bond terminal (the carbons are numbered according to IUPAC).

The terminal carbon in the double bond is numbered 1.

There are two geometric isomers of butene (i.e., *cis*-butene and *trans*-butene).

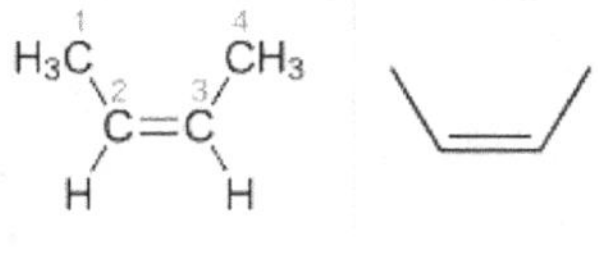

cis-2-butene

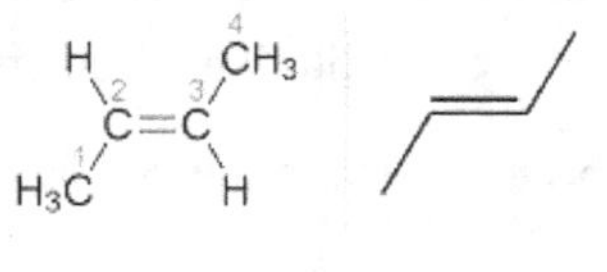

trans-2-butene

Geometric isomers are a subset of structural isomers. Geometric requires that the substituents from the double bond can be the same (i.e., *cis* and *trans*) or different (i.e., *E* and *Z*).

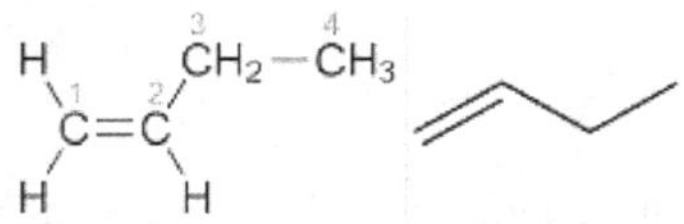

1-butene or butene (the position 1 is implied by IUPAC)

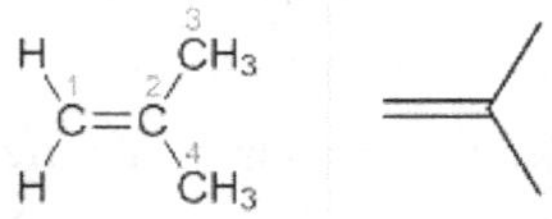

Isobutylene (not 2-methyl propene) according to IUPAC

Therefore, there are four structural isomers of butene.

2. D is correct.

An *asymmetric carbon* refers to a chiral carbon: carbon bonded to four different substituents. The other compounds do not contain asymmetric carbons because at least two of the three atoms or groups bonded to each carbon atom are the same.

3. D is correct.

The number of stereoisomers for a given molecule depends on the number of asymmetric carbon atoms (or chiral centers) present.

There is a 2^N number of possibilities, where N is the number of chiral centers present.

The molecule has 4 chiral centers: $2^4 = (2 \times 2 \times 2 \times 2) = 16$ stereoisomers.

For some molecules, symmetry elements may be redundant structures, so the 2^N calculation gives the maximum number of stereoisomers and not necessarily the actual number that exists.

College Organic Chemistry Practice Questions with Detailed Explanations

4. A is correct.

Remember: if an alkene has geminal disubstitution (i.e., two identical moieties bonded to the same alkene carbon), then *cis* and *trans* isomerism is not possible.

5. C is correct.

The naming reveals that the compounds are the same: 7-ethyl-4-isopropyl-3,6-dimethyldecane

6. C is correct.

The bromine substitution changes from vicinal (on adjacent carbons) to geminal (on the same carbon). Because the connectivity changes, the structures are constitutional isomers.

The first molecules cannot be chiral because it has a mirror plane of symmetry.

7. E is correct.

The molecules shown above are mirror images of each other.

Assign *R* and *S* at each chiral center to evaluate the relationship between the two chiral molecules.

8. E is correct.

In *polarimeters* (i.e., plane-polarized light), enantiomers have the same magnitude of specific rotation but opposite sign.

9. B is correct.

Isomers are compounds with the same molecular formula but a different structure.

A: *hydrocarbons* are organic molecules containing only carbon and hydrogen (i.e., no heteroatoms such as oxygen or nitrogen).

C: *homologs* are a series of compounds with the same general formula, usually varying by a single parameter (e.g., length of the carbon chain).

D: *isotopes* have different mass numbers due to differences in the number of neutrons. The number of protons determines the identity of the element.

E: *allotropes* are each of two or more different physical forms in which an element can exist (e.g., carbon exists as graphite, charcoal, and diamond).

10. B is correct.

(Z)-1-bromo-1-chloropropene   (E)-1-bromo-1-chloropropene

These molecules can only adopt one conformation because of the rotational barrier of the alkene.

continued…

Configurational isomers involve a double bond. If the molecules have optical activity, they may be enantiomers (non-superimposable mirror images) or diastereomers. If they lack optical activity, they are *geometric isomers*.

A: *constitutional isomers* have the same molecular formula but different connectivity. The molecules are not the same, but the connectivity is the same, so they are not constitutional isomers.

C: *identical molecules* may be drawn with different orientations on the paper but are the same.

D: *conformational isomers* refer to different molecules due to free rotation around single bonds (e.g., Newman projections, chair flips for cyclohexane).

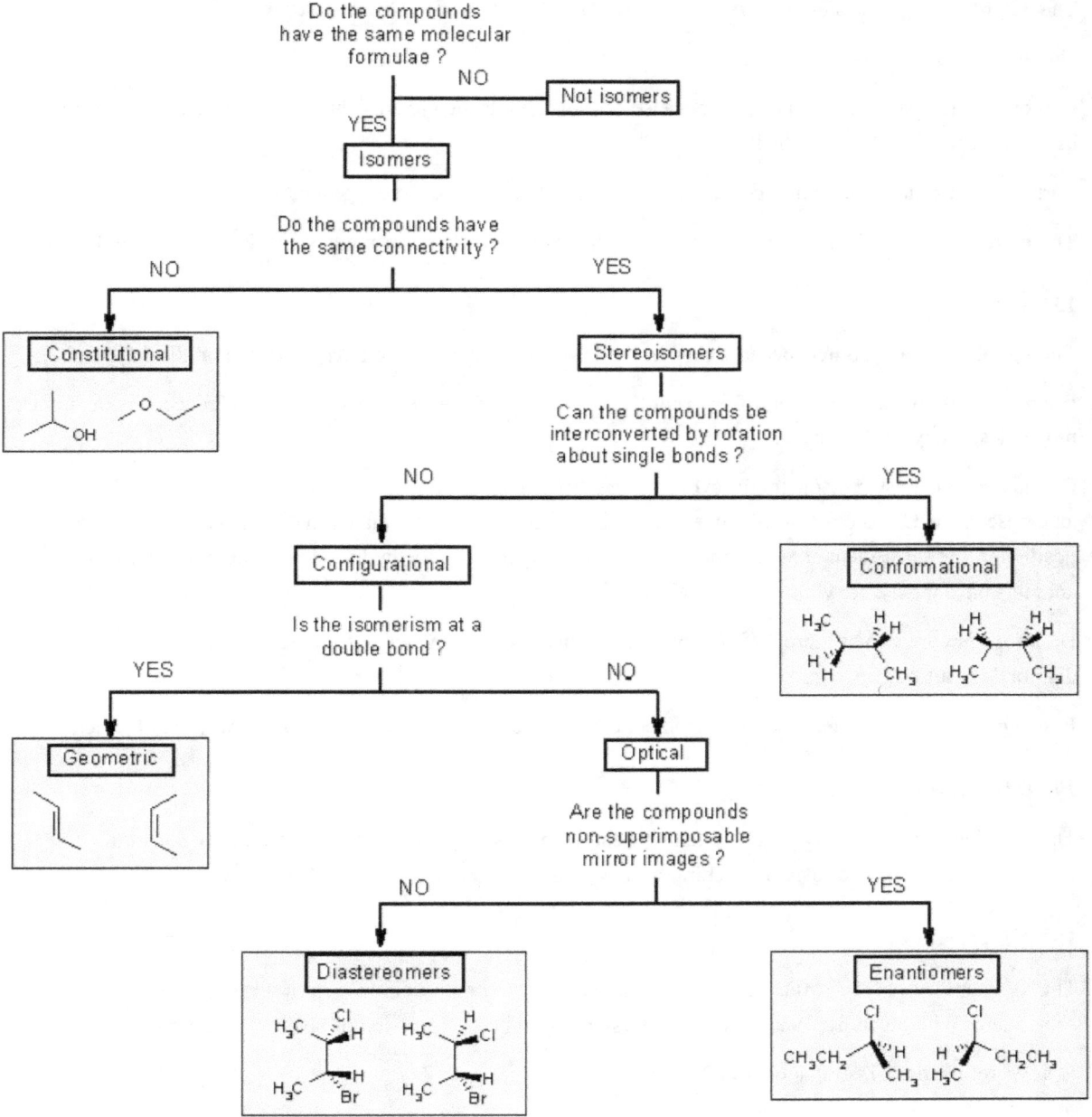

11. B is correct.

Chiral carbons refer to asymmetric carbons typically bonded to three other atoms.

Carbon 2 cannot be an asymmetric carbon (or chiral carbon) because carbon atoms with trigonal planar configuration possess a plane of symmetry and are achiral.

12. D is correct.

The number of stereoisomers depends on chiral centers, defined as a carbon bonded to 4 different substituents.

The number of possible stereoisomers is 2^n, where n is the number of chiral centers.

Number the carbons with the carbonyl carbon as #1.

Carbons 1 and 6 are not chiral centers because carbon 1 is attached to only three substituents, while carbon 5 has two identical hydrogen substituents.

Carbons 2, 3, 4, and 5 are chiral centers since each is bonded to 4 different groups.

Therefore, the molecule has 4 chiral centers, and the number of stereoisomers is $2^4 = 2 \times 2 \times 2 \times 2 = 16$.

13. A is correct.

These molecules are isomers because they have the same number and atoms, but the bonding is different.

B: *epimers* are two isomers that differ in configuration at only one stereogenic center. Stereocenters in epimer molecules, if any, are the same.

C: *anomers* are diastereoisomers for cyclic sugars differing in the anomeric carbon configuration (C–1 atom of an aldose or the C–2 atom of a 2-ketose). The cyclic forms of carbohydrates can exist in α– or β–, based on the position of the substituent at the anomeric center. α–anomers have the hydroxyl pointing down, while β–anomers have the hydroxyl pointing up.

D: *allotropes* are two or more different physical forms in which an element can exist. Graphite, charcoal, and diamond are allotropes of carbon.

E: *geometric isomers* differ in the arrangement of groups about a double bond, ring, or other rigid structure.

14. D is correct.

The bromine and hydrogen atoms alternate the carbon atoms they are bonded to, so they are constitutional isomers. These molecules cannot be chiral because they have a mirror plane of symmetry.

15. A is correct.

The *anti*-conformer (180° offset) has the lowest energy because it minimizes steric strain (i.e., bulky substituents within van der Waals radii) and torsional strain (i.e., repulsion of bonding electrons when eclipsed).

Gauche refers to a dihedral angle of 60°.

diagrams below…

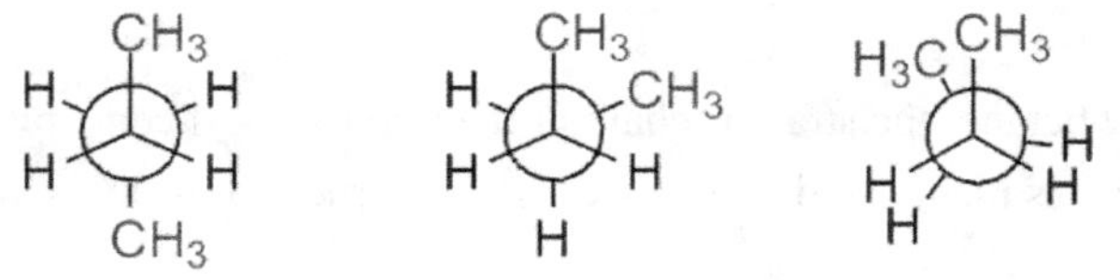

anti   gauche   eclipsed

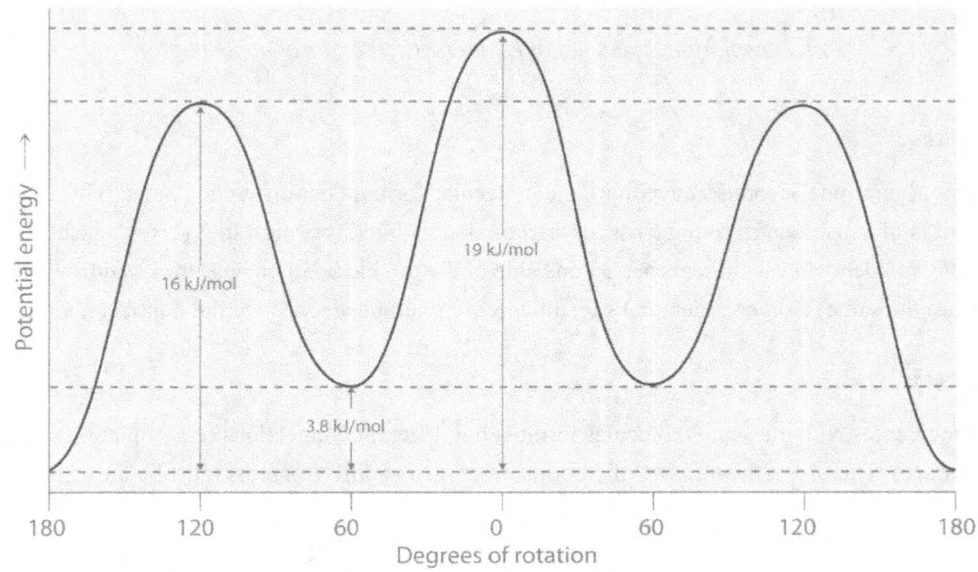

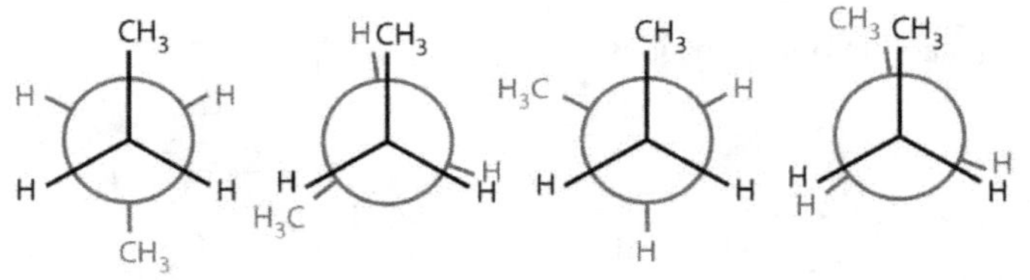

(Most stable)   (Least stable)

16. D is correct.

Draw each isomer and be cognizant of equivalent structures.

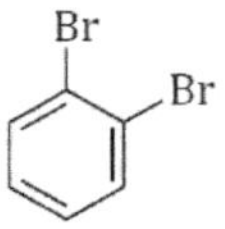

ortho-dibromobenzene 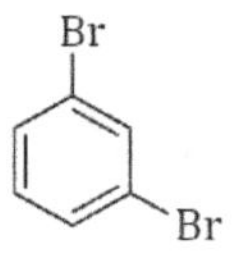   *meta*-dibromobenzene 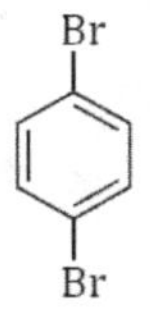   *para*-dibromobenzene

There are only three isomers for the given molecular formula.

17. D is correct.

There are two ways to draw butane; the straight-chain or the tertiary alkyl constitutional isomer. Note that the term "constitutional isomers" is recognized by IUPAC, while "structural isomers" was common historically.

Two constitutional isomers of n-butane (left) and isobutene (right)

18. C is correct.

Geometric (configurational) isomers have the same molecular formula but different connectivity of atoms due to the orientation of substituents around a carbon-carbon double bond (or ring). In Z (*cis* when substituents are the same) isomers, identical substituents are on one side of the double bond or ring. In E (*trans* when substituents are the same) isomers, identical substituents are on opposite sides of the double bond or ring.

19. A is correct.

Isomers are molecules with the same molecular formula but different connections (e.g., functional groups) between the atoms. Therefore, the number of carbons, hydrogen and heteroatoms must be the same.

20. C is correct.

Cis means that the groups are on the same side of the alkene (or face of a cyclic structure).

Trans means that the groups are on opposite sides.

Cis and trans isomerism can occur with disubstituted cycloalkanes.

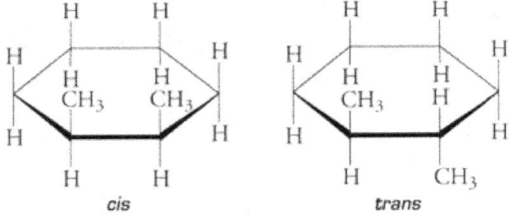

cis-2-butene   trans-2-butene

Cis and trans isomerism can occur with alkenes (or for a ring).

21. D is correct.

The number of each atom must be consistent among a pair or group of isomers, meaning that the oxidation state of the carbon atoms of the compound is consistent.

A change in the oxidation state of the compound suggests that the hydrogen count is different.

For instance, the isomer of the given alcohol is another alcohol and not an aldehyde or a ketone.

22. B is correct.

There are two stereoisomers for the vicinal disubstituted alkene (1,1-chlorofluoroethene), and there is only one isomer of the 1,2 disubstituted alkene (*cis* and *trans*).

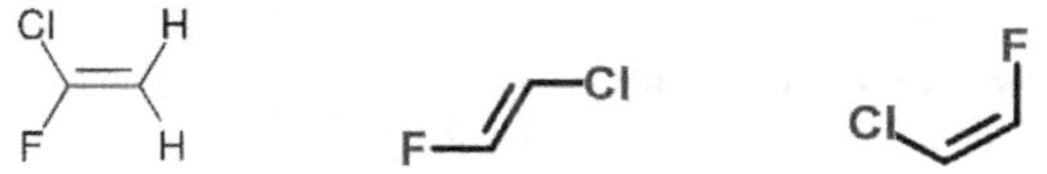

23. D is correct.

Use the rotation rule (clockwise = *R*, counterclockwise = *S*) to determine the configuration after assigning priorities. When the lowest priority group is on the horizontal bond, the assigned configuration is reversed.

24. C is correct.

Enantiomers are chiral molecules (i.e., attached to four different groups) and non-superimposable mirror images (i.e., *R* and *S*).

The two molecules are mirror images of each other, so they are enantiomers.

25. A is correct.

There are two stereoisomers for the vicinal disubstituted alkene (i.e., *cis* and *trans*).

cis- and *trans*-dichloroethylene

There is only one isomer of the geminal (i.e., on the same atom) disubstituted alkene.

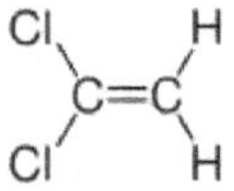

1,1-dichloroethene

26. B is correct.

To determine the enantiomer of the compound, draw the mirror image of the compound.

This requires inverting the two stereocenters, as inverting only one results in a diastereomer.

27. B is correct.

Meso compounds are molecules with chiral centers and require an inversion center or a plane of symmetry. These elements of symmetry make molecules with asymmetric carbon atoms achiral overall.

28. E is correct.

Any molecule that has a non-identical mirror image is chiral and has an enantiomer.

Draw the structure of the cyclic compound:

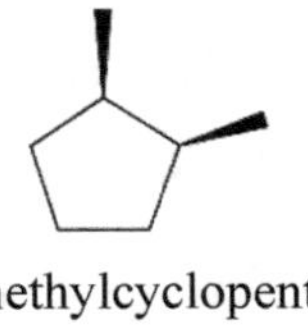

cis-1,2-dimethylcyclopentane

Comparison of the molecule reveals an internal plane of symmetry.

A *meso* compound and its enantiomer (i.e., a non-superimposable mirror image of a chiral molecule) are the same. Therefore, a *meso* compound does not have an enantiomer.

29. D is correct.

Structural isomers have the same molecular formula but a different connection to the atoms. With molecular formula $C_4H_8Cl_2$, the carbon skeleton is butane.

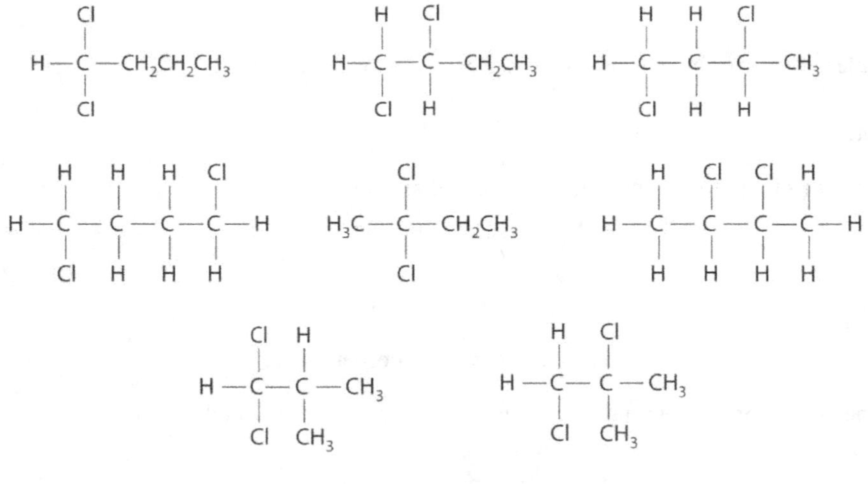

Eight structural isomers exist: three isomers have chiral carbons.

From top left and moving to the right: molecules 2, 3 and 6 are chiral (i.e., a carbon attached to four different substituents). Molecule 2 has one chiral carbon; molecule 3 has one chiral carbon. Structural isomer 6 contains two chiral carbons. One of the stereoisomers of 2,3-dichlorobutane has an internal plane of symmetry, making it the *meso* compound of R/S (or S/R, which is the same molecule) and therefore is achiral and optically inactive.

The other stereoisomer is R/R (or S/S, which is the same molecule) and therefore is chiral and exhibits optical activity.

Therefore, there are three optically active isomers of $C_4H_8Cl_2$.

Detailed Explanations: Molecular Structure and Spectra

30. C is correct.

Chiral carbon atoms have four different groups bonded to a carbon atom.

The molecule is named with the alcohol as the highest priority group and designated carbon 1.

The carbon atoms with two or more hydrogen atoms (i.e., carbons 1 and 5) are achiral because at least two groups are the same.

31. A is correct.

Achiral compounds cannot rotate the plane of polarized light.

The solutions of achiral compounds are optically inactive.

B: no relationship between absolute configuration (*R/S*) and specific rotation (+/−) of light in the polarimeter.

C: *meso* compounds are achiral, but not all achiral molecules are *meso*.

D and E: *meso* compounds are achiral, contain two or more chiral centers, and an internal plane of symmetry.

32. A is correct.

The stereochemistry of the alkene is '*E*' (highest priority groups are on opposite sides of the alkene).

The priority groups are ranked according to the Cahn-Ingold-Prelog rules for prioritization based on the atomic number of atoms attached to the alkene.

The chain possessing the heaviest atoms proximal to the alkene generally has higher priority. In this example, the bromomethyl group has a higher priority.

The alcohol and methoxy groups contain oxygen atoms, but the methoxy oxygen is closer to the double bond.

33. C is correct.

For Fischer projections, horizontal lines represent bonds projecting outward (i.e., wedges), whereas vertical lines represent bonds going back (i.e., dashed lines). A Fischer projection does not include a carbon specified at the cross of the vertical and horizontal lines (i.e., a C is implied, but no C is written on the structure).

For assigning *R/S* in Fischer projections, read the ranked (1 → 3) priorities as clockwise (*R*) or counterclockwise (*S*). If the lowest priority is vertical (i.e., points into the page), then assign *R/S*. If the lowest priority is horizontal (i.e., points out of the page), reverse (*R* → *S*, *S* → *R*).

Compound I: the order of priority is hydroxyl, carboxyl, methyl, and hydrogen. The order of increasing priority is counterclockwise, and the configuration appears S. However, the lowest priority (group 4 is H) is horizontal (pointing outward), so the absolute configuration is *R*.

Compound II: the order of priority is nitrogen, carboxyl, methyl, and hydrogen. The order of increasing priority is counterclockwise. The lowest priority (group 4 is H) is vertical and therefore points away. The absolute configuration is *S*.

continued…

Compound III is achiral because the carbon is not attached to four different groups, and therefore the molecule is neither *R* nor *S*.

Compound IV: the order of priority is hydroxyl, carbonyl (aldehyde), methyl (methanol), and hydrogen. The order of increasing priority is counterclockwise. The lowest priority (group 4 is H) is horizontal and therefore points towards the viewer. The absolute configuration is *R*.

Compounds I and IV have the same absolute configuration.

34. A is correct.

The root name of the compound is cyclopentane because the ring possesses 5 carbons.

The Cl substituents are adjacent (i.e., position 1,2) on the ring and must be oriented on opposite sides.

35. B is correct.

Chiral molecules include carbons bonded to four different substituents. This molecule contains no stereogenic centers (i.e., chiral centers), and therefore the molecule cannot be chiral.

36. A is correct. The observed rotation is half the value of the specific rotation for the pure enantiomeric substance. While the effect of the combined opposite enantiomers is canceled, the mixture should have a 50% excess of the pure substance.

The mixture must have 75% of the pure enantiomeric substance, where 25% of the rotation cancels the effect of the 25% opposite rotation of its enantiomer.

37. A is correct.

The carbon attached to the leaving group is tertiary (bonded to 3 carbons) and chiral because it is bonded to 4 different substituents, and the molecule is optically active. Tertiary alkyl halides undergo S_N1 reactions (forming a trigonal planar carbocation) but do not undergo S_N2 reactions because of steric hindrance.

In the first step of the S_N1 reaction, the bromine dissociates to form a stable tertiary carbocation, which results in the loss of optical activity. A positively charged carbon is sp^2 hybridized (trigonal planar) and achiral because it has only three substituents.

In the second step, the HCN nucleophile attacks the trigonal planar (flat) carbocation from either side of the plane (i.e., top or bottom) with approximately equal probability. As a result, the reaction yields approximately equal amounts of two chiral products.

The products are enantiomers (i.e., chiral molecules are non-superimposable mirror images), and each enantiomer rotates the plane of polarized light to the same extent but in opposite directions.

Therefore, the product is an optically inactive racemic mixture (i.e., both enantiomers in the solution), and there is a loss of optical activity in the solution.

Racemization means loss of optical activity and often involves a carbocation intermediate (S_N1 reaction), whereby the incoming nucleophile attacks from either side of the trigonal planar carbocation.

continued…

B: *mutarotation* occurs in monosaccharides (i.e., sugars) and involves the equilibrium between open-chain forms and cyclic hemiacetal forms (e.g., Haworth projections) in aqueous solutions.

D: *inversion* of absolute configuration only occurs in S_N2 reactions, whereby a nucleophile attacks the substrate from the side opposite the leaving group (backside) in a one-step reaction.

From the concerted S_N2 reaction, the products have the same absolute configuration (often inverted from the backside attack), and the product is considered chiral.

38. E is correct.

Draw the structure of each of the possibilities and count the number of isomers.

Two isomers can be formed from the geminal substitution of the chlorine atoms; three isomers result from the (1,2), (1,3), and (1,4) disubstitution.

The last isomer involves a (2,3) dichloro substitution.

The (2,3) disubstitution can exist as a pair of diastereomers.

39. C is correct.

An asymmetric (i.e., chiral) carbon is bonded to four different substituents.

There are three asymmetric carbons in this molecule.

The methylene is symmetrical; the isopropyl and the geminal dimethyl groups have symmetrical carbons.

40. C is correct.

Because one of the stereocenters has a different *R/S* configuration, the molecules are diastereomers.

41. D is correct.

A *meso* compound has chiral centers but is not chiral because it has an internal plane of symmetry.

Tartaric acid is a four-carbon polyol with two carboxylic acids and two chiral centers.

There are three stereoisomers: the (+) form, the (−) form, and the *meso* form.

Meso compounds are identical to their mirror images.

Each of the two stereoisomers rotates plane-polarized light as indicated by the notation of (+) or (−), but *meso*-tartaric acid is achiral (i.e., no net rotation of plane-polarized light).

III: *racemic mixtures* are solutions with enantiomers (i.e., chiral molecules that are mirror images).

A *meso* compound is a single molecule that contains two enantiomers joined.

42. A is correct.

A racemic mixture contains equal quantities of two enantiomers (i.e., isomers that are non-superimposable mirror images).

Compound I is D-fructose in a Fischer projection.

Compound II is D-fructose in a straight chain.

Compound III is D-glucose in a straight chain.

43. B is correct.

The molecule that contains the chiral carbon is the one that has a central carbon with four different groups as substituents. These carbons are described as asymmetric (i.e., stereogenic center or chiral carbons).

44. B is correct.

The chlorine atom occupies the internal (i.e., 2nd) carbon on the first molecule, and the chlorine atom is bonded to a terminal (i.e., 1st) carbon on the second molecule.

Constitutional (i.e., structural or configurational) isomers have the same molecular formula but different connectivity of the atoms.

A: *conformational* isomers involve free rotation around a single bond (e.g., Newman projections).

C: *diastereomers* are chiral molecules (i.e., attached to four different groups) with two or more chiral centers and are non-superimposable non-mirror images (i.e., *R,R* and *S,R*).

D: *enantiomers* are chiral molecules (i.e., attached to four different groups) and are non-superimposable mirror images (i.e., *R* and *S*).

The exception of the two or more chiral center requirements for diastereomers is geometric isomers that contain double bonds (i.e., *cis* and *trans*).

45. A is correct.

For an alkene to experience *cis-trans* isomerization, the *pi* bond of the double bond is broken.

The *pi* bond of the alkene makes the double bond rigid. Heating the alkene at high temperatures or exposure to electromagnetic radiation (e.g., UV radiation) may cause the *pi* bond to homolytically cleave to 1,2-diradical and rotate about the *sigma* bond to form the diastereomers.

E / Z and *cis / trans* isomers are geometric isomers and classified as diastereomers.

46. B is correct.

When determining whether an alkene is the *Z / E* (or *cis / trans*) stereoisomer, it is important to identify the higher priority substituent group at each of the two carbon atoms of the alkene.

The priority groups are ranked by the Cahn-Ingold-Prelog rules for prioritization based on the atomic number of atoms attached to the alkene.

If the higher priority groups are positioned on the same side of the double bond, the molecule is *Z* (*cis* notation can be used if the substituents are the same).

If the higher priority groups are positioned on the opposite sides of the double bond, the molecule is *E* (*trans* notation can be used if the substituents are the same).

The stereochemistry about the alkene is 'Z.' The highest priority groups – Br and Cl across the double bond – are on the same side of the alkene.

Cis–trans relationship cannot be used to describe the molecule because the substituents across the double bond are different.

47. D is correct.

Determine the molecular formula of the given molecule. 2-methylbutane has 5 carbon atoms and 12 hydrogen atoms.

Only *n*-pentane has the same molecular formula.

The notation *n*– represents normal (or straight chain).

Notes for active learning

Molecular Structure and Spectra – Detailed Explanations

1. D is correct.

The local magnetic field generated by the circulating current of the benzene ring causes the protons attached to the ring to be further deshielded.

Electron-donating and withdrawing groups attached to the ring shift the resonances of protons up or downfield.

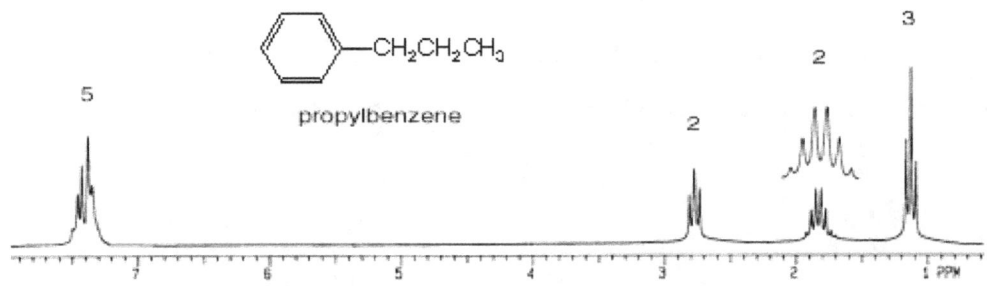

^{1}H NMR spectra with characteristic absorption for the phenyl ring between 6.0-8.0 ppm

2. D is correct.

The shielding effect arises from the electron density associated with the nuclei.

Hydrogen nuclei with more electron density are shielded, and their chemical shifts appear more upfield in the spectrum.

3. D is correct.

2-methylpropanoic acid

Carbonyl carbons show an IR absorption between approximately 1630 and 1780 cm^{-1}. The carbonyl of a carboxylic acid is reported to be between 1710 and 1780 cm^{-1}.

Because carboxylic acids contain a hydroxyl (~OH) group, another signal around 3300 to 3400 cm^{-1} should be expected in the spectrum for this compound.

4. A is correct.

5. A is correct.

The n to *pi** transition for ketones is the electron transition requiring the least energy.

Antibonding orbitals are vacant and serve as the acceptor orbitals for the electron excitation. The orbitals of nonbonding electrons have more energy than the orbitals for *pi* bonds.

Electron transitions from *sigma* bonds are difficult because electron energy is low and requires high energy to be promoted.

6. E is correct.

The peak area ratios of the spin states correspond to the values derived from Pascal's triangle.

The heights of the peaks may not correspond to the same ratio, but the area does.

n	2^n	multiplet intensities	
0	1	1	Singlet (s)
1	2	1 1	Doublet (d)
2	4	1 2 1	Triplet (t)
3	8	1 3 3 1	Quartet (q)
4	16	1 4 6 4 1	Pentet
5	32	1 5 10 10 5 1	Sextet
6	64	1 6 15 20 15 6 1	Septet
7	128	1 7 21 35 35 21 7 1	Octet
8	256	1 8 28 56 70 56 28 8 1	Nonet

Pascal's triangle

7. D is correct.

The most deshielded protons in the NMR spectrum have the largest δ shift (i.e., downfield).

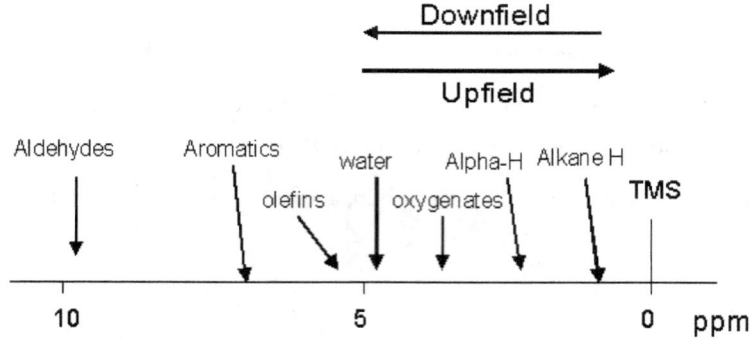

NMR with an approximate δ shift of some functional groups

8. E is correct.

The M–18 peak corresponds to the loss of water. The loss of water in a substrate containing alcohol may suggest that an alkene intermediate is generated during the ionization process.

Detailed Explanations: Molecular Structure and Spectra

9. B is correct.

The *alpha nitrogen* atom makes the neighboring C–H bonds less shielded because the nitrogen atom is electronegative and withdraws electron density through an inductive effect.

Terminal or substituted alkyl C–H bonds (away from electronegative atoms) resonate between 1 and 2 ppm.

10. C is correct.

3,3-dibromoheptane (below) produces 6 NMR signals.

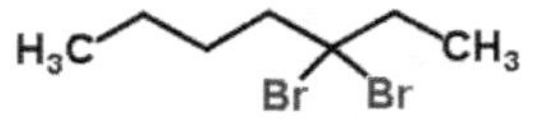

A: 1,1,2-tribromobutane (below) produces 4 NMR signals.

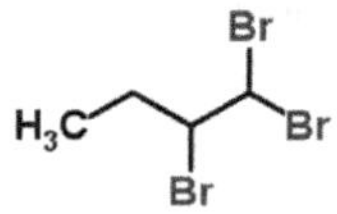

B: bromobutane (below) produces 4 NMR signals.

D: dibutyl ether (below) produces 4 NMR signals.

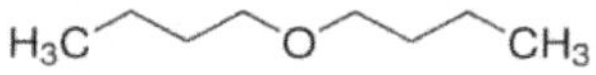

11. C is correct.

Molecules containing the carbonyl functional groups, such as ketones, typically have C=O resonance frequencies at 1710 cm^{-1}.

However, the stretching frequency is lower if the carbonyl group is in conjugation with another group (in this case, the alkene).

Resonance contributes to the carbon-oxygen single bond character of the carbonyl, and single bonds are weaker than double bonds.

12. A is correct.

The *topicity* of the protons can be determined by labeling them as H_a and H_b to produce a "chiral center." If this causes the molecule to have two or more chiral centers, then the protons are *diastereotopic*.

If the labeling causes the molecule to have only one stereocenter, the labeled protons are *enantiotopic*.

If labeling the protons as H_a and H_b does not lead to the formation of a chiral center (as in methane), then the protons are *homotopic*.

13. C is correct.

1-chlorobutane (below) has a chiral center and produces 4 NMR peaks.

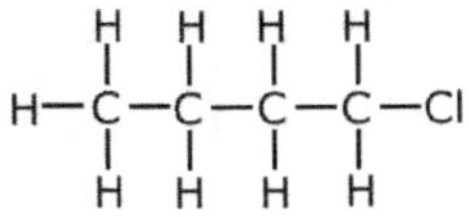

A: 3,3-dichloropentane (below) is a symmetrical molecule that produces 2 NMR proton peaks.

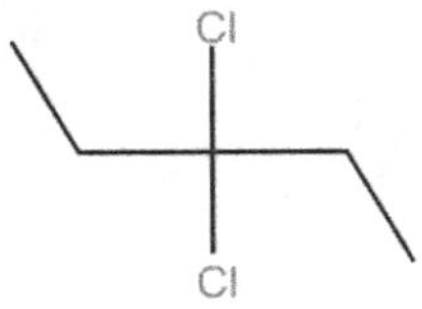

B: 4,4-dichloroheptane (below) is a symmetrical molecule that produces 3 NMR proton peaks.

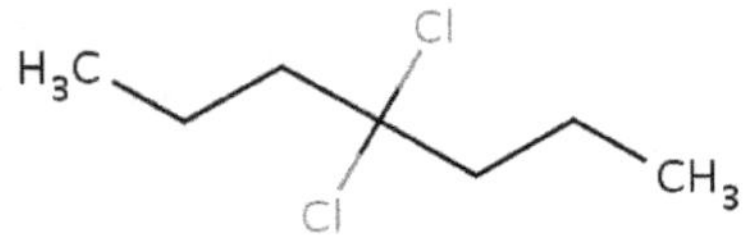

D: 1,4-dichlorobutane (below) produces 2 NMR proton peaks.

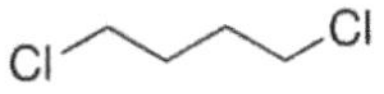

E: dichloromethane (below) produces 1 NMR proton peak.

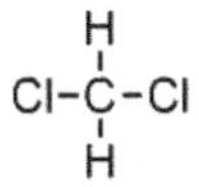

14. B is correct.

Carboxylic acid protons are some of the most deshielded protons.

These functional groups can partially ionize, causing the hydrogen to develop a high degree of positive (cationic) character.

15. C is correct.

The stretching frequency is much higher than a typical carbonyl group.

The chlorine atom is a poor electron donor to the carbonyl through lone pair conjugation, and it tends to withdraw more electron density through induction.

The lone pair of electrons on the oxygen atom can donate to carbon-chlorine *sigma* orbital, shortening the bond.

The shorter (or stronger) a bond, the higher is its stretching frequency in the IR spectrum.

If the lone pair conjugation with the carbonyl is the more dominant effect (e.g., amides), the carbonyl stretching frequency is lower than 1710 cm^{-1}.

If the inductive effect of the heteroatom outweighs the conjugation effect into the carbonyl group, the carbonyl stretching frequency is higher than 1710 cm^{-1}.

16. B is correct.

The various peaks between 900 and 1500 cm^{-1} correspond to the unique fingerprint region in the IR spectrum.

The prominent peak at 1710 cm^{-1} indicates the carbon-oxygen double bond of a carbonyl (C=O) group (aldehyde, ketone, acyl halide, anhydride, carboxylic acid, ester, or amide).

A similar (prominent and broad peak) between 3300 and 3500 cm^{-1} is characteristic of alcohol.

The alcohol of a carboxylic acid would show an IR absorption peak around 2800 and 3200 cm^{-1}.

17. B is correct.

Spectroscopy (e.g., NMR) generally is used in the identification of compounds.

Distillation, crystallization, and extraction are common techniques to isolate and purify compounds.

18. C is correct.

For carboxylic acid derivatives, the heteroatom either increases or decreases the stretching frequency of the carbonyl group.

If the lone pair conjugation with the carbonyl is the more dominant effect (e.g., amides), the carbonyl stretching frequency is lower than 1710 cm^{-1}.

If the inductive effect of the heteroatom outweighs the conjugation effect into the carbonyl group (e.g., esters and acid chlorides), then the stretching frequency is higher than 1710 cm^{-1}.

19. A is correct.

NMR spectroscopy provides information about the local environment of the proton.

Equivalent hydrogens (i.e., hydrogens in identical locations about other atoms) produce a single NMR signal.

Nonequivalent hydrogens give separate NMR signals on the spectrum.

Since the molecule produces one signal for NMR, all hydrogens are equivalent. In $(CH_3)_3CCCl_2C(CH_3)_3$, the methyl groups are equivalent, and this molecule produces only one signal in NMR.

B: $(CH_3)_2CHCH_2CH_2CH(CH_3)CH_2CH_3$ produces eight signals. Additionally, the splitting produces a complex NMR pattern indicating the number of adjacent Hs.

C: $(CH_3)_2CHCH_2(CH_2)_4CH_3$ produces eight signals. Additionally, the splitting produces a complex NMR pattern indicating the number of adjacent Hs.

D: $CH_3(CH_2)_7CH_3$ produces five signals. For symmetrical molecules, one signal is for the terminal $CH_3(1,9)$, one for the $CH_2(2,8)$, one for the $CH_2(3,7)$, one for the $CH_2(4,6)$, and one for the CH_2 at 5.

20. A is correct.

Note: the displayed masses on the mass spectrum corresponds to molecular fragments with a charge of +1. It is possible to generate dications from the ionization in the mass spectrometer, and therefore an additional calculation may be required to determine the true mass of the fragment.

21. B is correct.

The two nuclear spin states for protons are *alpha* and *beta*.

The *alpha* spin state has less energy than the *beta* spin state because the *alpha* spin state has the same direction as the applied external field.

22. C is correct.

IR spectroscopy provides information about functional groups.

A: *mass spectrometry* (MS) provides information about molecular weight.

B: *nuclear magnetic resonance* (NMR) spectroscopy provides information about protons.

D: *UV spectroscopy* provides information about conjugated (i.e., sp^2 hybridization) double bonds.

E: *polarity* refers to the difference in electron density due to electronegative atoms.

23. A is correct.

The *fragmentation pattern* of the spectra provides structural information and determination of the molar weight of an unknown compound.

Cleavage occurs at alkyl-substituted carbons reflecting the order generally observed in carbocations.

3,3-dimethyl-2-butanone

The *base peak* for this molecule is the acetyl intermediate. This intermediate results from the ionization of the carbonyl oxygen atom to form an oxygen-centered radical cation.

The carbon-carbon bond between the *tert*-butyl group and the carbonyl can homolytically cleave to give the acylium cation.

For example:

acetone sample
MW = 58

radical cation
(molecular ion)
m/z = 58

acylium ion
m/z = 43

methyl radical
(not detected)

24. C is correct.

Topicity is the stereochemical relationship between substituents.

These groups, depending on the relationship, can be *heterotopic, homotopic, enantiotopic,* or *diastereotopic*.

The protons are chemically equivalent or homotopic because the groups are equivalent.

If labeling the protons of methylene (~CH_2~) group as H_a and H_b does not form *enantiomers*, the molecule is *homotopic*.

25. A is correct.

IR active molecules must have polarized covalent bonds to absorb IR.

When a Cl–Cl bond with atoms of the same electronegativity stretches or bends, no dipole is created, and therefore the molecule is IR inactive.

B: CO (C≡O) contains covalent bonds whereby carbon is attached to the electronegative oxygen, which creates a dipole generating an IR signal.

C: $CH_3CH_2CH_2OH$ contains covalent bonds attached to an electronegative oxygen, which creates a dipole generating an IR signal.

D: CH_3Br contains covalent bonds attached to electronegative bromines, creating a dipole with an IR signal.

E: HCN contains covalent bonds attached to the electronegative nitrogen, creating a dipole with an IR signal.

26. A is correct.

Electromagnetic spectrum:

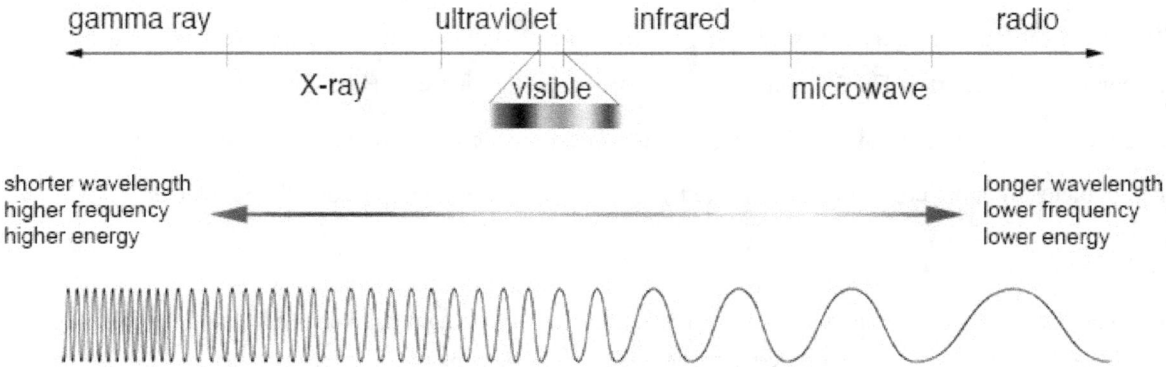

Radio waves have the lowest energy on the electromagnetic spectrum, infrared radiation has more energy than radio waves, and ultraviolet light has more energy than infrared radiation.

27. E is correct.

The type of electromagnetic radiation (EMR) needed to excite an electron in the molecule from the highest energy occupied molecular orbital (*HOMO*) to the lowest energy unoccupied molecular orbital (*LUMO*) corresponds to UV-visible light.

Exciting an electron from the *sigma* bond is more difficult because the *sigma* bond has very low energy and requires higher energy radiation to be promoted to a vacant orbital.

Furthermore, the *sigma** orbital has very high energy; promoting electrons to this vacant orbital requires stronger radiation.

28. A is correct.

UV spectroscopy is useful for identifying compounds with conjugated double bonds. Neither dimethyl ether nor bromoethane has conjugated double bonds, so UV is not a good analytical technique to distinguish them.

B: *mass spectrometry* (MS) provides information about the molecular weight, the number and size of molecular fragments, and a unique fingerprint pattern.

C: *infrared* (IR) spectroscopy determines if the molecule contains certain functional groups and gives a unique fingerprint pattern for a molecule.

D: *proton nuclear magnetic resonance* (NMR) examines the molecular environment of the hydrogens and is related to where the signal is located on the spectrum.

NMR is useful in determining the connectivity of the atoms and identifies the relative numbers of each kind of hydrogen (i.e., integration number given by the area under each signal) and the number of hydrogen atoms on adjacent atoms (i.e., splitting pattern as determined by n + 1, where n = number of adjacent Hs).

29. D is correct.

The compound is an ester. The 3.8 ppm septet corresponds to the single C–H bond near the oxygen of the ester.

The singlet at 2.2 ppm suggests that the ~CH$_3$ group is near the carbonyl group.

The doublet at 1.0 ppm corresponds to the methyl groups of the isopropyl portion of the molecule.

30. E is correct.

Conjugated polyenes absorb light at longer wavelengths than unconjugated alkenes because the additional *p* orbital overlap in larger *pi* systems decreases the energy difference between the highest occupied molecular orbital (*HOMO*) and the lowest unoccupied molecular orbital (*LUMO*).

The *LUMO* is an antibonding orbital and has more energy than the *HOMO*, a bonding orbital.

The longer the conjugated system, the smaller the energy gap between the two molecular orbitals (MO); this requires radiation of less energy (and longer wavelength) for electron transitions.

When a molecule absorbs UV/visible radiation, electrons are promoted from one orbital to a higher energy orbital.

31. B is correct.

For NMR, the position that hydrogen absorbs is determined by the chemical environment.

If the chemical environment of two hydrogens is identical, only one signal is produced.

Therefore, equivalent hydrogens could be replaced by another group to yield the same molecule.

The challenge is in determining whether hydrogens are in identical environments (i.e., symmetric molecules).

1,2-dibromoethane has one absorbance for the NMR spectrum because a molecule has one type of H atom.

A: *tert*-butyl alcohol produces 2 signals.

C: *toluene* produces 2 signals.

D: *methanol* produces 2 signals.

E: *phenol* produces 4 signals.

32. A is correct.

IR absorption between 1630 cm^{-1} and 1740 cm^{-1} is characteristic of carbonyls (e.g., aldehydes, ketones, anhydrides, carboxylic acids, esters, and amides).

An IR absorption of 1735 cm^{-1} is characteristic of an ester.

33. D is correct.

UV light has enough energy to excite electrons to higher energy vacant orbitals as photoactivated atoms.

This radiation generally is not strong enough to eject electrons from atoms of most elements to form ions.

34. B is correct.

Infrared radiation (IR) is useful in 1500 to 3500 cm^{-1} (called wavenumbers). In this range, the molecular vibrations of molecules are active, and characteristic absorbance frequencies identify functional groups.

Below 1500 cm^{-1} is the fingerprint region used for more detailed analysis once the target molecule(s) have been identified.

A: *nuclear magnetic resonance* (NMR) spectroscopy involves a sample containing the compound being subjected to a high-intensity magnetic field and scanning through the radio-frequency range of the electromagnetic spectrum for particular absorptions.

NMR relies on the magnetic properties of specific atomic nuclei and determines the physical and chemical properties of atoms within molecules. NMR can be used to deduce the structure (connectivity) of the atoms within the molecule.

C: the *UV range* (not the IR range) of wavelengths is between 200–400 nm and corresponds to the energy required for electronic transitions between the bonding or nonbonding molecular orbitals and antibonding molecular orbitals.

continued…

UV spectroscopy is useful for studying compounds with double bonds, especially in conjugated (i.e., alternating double and single bond) molecules.

When molecules containing π-electrons (or non-bonding electrons) absorb UV energy, these electrons are promoted (i.e., excited) to higher antibonding molecular orbitals.

The more easily excited the electrons (i.e., lower energy gap between the *HOMO* and the *LUMO*), the longer the wavelength of UV light it absorbs.

D: *mass spectrometry* (MS) studies compounds through the fragmentation of molecules, although the unfragmented parent peak provides useful information.

Note: mass spectrometry destroys (fragments) the sample and is not preferred for rare/limited samples.

35. B is correct.

The IR absorbance at 1710 cm^{-1} indicates the carbonyl group of a ketone or an aldehyde.

The carbonyl (C=O) is present in the derivatives of carboxylic acid: acyl halide, anhydride, carboxylic acid, esters, and amide.

NMR can distinguish between an aldehyde (NMR δ 9–10) and a ketone (no characteristic NMR signals).

Aldehydes have additional peaks at 2700–2800 cm^{-1}, while ketones do not.

36. B is correct.

The broad, deep absorption between 3000 cm^{-1} to 3500 cm^{-1} is characteristic for alcohol (~OH).

The carbonyl (C=O) is also present in carboxylic acid derivatives: acyl halide, anhydride, carboxylic acid, esters, and amide.

NMR distinguishes between an aldehyde (NMR δ 9–10) and a ketone (no characteristic NMR signals).

Aldehydes have additional peaks at 2700–2800 cm^{-1}, while ketones do not.

Notes for active learning

Notes for active learning

Alkanes and Alkyl Halides – Detailed Explanations

1. C is correct.

A nucleophile donates lone pairs of electrons; therefore, it is a Lewis base.

2. D is correct.

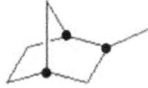 There are three tertiary alkyl positions (i.e., carbon bonded to three other carbons).

3. B is correct.

S_N2 is bimolecular; rate = k [substrate] × [nucleophile].

Therefore, if the nucleophile ($^-$OH) concentration is doubled, then the reaction rate doubles.

Since the alkyl halide is primary, a unimolecular (S_N1) reaction occurs.

Water stabilizes the carbocation intermediate and thereby increases the rate of the reaction.

4. D is correct.

Only the electrophile concentration controls the reaction rate because the rate-determining step is the unimolecular formation of the carbocation.

S_N1 reactions proceed *via* a carbocation in the first step of the mechanism.

In the second step, the nucleophile forms a new bond by attacking the carbocation.

S_N1 undergoes first-order kinetics, whereby:

  rate = k[substrate]

5. E is correct.

Alkanes are only composed of hydrogen and carbon atoms. Because the electronegativity of these atoms is quite similar, large bond dipoles are not expected.

Hydrogen bonding is a force that acts through highly polarized bonds to form intermolecular bonds to partial positive hydrogen atoms (H attached to F, O or N).

6. A is correct.

The sodium methoxide in this reaction acts as a strong base and abstracts a proton on the carbon atom that neighbors the chloride. The mechanism for this process is E_2.

The reaction is not S_N1 because heat is needed to generate the carbocation, not included in the rxn conditions.

7. B is correct. The order of stability of carbocations is 3° > 2° > 1°.

Therefore, more substituted carbocations are more stable.

8. D is correct.

1-bromopropane is a primary halogen and a strong nucleophile, leading to S_N2, preferred without steric hindrance.

9. C is correct.

Molecules with double or triple bonds cannot undergo free rotation because the *pi* bond rigidifies the structure of the compound. To achieve the bond rotation, the *pi* bond(s) must be broken.

Furthermore, for cyclic compounds, the smaller the ring size, the fewer degrees of freedom it has.

Due to strain energy, cyclopropane is unable to rotate freely.

10. B is correct.

The concentration of substrate and nucleophile control the reaction rate because the rate-determining step is bimolecular.

S_N2 undergoes *second-order kinetics* whereby:

$$\text{rate} = k\,[\text{substrate}] \times [\text{nucleophile}]$$

11. D is correct.

The heat of combustion ($\Delta H_c°$) is the energy released (as heat) when a compound undergoes complete combustion with oxygen.

The chemical reaction is typically a hydrocarbon reacting with oxygen to form carbon dioxide, water, and heat. General formula:

$$C_nH_{2n+2} + ((3n+1)/2)O_2 \rightarrow (n+1)H_2O + nCO_2 + \text{energy}$$

The heat of combustion of a compound depends on three main factors: molecular weight, angle strain, and degree of branching.

In most cases, the compound with a higher molecular weight (i.e., more C–C and C–H bonds) has the larger heat of combustion.

For straight-chain alkanes, each addition of methylene (~CH_2~) groups adds approximately –157 kcal/mole to the heat of combustion.

For cycloalkanes, the heat of combustion increases with increasing angle strain.

Reference: $1\,kJ\cdot mol^{-1}$ is equal to $0.239\,kcal\cdot mol^{-1}$

$\Delta H_c°$ for alkanes increase by about 657 kJ/mol (157 kcal/mol) per ~CH_2~ group.

For example, heptane has 4 more ~CH_2~ groups than propane:

$$4 \times -157\text{ kcal per mole} = -628\text{ kcal/mol}$$

Yields:

$$-530\text{ (propane)} + -628 = -1{,}094\text{ kcal/mol (heptane)}$$

12. D is correct.

Identify the molecules that contain nitrogen or oxygen heteroatoms because these atoms enable molecules to participate in hydrogen bonding and contribute to dipolar interactions.

Cyclopentane is only composed of hydrogen and carbon, so it is the least water-soluble.

13. D is correct.

Unimolecular elimination occurs *via* E_1 and forms a carbocation in the slow (rate-determining) step.

A: a single step (concerted) process describes bimolecular (E_2), not E_1 elimination.

B: homolytic (compared to the more common heterolytic) cleavage of a covalent bond yields free radicals.

C: free radicals occur with peroxides (H_2O_2) or dihalides / UV, and radicals are an intermediate in unimolecular (E_1) elimination.

14. C is correct.

Since an S_N1 reaction proceeds through a planar, sp^2 hybridized carbocation intermediate (which can be attacked from either side), it forms a racemic mixture (i.e., both *R* / *S* stereoisomers are present).

15. B is correct.

S_N1 reactions favor substituted alkyl halides because of the stability of the carbocation intermediates, whereas S_N2 reactions favor unsubstituted reactants due to minimal steric hindrance for the approaching nucleophile.

A: S_N1 reaction rates are greatly affected by electronic factors (degree of substitution and inductive influence of electronegative atoms), while S_N2 reaction rates are greatly affected by steric (hindrance) factors.

C: S_N1 reactions proceed *via* a carbocation intermediate, but S_N2 reactions (*bimolecular*) proceed *via* a single-step reaction involving a concerted mechanism with a transition state instead of a carbocation.

D: S_N2 reactions are bimolecular reactions that proceed *via* a single-step reaction with a transition state (i.e., bond making and breaking events) rather than *via* an intermediate.

E: S_N2 reactions are favored by polar aprotic solvents (i.e., cannot dissociate a proton into the solution), such as THF, DMSO, EtOAc, which do not stabilize the strong (i.e., negatively charged anion) nucleophile, so it remains more reactive to drive the bimolecular reaction of S_N2.

S_N1 reactions proceed *via* polar protic solvents (e.g., H_2O, methanol), which tend to stabilize the carbocation (positive charge) by electrostatic attraction to the resulting anion of the dissociated solvent.

16. D is correct.

Structural isomers have the same molecular formula (C_7H_{16}) but different atomic connections.

Molecular mass would be a consideration if the molecules were not isomers (same molecular formula).

Branching in alkanes lowers the boiling point because branched molecules cannot interact as effectively as unbranched molecules and have less surface area. 2,2,4-trimethylpentane has the lowest boiling point because it is the most highly branched.

Hydrogen bonding is a strong attractive force between molecules (intermolecular force) but requires hydrogen to be attached to an electronegative atom (fluorine, oxygen, or nitrogen).

The next most attractive intermolecular force is dipole-dipole interaction.

17. D is correct.

S_N2 reactions are concerted reactions with a single step and do not form charged intermediates.

Carbocation intermediates are a feature of S_N1 and E_1 mechanisms.

18. B is correct.

The twist-boat is located at a local energy minimum (or trough) for the conformers of cyclohexane.

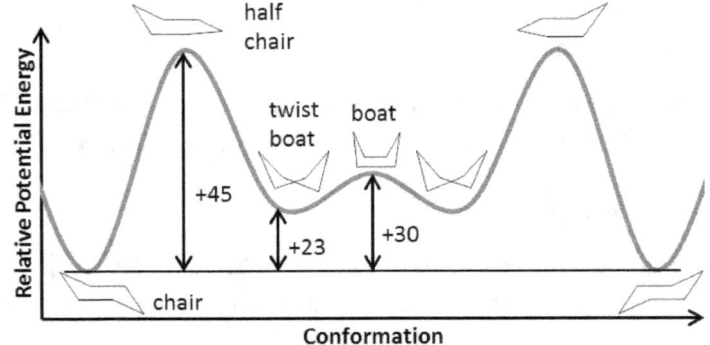

Relative energy diagram for conformers (i.e., chair flips) of cyclohexane

19. B is correct.

To form propyl chloride, one of six hydrogens (3 on each methyl group) can be substituted. To form isopropyl chloride, the hydrogen extracted must be one of the two on the middle carbon.

Based on statistical probability, propyl chloride forms in a 3:1 ratio compared to isopropyl chloride.

When terminal hydrogen is extracted (to form propyl chloride), a primary radical intermediate is formed; when internal hydrogen is extracted (to form isopropyl chloride), a secondary radical intermediate forms.

Carbon radicals are electron-deficient because they do not have a full octet.

Alkyl groups are electron-donating (*via* hyperconjugation), so a secondary radical is more stable than a primary radical, and therefore the formation of a secondary radical is more likely.

The reaction proceeding through the secondary radical intermediate is faster. Therefore, the reaction leading to the formation of isopropyl chloride proceeds more readily. This is contrary to strictly statistical considerations, and more isopropyl chloride is formed than the predicted 25%.

Note: it is impossible to predict what percentages form because other factors, such as temperature, are important for the empirical yield determination for each product.

20. D is correct.

S_N1 reactions proceed *via* a carbocation in the first step of the mechanism.

In the second step, the nucleophile forms a new bond by attacking the carbocation.

S_N1 undergoes first-order kinetics, whereby:

    rate = k[substrate]

where k is a rate constant determined experimentally.

21. C is correct.

Conformations in which the single bonds are staggered are more stable than those in which they are eclipsed due to steric repulsion and torsional strain (i.e., bonding electron between adjacent atoms – such as observed in eclipsed Newman projections).

Torsional strain is present in Newman projections when the substituents are eclipsed. It originates from the repulsion of the bonding electrons (i.e., not the steric interactions of the substituents).

Conformations that put the largest substituents in an *anti* (180° offset) arrangement are more stable than those in a *gauche* (60° offset) arrangement.

1,2-dibromoethane shown anti in the Newman projection

22. D is correct.

S_N2 reactions are favored with primary substrates, strong nucleophiles, good leaving groups, and polar aprotic solvents. 1-bromobutane is a primary alkyl halide, ⁻CN is an extremely strong nucleophile, and bromine is a good leaving group.

S_N2 occurs exclusively over elimination E_2 because ⁻CN is a strong, linear nucleophile; ⁻CN is not sterically hindered as a base.

Only when strong, bulky bases (e.g., *tert*-butoxide) are used, elimination (e.g., E_2) is favored over substitution (e.g., S_N2).

S_N1 and E_1 reaction mechanisms are not favored with primary alkyl halides because a primary carbocation is unstable.

23. B is correct.

S_N1 proceeds when the substrate forms a carbocation.

Iodide is the best leaving group (i.e., most stable anion) and forms the cation the fastest.

24. C is correct.

Acetate anion

Dimethyl sulfoxide (DMSO) is an organosulfur compound with the formula $(CH_3)_2SO$.

The colorless liquid is an important polar aprotic solvent that dissolves polar and nonpolar compounds and is miscible with water and a wide range of organic solvents.

25. C is correct.

Strong nucleophiles have a negative formal charge (i.e., lone pairs excess), while weak nucleophiles are neutral species with a lone pair of electrons.

26. A is correct.

E_1 is a unimolecular elimination that proceeds *via* a carbocation intermediate in a two-step reaction.

Substitution of the substrate (3° > 2° > 1° >> methyl) increases the rate of E_1 (and S_N1) reactions because highly branched carbon chains (with more substituted carbons) form more stable carbocations.

S_N2 is a bimolecular nucleophilic substitution that proceeds *via* a one-step (concerted mechanisms) displacement of a leaving group by a nucleophile.

S_N2 is favored by unbranched carbon chains because the nucleophilic displacement of a leaving group by a nucleophile is favored due to less steric hindrance.

Secondary carbons could undergo both E_1 and S_N2. The cyanide group is a very poor leaving group because it is an unstable anion and is unlikely to dissociate *via* E_1 or be displaced *via* S_N2.

Br^- forms a stable anion and can dissociate *via* E_1 or be displaced by S_N2.

B: $(CH_3CH_2CH_2)_3CBr$ is a tertiary alkyl halide and is not favored.

C: $(CH_3CH_2CH_2)_3CCH_2Cl$ is a primary alkyl halide that lacks β-hydrogens and therefore cannot form a double bond by elimination (E_1 or E_2).

D: $(CH_3CH_2CH_2)_2CHCN$ has the leaving group (cyanide) bonded to a secondary carbon.

E: $CH_3CH_2CH_2CH_2Br$ has the leaving group (bromine) bonded to a primary carbon.

27. D is correct.

Boiling requires the molecules in the liquid phase to overcome the attractive intermolecular forces (e.g., hydrogen bonding, dipole-dipole & London dispersion forces) and move into the gas phase.

continued…

The stronger these interactions are, the more energy (i.e., heat) is needed for the molecules to separate from their neighbors and migrate into the gaseous state.

Molecules have lower boiling points when branching increases because branching disrupts the spatial packing of molecules in the solid and liquid phase, and therefore branching reduces intermolecular attractions.

A: *cis*-2-pentene has a slightly higher boiling point because unsaturation establishes a dipole moment that raises their relative boiling point compared to alkanes.

B: 2-pentyne has a slightly higher boiling point because unsaturation establishes a dipole moment that raises their relative boiling point compared with alkanes.

C: pentane is a hydrocarbon and only experiences weak London dispersion forces.

E: 3-pentanol exhibits hydrogen bonding and has a relatively high boiling point compared with molecules of comparable molecular weight that lack hydrogen bonding.

28. C is correct.

A *trans* isomer requires that the substituents point in opposite directions (up and down).

Therefore, one substituent is located axial, while the other substituent is equatorial.

In this substituted cyclohexane, the molecule is more stable when the larger substituents are equatorial.

When comparing methyl and isopropyl, the isopropyl is larger and therefore is located equatorially.

29. A is correct.

E_1 and S_N1 reactions are strongly favored by highly branched carbon chains and good leaving groups.

E_2 reactions are largely independent of the structure of carbon chains and are favored by good leaving groups, which can easily be eliminated by basic conditions.

S_N2 reactions are strongly favored by substrates with unbranched carbon chains.

$(CH_3CH_2CH_2)_3CBr$ is a tertiary alkyl halide.

B: $CH_3CH_2CH_2CH_3$ is a hydrocarbon where alkanes do not undergo elimination or nucleophilic substitution. Alkanes are unreactive to most organic chemistry reagents and can either undergo combustion (i.e., burning of propane) or free radical halogenation to introduce a halogen as a leaving group.

C: $(CH_3CH_2)_3COH$ is a highly branched tertiary alcohol carbon chain, so it cannot undergo an S_N2 reaction. Also, ^-OH is a very poor leaving group, so it does not readily undergo substitution. With heat, alcohols undergo elimination *via* dehydration (i.e., removal of water).

D: $CH_3CH_2CH_2CH_2Br$ is a primary alkyl halide that undergoes S_N2 and E_2, but neither S_N1 nor E_1.

30. A is correct.

In a complete combustion reaction, a compound reacts with an oxidizing element (e.g., oxygen), and the products are compounds of each element in the fuel combined with the oxidizing element.

General formula:

$$C_nH_{2n+2} + [(3n + 1) / 2]O_2 \rightarrow (n+1)H_2O + nCO_2 + \text{energy}$$

For example, methane yields:

$$CH_4 + 2\,O_2 \rightarrow CO_2 + 2\,H_2O + \text{energy}$$

Nonane:

$$C_9H_{20} + [(3 \times 9 + 1) / 2]O_2 \rightarrow (9 + 1)H_2O + 9\,CO_2 + \text{energy}$$

$$C_9H_{20} + 14\,O_2 \rightarrow 10\,H_2O + 9\,CO_2 + \text{energy}$$

Since nonane has the molecular formula C_9H_{20}, the combustion of 1 mole of neopentane produces 9 moles of CO_2 and 10 moles of H_2O.

31. D is correct.

Free radical halogenation is a few reactions (along with combustion) that alkanes undergo and occur *via* a highly reactive halogen radical.

A radical is a single, neutrally charged atom with an unhybridized *p* orbital with a single unpaired electron that causes it to be reactive.

Steps for free radical halogenation:

> Step I: initiation forms the free radical from a diatomic molecule (X_2) by homolytic bond cleavage. Initiation is shown in the first reaction equation and is usually catalyzed by ultraviolet light (hv), heat, or by an attack by another free radical.
>
> Step II: propagation involves a radical and a neutral molecule, and the products are a new neutral molecule and a new radical. The halogen-free radical attacks the neutral alkane and, *via* homolytic bond cleavage, produces a new H–X bond and another highly reactive free radical as the alkyl radical.
>
> The alkyl radical reacts with Br_2 to form the alkyl halide and another halogen radical. This new halogen radical then starts the process again, thus causing a chain reaction (propagation).
>
> Step III: termination joins the two radicals. For example, an alkyl radical attacks a halogen radical to produce a new neutral molecule.

I: $Br_2 + hv \rightarrow 2\,Br\bullet$ is UV light-induced generation of two free radicals and is the chain initiating step.

II: $Br\bullet + RH \rightarrow HBr + R\bullet$ is a chain propagating step because the reaction advances the chain onward by generating a new neutral product plus a new free radical.

III: $R\bullet + Br_2 \rightarrow RBr + Br\bullet$ is a chain propagating step because the reaction advances the chain onward by generating a new neutral product plus a new free radical.

32. B is correct.

As a concerted mechanism, the single step of the S_N2 reaction is a simultaneous substitution occurring with the formation of a new bond while the original bond breaks.

S_N2 undergoes second-order kinetics.

$$\text{rate} = k \, [\text{substrate}] \times [\text{nucleophile}]$$

33. A is correct.

Alkanes undergo free radical halogenation to substitute a C–H bond with a C–X bond, where X is a halogen.

Alkenes and alkynes generally undergo addition reactions instead, where an electrophile may add across a carbon-carbon *pi* bond.

34. B is correct.

Bimolecular nucleophilic substitution (S_N2) occurs at the fastest rate when the substrate is the least hindered.

The relative reactivity of the alkyl halides for S_N2 is methyl > primary > secondary >> tertiary.

S_N2 reactions do not occur on sterically hindered tertiary substrates.

A: 1-chloro-2,2-diethylcyclopentane is a secondary alkyl halide.

This substrate does not undergo S_N2 substitution as rapidly as 1-chlorocyclopentane because branching adjacent to the carbon-containing leaving group reduces the rate of S_N2 reactions since the approach of the nucleophile is impeded compared to a straight-chain molecule.

C: 1-chlorocyclopentene does not undergo nucleophilic substitution because it has a halogen attached to a vinylic (i.e., on a double bond) carbon.

D: 1-chloro-1-ethylcyclopentane is a tertiary halide and does not undergo S_N2 nucleophilic substitution.

The mechanism is S_N1 (with a carbocation intermediate).

E: *tert*-butylchloride is a tertiary halide and does not undergo S_N2 nucleophilic substitution.

The mechanism is S_N1 (with a carbocation intermediate).

35. D is correct.

Stable molecules are the best leaving groups.

If the leaving groups are charged, then the bromide ion is the best leaving group because the anion is stable (due to atomic size).

The order of the halogens as leaving groups: $I^- > Br^- > Cl^- > F^-$

The order of the halogens as nucleophiles: $I^- > Br^- > Cl^- > F^-$

36. E is correct.

Radical termination steps involve an overall decrease in the number of radicals when comparing the starting materials to the products.

The correct answer involves a decrease in the number of radicals (from two to zero).

A: the number of radicals increases from zero to two, which is an initiation step.

B: the number of radicals (one) does not change, which are propagation steps.

C and D: the number of radicals (one) does not change, which are propagation steps.

E: the number of radicals decreases from two to zero, which is a termination step.

Notes for active learning

Notes for active learning

Alkenes – Detailed Explanations

1. C is correct.

Although these reactions produce the same products and in the same quantities, what is different is the rate of the borane addition for each alkene.

(E)-3-heptene   (Z)-3-hexene

The E has the highest priority substituents on opposite sides, while the Z alkene has the alkyl substituents on the same side.

Therefore, because Z alkenes are less thermodynamically stable than E alkenes, Z alkenes are more reactive.

Furthermore, the approach of the borane to the Z alkene is less sterically demanding because the Z alkene is more open to an attack (alkyl groups are on the same side).

2. B is correct.

The addition of deuterium (an isotope of hydrogen) across a C=C double bond is metal-catalyzed. It proceeds with *syn* stereoselectivity, adding both deuterium to the same face of the double bond.

3. D is correct.

The trapping of mercury-alkene cations typically opens on the more substituted side, but the *tert*-butyl group sufficiently blocks access to this side. Therefore, the cation is opened on the more terminal side.

4. D is correct.

HI is the strongest acid of the given molecules, so it protonates the fastest (followed by HBr, then HCl).

Also, I^- is the best nucleophile, so it adds to the carbocation most rapidly.

5. D is correct.

In 1,3-butadiene, C–2 and C–3 are sp^2 hybridized with unhybridized p orbitals involved in π bonding. The bond between C–2 and C–3 results from the overlap of two sp^2 hybridized orbitals.

There cannot be a partial double-bond character due to σ electrons; double bonds result only from π electrons.

6. A is correct.

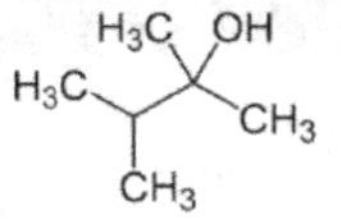

In the first step of the reaction, the H⁺ adds to the double bond creating a carbocation.

The nucleophilic water attacks the carbocation, and the acid is regenerated in the last step of the reaction.

2,3-dimethyl-2-butanol

7. A is correct.

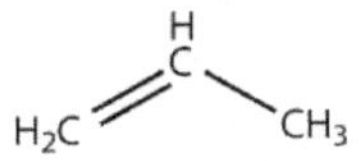

The rate law may not reveal the steps in a reaction mechanism.

Furthermore, the bromination of alkenes proceeds in two steps, not one.

8. D is correct.

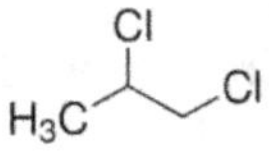

Draw the structures of the starting material and product.

Two halogen atoms are added to the molecule, and an efficient method to achieve this is by exposing the alkene to a diatomic halogen species (e.g., Cl₂ or Br₂).

The chlorine atoms add in an *anti*-orientation from the 3-membered ring of the chloronium ion.

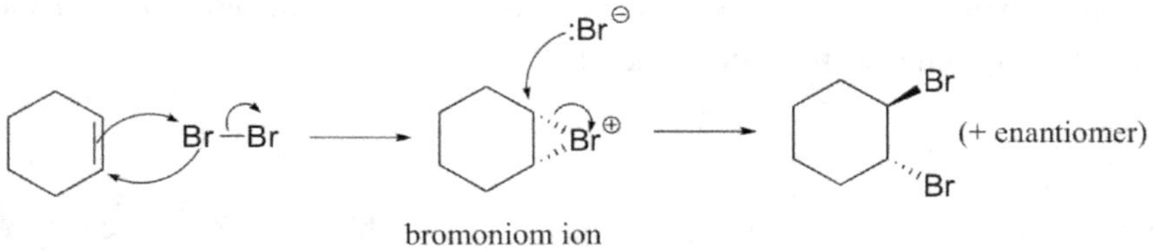

bromoniom ion

Bromonium (like chloronium) ions undergo the same reaction mechanism.

9. A is correct.

The Zaitsev product is the most thermodynamically stable.

Thermodynamically stable alkenes are most substituted (3° > 2° > 1°).

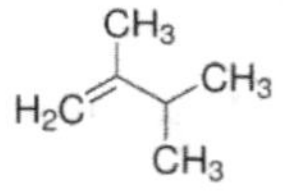

2,3-dimethyl-1-butene has a 1° alkene carbon (terminal) and a 3° carbon (at position 2).

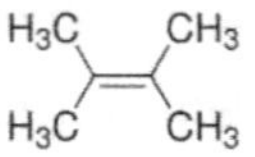

2,3-dimethyl-2-butene has two 3° alkene carbons at positions 2 and 3.

Hyperconjugation of the neighboring C–H bonds into the *pi** orbital of the alkene lowers the alkene's energy.

Furthermore, the sp^3 hybridized alkyl groups help to inductively donate electron density to the more electronegative sp^2 hybridized carbon atoms.

10. C is correct.

Conjugation is the alternation of double and single bonds, which results in the delocalization of electrons *via* resonance through the *sp²* hybridized carbons, resulting in increased stability for the molecule.

1,3,5-heptatriene has three conjugated double (π) bonds.

A: 1,2-hexadiene has two cumulated, rather than conjugated, π bonds.

Cumulated molecules have a *sp* carbon in the center of the two double bonds – this is unstable and increases the overall energy of the molecule.

B: 1,3-hexadiene has two conjugated double bonds.

D: 1,5-hexadiene has two isolated π bonds.

E: 1,2,3-heptatriene is cumulated. Cumulated molecules have a *sp* carbon in the center of the two double bonds – this is unstable and increases the overall energy of the molecule.

11. A is correct.

Water, methanol, and acetic acid are poor solvents because they can react as nucleophiles and add to alkenes.

12. C is correct.

The hydration proceeds with Markovnikov addition.

The first step in the reaction is protonation to form the tertiary carbocation trapped by water.

13. A is correct.

A conjugated system has two or more double or triple bonds separated by a single bond (i.e., *sp²* hybridization at each carbon in the conjugated system).

$H_2C=C=C(CH_3)_2$

1,2-butadiene is not conjugated because the double bonds are adjacent (not separated by a single bond). Adjacent double bonds are cumulated and are unstable.

B: cyclobutadiene is conjugated because the molecule has two double bonds separated by single bonds.

C: benzene is conjugated because a single bond separates each double bond.

D: 1,3-cyclohexadiene is conjugated because a single bond separates the double bonds between carbons 1-2 and carbons 3-4.

E: 2,4-pentadiene is conjugated because the molecule has two double bonds separated by a single bond.

14. A is correct.

In $(CH_3)_2C=C(CH_3)CH_2CH_3$, both carbons of the alkene are equally substituted (i.e., tertiary), so the two putative carbocations are approximately equally stable.

The addition of a hydrogen halide across a double bond can create two chiral centers, one at each of the former sp^2 carbons. The carbocation forms preferentially at the more substituted carbon.

15. C is correct. Oxymercuration-reduction of an alkene yields Markovnikov orientation and *anti*-addition.

By comparison, hydroboration-oxidation of an alkene is consistent with *anti*-Markovnikov and *syn* addition.

Step one uses BH_3 to add BH_2 and H as *syn* addition across the double bond of the alkene.

Step two uses peroxides and nucleophilic oxygen for *syn* addition of a hydroxyl. Other nucleophiles (e.g., methanol) can add *syn* addition as an ether substituent across the double bond.

16. D is correct.

Geometric isomers have the same molecular formula but different connectivity between the atoms due to the orientation of substituents around a carbon-carbon double bond (or ring).

In *cis* isomers, the same substituents are on one side of the double bond or ring, while in *trans* isomers, the same substituents are on opposite sides of the double bond or ring.

Isomers (same molecular formula, but different molecules) with no double bonds cannot be geometric isomers.

Isomers include:

    Constitutional isomers – different connectivity of the backbone or containing different functional groups;

    Enantiomers – mirror images of chiral molecules;

    Diastereomers – non-mirror chiral molecules with 2 or more chiral centers or geometric isomers containing double bonds (i.e., designated as *cis*/*trans* or *E*/*Z*).

17. A is correct.

The methyl group adjacent to the carbocation migrates *via* a methide shift to give a tertiary carbocation, which then loses a proton to form the most substituted alkene (i.e., both carbons of the alkene are tertiary).

18. B is correct.

The reaction conditions are for the halohydration of an alkene. The bromonium ion forms first; this is attacked (ring opens) by water as a nucleophile on the more substituted side.

19. D is correct.

A *bromonium ion* is a cyclic (three-membered ring) structure of a bromine atom attached to two unsaturated carbons (i.e., an alkene).

The bromonium ion is formed when the nucleophilic double bond adds to a bromine atom of Br_2, releasing Br^-.

The resulting bromine anion (or solvent if it is a nucleophile – contains lone pairs of electrons: H_2O, NH_3 or CH_3OH) bonds to the more substituted carbon of the bromonium structure dibrominated product. Cl_2 follows the same reaction mechanism.

An *anti*-product is formed because the Br^- (or nucleophilic solvent with lone pairs) must approach the three-membered ring of the bromonium ion from the side opposite the bromonium ion.

$CH_3CH_2CH=CH_2 + HBr \rightarrow CH_3CH_2CHBrCH_3$: HBr adds to alkenes *via* a Markovnikov mechanism. The double bond attacks the proton, and hydrogen adds to the carbon of the double bond with fewer alkyl substituents (i.e., less substituted carbon).

As a result, a carbocation intermediate forms on the more substituted carbon.

Second, Br^- adds to the carbocation (more substituted carbon) to form the alkyl halide. The mechanism involves a carbocation and not a bromonium ion.

Unlike a bromonium ion, which yields the *anti*-stereochemistry, the trigonal planar carbocation undergoes nucleophilic attack from either face (e.g., the top and bottom). It produces enantiomers (a racemic mixture) if the product contains a chiral center on the carbocation carbon.

20. C is correct.

cis-3-methyl-2-hexene undergoes Markovnikov addition because the first step is protonation to give the more stable (more substituted) carbonium ion. It gives *syn*- and *anti*-addition products because, in the second step, the bromide ion attacks the top and bottom faces of the carbonium ion.

Adding a hydrogen halide to an asymmetrical alkene leads to either halogenated alkene. The halide is on the more substituted carbon or a product in which the halide is on the less substituted carbon.

The former addition follows Markovnikov's rule, and the latter is an example of an *anti*-Markovnikov addition. If a mixture of the two products is formed, with a predominance of one product, the reaction is *regioselective*.

21. E is correct.

Addition reactions require two or more reagents to combine to form a single product.

For the oxidation of an alkene (e.g., epoxidation or dihydroxylation), only the oxygen atoms of a reagent are transferred to the alkene. A reagent byproduct is typically given off in the reaction, such as the carboxylic acid from *m*CPBA oxidation (i.e., peroxyacid) or the reduced metal from potassium permanganate ($KMnO_4$) or osmium tetraoxide (OsO_4) oxidation.

Ozonolysis (O_3 or Cr_2O_7) is a type of oxidation of alkenes splitting the alkene into two separate carbonyl-containing products (i.e., cleavage reaction) and is not an addition reaction.

22. A is correct.

Hydrogen bromide will not substitute an alkane because alkanes are unreactive. Alkanes undergo substitution only under extreme conditions (e.g., *hv* as UV light) or yield combustion products CO_2 and H_2O.

N-bromosuccinimide (NBS) adds bromine to the allylic position (i.e., one away from a double bond). The product is: butene + NBS → 3-bromobutene

Reaction 2: at high temperatures, alkanes undergo combustion to form CO_2 and H_2O.

Reaction 3: free radical substitution is initiated for highly reactive free radicals (e.g., Br_2 or Cl_2), with Br_2 being more selective, whereby the Br radical adds to the more substituted carbon of the alkene.

Reaction 4: Br_2 in CCl_4 adds bromines to an alkene as a bromonium intermediate, with the second bromide adding to the more substituted carbon of the alkene.

23. C is correct.

Use the following formulae to calculate the *degrees of unsaturation*:

C_nH_{2n+2}: for an alkane (0 degrees of unsaturation)

C_nH_{2n}: for an alkene or a ring (1 degree of unsaturation)

C_nH_{2n-2}: for an alkyne, 2 double bonds, 2 rings or 1 ring and 1 double bond (2 degrees of unsaturation)

There are two degrees of unsaturation for the compound C_6H_{10}.

24. C is correct.

Conjugated (alternating double and single) bonds are more thermodynamically stable than unconjugated double bonds.

B: adjacent double bonds of allenes are not conjugated but cumulated because the *pi* bonds are oriented 90 degrees apart. Allenes tend to be less stable, especially when confined to cyclic structures.

The other structures are isolated double bonds with (one or more) intervening sp^3 hybridized carbons between the sp^2 carbons of the double bonds.

25. A is correct.

Product A is an example of Markovnikov addition, whereby the hydrogen adds to the least substituted carbon because the most stable carbocation is formed. The bromine then adds to the (most stable) carbocation. In this example, the hydrogen adds to the secondary carbon, and the bromine adds to the tertiary carbon.

Product B involves free radical intermediates because of the hydrogen peroxide (H_2O_2).

Hydrogen peroxide causes the reaction to proceed *via* a radical intermediate (not carbocation), and the regiochemistry (where the substituents add) is *anti*-Markovnikov. The bromine adds to the least substituted carbon, and the H adds to the most substituted carbon radical.

Product C yields 2-methyl-2-butanol, according to Markovnikov addition.

26. A is correct.

Vinyl refers to an atom attached to carbon on a double bond.

Allylic refers to an atom attached to a carbon adjacent (β) to the double bond.

The chlorine substituent is directly attached to the alkene carbon atoms in vinyl chloride.

27. E is correct.

An *anti*-Markovnikov addition of water across the alkene double bond is needed.

Peroxides (H_2O_2) are a characteristic reagent for *anti*-Markovnikov regiospecificity.

28. D is correct.

Carbon-carbon *pi* bonds are elements of unsaturation, and unsaturated compounds can be reduced to give more reduced molecules (e.g., alkanes).

Alkenes have a higher oxidation state than the alkane and are equivalent to a C–O or C–X bond, where X is a more electronegative atom.

29. A is correct.

An *alkoxide* is the conjugate base of an alcohol and therefore consists of an organic moiety (i.e., group) bonded to a deprotonated (i.e., negatively charged) oxygen atom.

Secondary halides undergo bimolecular elimination (E_2) with strong bases, especially hindered ones like potassium *tert*-butoxide, $KOC(CH_3)_3$.

B: E_1 designates unimolecular elimination, generally observed in protic (i.e., H^+ donating) solvents (e.g., water or alcohols) and not when subjected to a strong alkoxide base.

C: S_N2 designates bimolecular nucleophilic substitution.

D: S_N1 designates unimolecular nucleophilic substitution.

30. B is correct.

The bromine adds to the internal position of the epoxide because the partial positive charge is greater at the more substituted position. The mechanism follows *anti*-addition stereochemistry and would be shown in the final product if the alcohol and halogen were attached to chiral carbons.

31. B is correct.

Protonation of the alkene forms the more stabilized carbocation and this cation forms at the secondary alkyl position. The cation is then trapped by water to form the alcohol.

32. A is correct.

The reaction shown is an acid-catalyzed dehydration reaction.

Alcohol in sulfuric acid and heat is characteristic of the E_1 mechanism to form an alkene.

The generated cation undergoes a ring expansion to give a more stable tertiary carbocation, which is eliminated to form the tertiary (i.e., trisubstituted) alkene.

From the rearranged carbocation, the alpha proton is eliminated by the conjugate base HSO_4^- to regenerate the sulfuric acid.

The ring (i.e., bond angle or Baeyer) strain energy drives ring expansion.

33. A is correct.

Alkanes and alkenes have *sigma* bonds.

The *pi* (i.e., double) bond in the alkenes stabilize a negative charge (as an anion) and, therefore, are more acidic.

The increased *s* character on the hybridization of an sp^2 orbital of an alkene ($pK_a = 45$) allows the orbital to accommodate the negative charge with more stability than the sp^3 of an alkane ($pK_a = 50$).

Furthermore, the increased *s* character on the hybridization of an sp orbital of an alkyne ($pK_a = 28$) allows the orbital to accommodate the negative charge with more stability than the sp^2 of an alkene.

34. A is correct.

Tertiary alkyl halides form alkenes with strong bases *via* E_2, such as sodium ethoxide ($NaOCH_2CH_3$).

The most substituted alkene is the major (Zaitsev) product that is internal or more substituted.

The least substituted alkene is the minor (Hofmann) product that is terminal or less substituted.

B: 2-methylpent-3-ene is a molecule whereby the alkene does not connect carbon with the bromine or adjacent carbon. Rearrangement does not occur in E_2 reactions because rearrangement requires a carbocation (S_N1 or E_1) intermediate.

C: 2-methyl-2-methoxypentane is the product of substitution. Tertiary alkyl halides do not undergo substitution with Lewis bases; elimination is the mechanism for product formation.

continued…

D: 2-methylpentene is a less substituted alkene and only occurs with a bulky base (e.g., tert-butyl oxide or LDA).

E: 1-methylpentene is not the product of the reaction because the carbon structure has changed, whereby the methyl group has migrated from the second carbon to the first carbon.

Additionally, the molecule would be named 2-hexene according to the longest chain.

35. C is correct.

An *anti*-Markovnikov addition of water is needed across the alkene double bond.

The reagents are borane, followed by hydrogen peroxide and sodium hydroxide.

Comparison of Markovnikov vs anti-Markovnikov addition reactions for alkenes.

Each reaction above occurs in separate steps as indicated by the vertical separation line.

A: the product of oxymercuration-demercuration: 1) Hg(OAc)$_2$, H$_2$O/THF; 2) NaBH$_4$

36. A is correct.

The *Cope elimination* is an intramolecular elimination reaction that occurs when the oxygen atom of the oxide of a tertiary amine removes a proton from an adjacent position.

This results in the formation of a *syn* alkene. The reaction requires heat.

The *Cope elimination* yields the same products as a Hofmann elimination (i.e., exhaustive methylation).

Notes for active learning

Alkynes – Detailed Explanations

1. E is correct.

In excess hydrogen gas, the triple bond of 3-heptyne is completely reduced to the alkane.

Platinum and palladium catalysts can be used to reduce alkynes to alkenes or alkanes.

2. C is correct.

Protonation of alkynes with acids generates the more substituted carbocation because this cation is the more stable intermediate. Because the cation exists on a vinyl carbon atom, this cation is a *vinyl cation*.

3. C is correct.

Use the following formulae to calculate the *degrees of unsaturation*:

C_nH_{2n+2}: for an alkane (0 degrees of unsaturation)

C_nH_{2n}: for an alkene or a ring (1 degree of unsaturation)

C_nH_{2n-2}: for an alkyne, 2 double bonds, 2 rings or 1 ring and 1 double bond (2 degrees of unsaturation).

The *general chemical formula for alkynes* is C_nH_{2n-2} because each *pi* bond of the alkyne represents one degree of unsaturation.

Cyclic alkynes have an additional degree of unsaturation because of the cyclic structure.

Therefore, C_9H_{16} is the molecular formula that describes an acyclic alkyne.

4. A is correct.

One *sigma* bond and two *pi* bonds are used to make the triple bond of alkynes.

Nitriles are another functional group that contains a triple bond.

5. A is correct.

Using a subscript of n for the number of carbons, the *degree of unsaturation* is:

Alkane: C_nH_{2n+2} = 0 degrees of unsaturation

Alkene: C_nH_{2n} = 1 degree of unsaturation (1 ring or 1 double bond)

Alkyne: C_nH_{2n-2} = 2 degrees of unsaturation (2 double bonds, 1 double bond and 1 ring or 2 rings)

6. D is correct.

Platinum and palladium are used to hydrogenate alkynes to alkenes (or alkanes) or reduce alkenes to alkanes.

$$H-C\equiv C-H \xrightarrow[Pt]{1\ eq.\ H_2} \underset{H\ \ \ \ H}{\overset{H\ \ \ \ H}{C=C}} \xrightarrow[Pt]{1\ eq.\ H_2} H-\underset{H\ \ H}{\overset{H\ \ H}{C-C}}-H$$

Two moles of hydrogen gas reduce the two pi bonds of the alkyne to an alkane.

7. B is correct.

Alkynes are triple bonded molecules made of hydrogen and carbon atoms.

The electronegativity difference between carbon and hydrogen is low, resulting in the molecule being less polar.

Molecules with little polarization may be more soluble in organic solvents as opposed to water.

8. A is correct.

Whenever a hydrocarbon is burned (adding O_2), the major byproducts of the reaction are water and carbon dioxide. Ash may form from the reaction, composed of carbon material that cannot undergo further oxidation.

9. C is correct.

$$H_3C-\equiv CH$$

Alkyne: C_nH_{2n-2} = 2 degrees of unsaturation due to the triple bond.

In the name of this molecule, the prefix is *pro–*, which indicates that the molecule is composed of three carbon atoms.

The *–yne* suffix indicates that a carbon-carbon triple bond exists in the molecule.

Therefore, propyne has three carbon atoms and four hydrogen atoms.

10. C is correct.

1-butyne is a gas at room temperature, and 1-propyne has an even lower boiling point, so it is a gas.

11. C is correct.

Two *pi* bonds use one mole of H_2 each, so 2 moles of hydrogen are consumed in the conversion.

12. D is correct.

Bromine atoms are added to both sides of the triple bond to produce a vicinal dibrominated alkene.

The addition product has the bromine atoms on opposite sides of the double bond; therefore, this addition proceeds through an *anti*-addition mechanism.

13. B is correct.

The substrate contains a carbon-carbon triple bond at the end of the chain. Therefore, to synthesize 2-hexanone from 1-hexyne, oxygen is introduced to the internal carbon atom of the alkene. This reaction requires aqueous acidic conditions so that addition proceeds as a Markovnikov addition.

$$\text{1-hexyne} + Hg^{2+}, H_2SO_4, H_2O \rightarrow \text{2-hexanone}$$

The protonation of the alkyne results in the formation of a secondary vinyl cation.

This cation is trapped by water to produce an enol that tautomerizes to form the ketone product.

14. A is correct.

Radical hydrogenation of a C≡C triple bond in the presence of sodium metal and liquid ammonia adds two hydrogen atoms (i.e., reduction) across the double bond with *anti*-stereoselectivity, producing an *E* alkene.

The mechanism for reducing an alkyne to a trans (E) alkene

Lindlar reagent (i.e., H_2, Pd, $CaCo_3$, quinolone, and hexane) reduces the alkyne to the *cis* (*Z*) alkene.

Reaction for reducing an alkyne to a cis (Z) alkene

The Lindlar reagent, unlike the reagents of Na and NH_3, can be used to reduce a terminal alkyne to an alkene. Na and NH_3 is a "poison catalyst" and are unreactive with a terminal alkyne.

15. C is correct.

When terminal alkynes are treated with Lewis acids, such as mercury salts in aqueous conditions, the ketone forms the major product instead of the aldehyde.

The mercury cation is electron-deficient and forms a complex with the alkyne. This coordination increases the electrophilicity of the carbon atoms of the alkyne, and water adds to the internal carbon atom because the partial positive charge is larger at this position. This addition forms an enol (i.e., hydroxyl attached to a carbon in a double bond) that tautomerizes to the ketone.

A reduction step with sodium borohydride ($NaBH_4$) is not necessary because the carbon-mercury bond could break (to give Hg^{2+} and the enolate) when the ketone forms.

16. B is correct.

Although hydroxide and high temperatures are employed in this reaction, the potassium hydroxide base is not strong enough to catalyze the isomerization of the triple bond, which is why the internal, not terminal, alkyne is recovered from the reaction.

17. D is correct.

The oxymercuration-demercuration [Hg(OAc)$_2$] of alkynes occurs with Markovnikov addition to generate a ketone enol. This enol tautomerizes to form a ketone. Sodium borohydride (NaBH$_4$) then reduces the ketone to the secondary alcohol.

18. B is correct.

Hydrogenation (i.e., reduction) involves the addition of hydrogens (H$_2$) to an unsaturated molecule.

Catalytic hydrogenation (H$_2$/Pd or Pt) of an alkyne is susceptible to a further reduction to yield an alkane.

Alkynes can be reduced to stereospecific products with special reagents.

An alkyne yields a *cis* alkene which requires the Lindlar catalyst (i.e., H$_2$, Pd, CaCo$_3$, quinolone, and hexane) and a *trans* alkene with Ni (or Li) metal over NH$_3$ (*l*).

A: *oxidation* is an increase in the number of bonds to oxygen. Increasing the number of bonds to oxygen often results from decreasing the number of bonds to hydrogens.

C: adding H$_2$ (i.e., hydrogenation) is an addition, not a substitution, reaction.

D: *hydration* reactions add water to an unsaturated (i.e., alkene or alkyne) molecule.

E: *elimination* reaction increases the degree of unsaturation in a molecule. Elimination describes the conversion of an alkane to an alkene (or alkyne) or converting an alkene to an alkyne.

19. C is correct.

Using a subscript of n for the number of carbons, the *degrees of unsaturation* uses the following formulae:

Alkane: C_nH_{2n+2} = 0 degrees of unsaturation

Alkene: C_nH_{2n} = 1 degree of unsaturation

Alkyne: C_nH_{2n-2} = 2 degrees of unsaturation

The formula C_nH_{2n-2} is the general formula for acyclic alkynes.

The general molecular formula for cyclic alkynes is C_nH_{2n-4}.

Cyclic alkynes (i.e., bond angle of 180°) are typically larger-sized rings with at least 8 carbon atoms in the ring.

20. C is correct.

The higher the pK_a of a compound, the less acidic the molecule is and the stronger the *conjugate base* is.

The pK_a of water is ≈ 15-16.

The pK_a of the terminal alkynes is ≈ 25.

The pK_a of the N-H bonds of neutral amines is ≈ 36-38.

Therefore, water and the terminal alkyne (but-1-yne or butyne) are more acidic.

21. B is correct.

When an alkyne is reduced by sodium in liquid ammonia, a single electron is transferred to the alkyne to produce an anion. The intermediate is a vinyl anion because it is on the atom that is in the double bond.

The anion protonates to give the vinyl radical.

The only cations produced during the reaction are the Na$^+$.

22. C is correct.

Use the following formulae to calculate the degrees of unsaturation:

C_nH_{2n+2}: for an alkane (0 degrees of unsaturation)

C_nH_{2n}: for an alkene or a ring (1 degree of unsaturation)

C_nH_{2n-2}: for an alkyne, 2 double bonds, 2 rings or 1 ring and 1 double bond (2 degrees of unsaturation)

Molecular formula $C_{10}H_{16}$ (C_nH_{2n-6}) has 3 degrees of unsaturation. It is consistent with an acyclic molecule that contains two alkyne functional groups or three alkenes or two alkenes and one ring, etc.

A molecule with two triple bonds has 4 degrees of unsaturation (C_nH_{2n-8}) or $C_{10}H_{14}$.

23. D is correct.

A catalytic system, which may produce alkenes from alkynes, is the Lindlar catalyst (i.e., H_2, Pd, $CaCo_3$, quinolone, and hexane).

An alkyne yields a *cis* alkene when subjected to the Lindlar catalyst.

Hydrogenation reactions catalyzed by platinum or palladium result in the formation of alkane products.

24. C is correct.

The acetylide anion has a pK_a ≈ 28. Due to the large differences in electronegativity between oxygen and carbon atoms, ions that possess negatively charged oxygen atoms are relatively more stable than carbon anions.

The sodium methoxide is the most stable conjugate base, and therefore the least basic.

The CH_3Li is the Gilman reagent, and CH_3MgBr is the Grignard; both are strong bases with a pK_a greater than 40.

25. D is correct.

The hydration of the terminal alkyne with BH₃ proceeds with *anti*-Markovnikov regioselectivity.

Enol is on the left and the keto on the right

The enol intermediate tautomerizes to the keto product, whereby the keto product (more stable) is over 99% of the observed product.

26. B is correct.

1-butyne is a terminal alkyne.

Terminal alkynes have additional chemical properties, such as their ability to form anions when exposed to strong bases (e.g., the acetylide anion that forms with NaNH₂).

27. D is correct.

The bond order for an alkyne is larger than for an alkene.

The larger the bond order, the shorter the bond. Therefore, the *pi* bond in an alkyne is shorter.

Furthermore, there is less *p* orbital overlap present in an alkyne than in an alkene.

Because the internuclear overlap is lower, the *pi* bond is weaker.

28. A is correct.

The Grignard reagent (CH₃CH₂MgBr), as a carbanion, is a strong base.

In the terminal alkyne shown, an acid-base reaction occurs by deprotonating the alkyne and producing ethane.

29. C is correct.

The hydration of the terminal alkyne proceeds with Markovnikov regioselectivity to produce a ketone.

A: the enol intermediate tautomerizes to the keto product.

B: CH₃CH₂CH₂CH=CHOH is the enol intermediate of the *anti*-Markovnikov reaction (hydroboration with BH₃).

D: CH₃CH₂CH₂CH₂CHO is the aldehyde product of the *anti*-Markovnikov reaction (hydroboration with BH₃).

E: CH₃CH₂CH₂CH(OH)CH₂OH is the germinal diol as would be formed from the treatment of an alkene with OsO₄ (osmium tetroxide to form the *syn*-diol).

30. B is correct.

Alkynes can undergo bromination to yield compounds with four bromine atoms incorporated in their structures.

The first halogenation is expected to proceed more quickly than the second halogenation.

Bromine atoms are quite large (about the size of a tertbutyl group), and the first bromination increases the steric bulk of the reactant to form the intermediate alkene.

Furthermore, the bromine atoms are more electronegative than carbon, so the *pi* bond of the alkene intermediate is less electron-rich and less nucleophilic than the alkyne *pi* bond.

31. D is correct.

Alkynes are oxidized by two mechanisms to yield the Markovnikov or *anti*-Markovnikov product.

$$CH_3C{\equiv}CH \xrightarrow[\text{HgSO}_4]{\text{H}_2\text{O, H}_2\text{SO}_4} CH_3\underset{\underset{\text{OH}}{|}}{C}{=}CH_2 \rightleftharpoons \underset{\text{a ketone}}{CH_3\overset{\overset{\text{O}}{\|}}{C}CH_3}$$

$$CH_3C{\equiv}CH \xrightarrow[\text{2. HO}^-,\text{ H}_2\text{O}_2,\text{ H}_2\text{O}]{\text{1. disiamylborane}} CH_3\underset{\underset{\text{OH}}{|}}{CH}{=}CH \rightleftharpoons \underset{\text{an aldehyde}}{CH_3CH_2\overset{\overset{\text{O}}{\|}}{C}H}$$

One of the *pi* bonds of the alkyne undergoes addition to yield an enol intermediate. The enol tautomerizes to generate the Markovnikov ketone or the *anti*-Markovnikov aldehyde.

32. A is correct.

The hydration of the alkyne forms a carbocation and proceeds through a Markovnikov-type mechanism.

The enol intermediate converts (i.e., tautomerizes) to the keto of the methyl phenyl ketone product.

33. C is correct.

The compound has five carbons, seven hydrogens, and one nitrogen.

Use the following formulae to calculate the *degrees of unsaturation*:

C_nH_{2n+2}: for an alkane (0 degrees of unsaturation)

C_nH_{2n}: for an alkene or a ring (1 degree of unsaturation)

C_nH_{2n-2}: for an alkyne, 2 double bonds, 2 rings or 1 ring and 1 double bond (2 degrees of unsaturation)

The molecule has 3 degrees of unsaturation.

Because one of the unsaturation elements is a ring, the molecule contains two *pi* bonds.

34. C is correct.

Alkynes can be oxidized to aldehydes or ketones as follows:

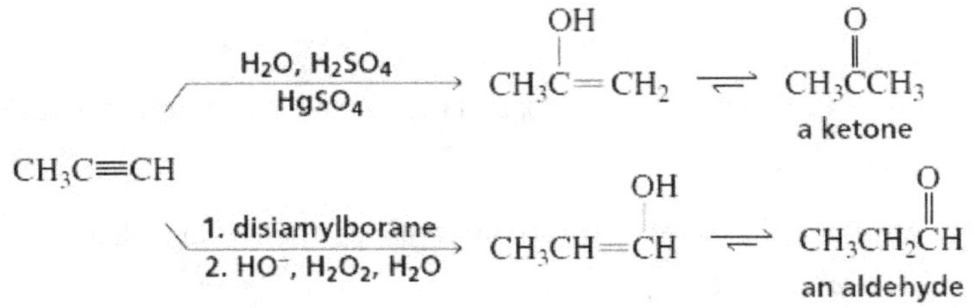

$$CH_3C\equiv CH \xrightarrow{\underset{HgSO_4}{H_2O, H_2SO_4}} CH_3\underset{OH}{C}=CH_2 \rightleftharpoons \underset{\text{a ketone}}{CH_3\overset{O}{\overset{\|}{C}}CH_3}$$

$$CH_3C\equiv CH \xrightarrow{\underset{2.\ HO^-,\ H_2O_2,\ H_2O}{1.\ \text{disiamylborane}}} CH_3CH=\underset{OH}{CH} \rightleftharpoons \underset{\text{an aldehyde}}{CH_3CH_2\overset{O}{\overset{\|}{C}}H}$$

Disiamylborane (BH$_3$) is used for the hydroboration of alkynes (and alkenes) and involves peroxides in step 2.

The hydration of the alkyne (or alkene) proceeds through an *anti*-Markovnikov addition; the preference for the *anti*-Markovnikov addition is due to the minimization of steric interactions.

The peroxide (H$_2$O$_2$) and $^-$OH converts the RBH$_2$ bond to an enol (C=C–OH) of the *anti*-Markovnikov product.

The *enol* (alkene and alcohol attached to the same carbon atom) is less stable and tautomerizes (i.e., migration of a proton) to the aldehyde (aldehyde or ketone are referred to as keto) and not the final product of the reaction.

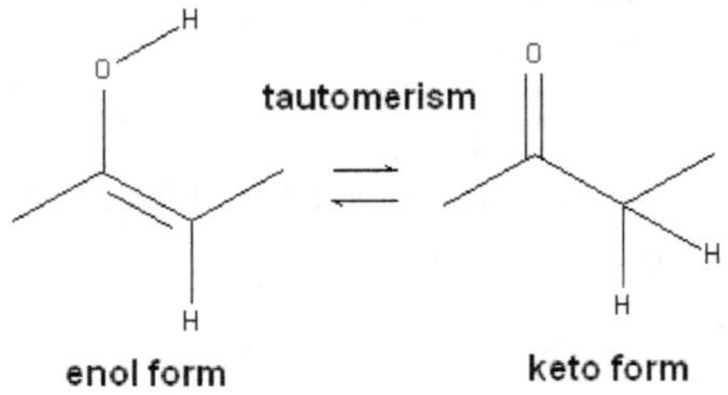

enol form ⇌ keto form (tautomerism)

The keto and enol form are structural isomers, with the keto form more than 99% of the final yield due to stability.

35. C is correct.

Like alkenes, alkynes are electron-rich functional groups and function as nucleophiles that donate electron density to electron-seeking electrophiles. The reactivity of alkenes and alkynes is similar, and they interact with electrophiles in analogous ways.

36. B is correct.

In this reaction, the bromine adds to the alkene group of C_2H_4 and forms 1,2–dibromoethane. The reaction proceeds *via anti*-addition from the 3-membered bridge structure of the bromonium (i.e., halonium) ion.

A: hydrogen gas cannot be generated as a product from this reaction.

Carbon-carbon multibonds do not form from the given reaction conditions because the reagents do not include a base to eliminate the bromine(s) to form an alkene or alkyne.

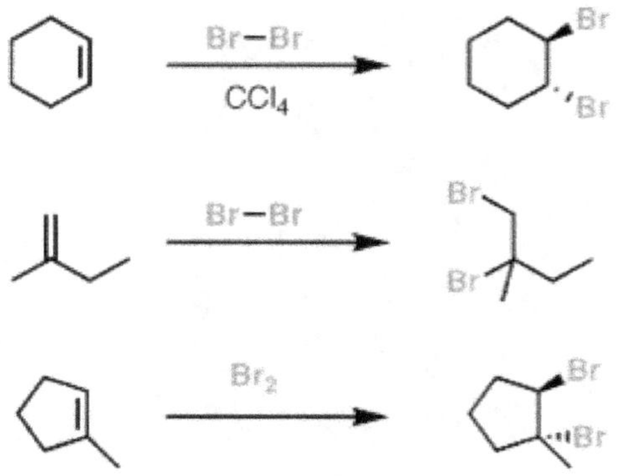

Sample *anti*-stereochemical products from the addition of bromine to an alkene. Each product has an enantiomer that is not shown.

Bromonium ion as an intermediate:

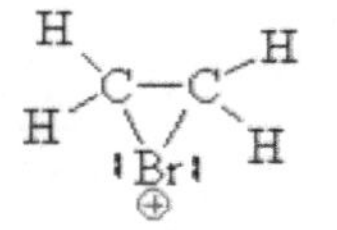

The reaction is regioselective (i.e., where) for the addition of the second nucleophile (i.e., Br⁻) to the halonium (i.e., bromonium) structure. The incoming nucleophile attacks the more substituted atom.

Mechanism of Br_2 addition to an alkene:

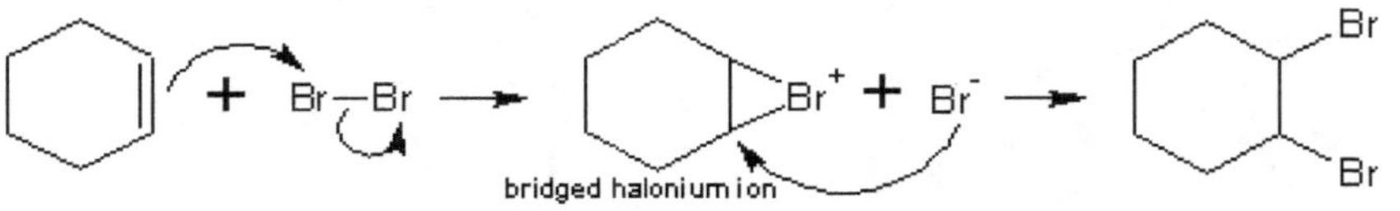

The incoming nucleophile attacks the bromonium bridged-structure ion at the most substituted position.

The bromonium ion (i.e., 3-membered bridged structure) undergoes S_N2 attack for *anti*-addition product formation. The stereochemical (i.e., *trans*) notation would include wedges and dashes.

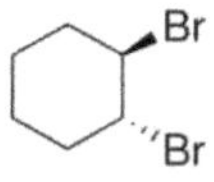

trans-1,2 dibromocyclohexane

Notes for active learning

Aromatic Compounds – Detailed Explanations

1. C is correct.

FeBr₃ acts as a catalyst to activate the alkyl bromide for electrophilic aromatic substitution (EAS) in this Friedel-Crafts alkylation. Since the ethyl substituent on ethylbenzene is slightly electron-donating, it functions as an *ortho*, *para* director for the substitution reaction.

Due to steric hindrance at the *ortho* position, *para* substitution is the major product.

2. B is correct.

The aromatic stabilization energy of benzene decreases the reactivity of its *pi* system relative to isolated or conjugated/nonaromatic alkenes. Because of this stabilization, arenes resist metal-catalyzed hydrogenation reactions and may require more specialized catalysts to facilitate their reductions.

3. E is correct.

Degenerate orbitals are with the same energy.

Bonding orbitals are below the horizontal plane, while the antibonding orbital is above.

In the molecular orbital (MO) diagram for benzene: MOs π2 and π3 are positioned in the second energy level, and π4 and π5 MOs are positioned on the third energy level.

Therefore, there are two pairs of degenerate MOs.

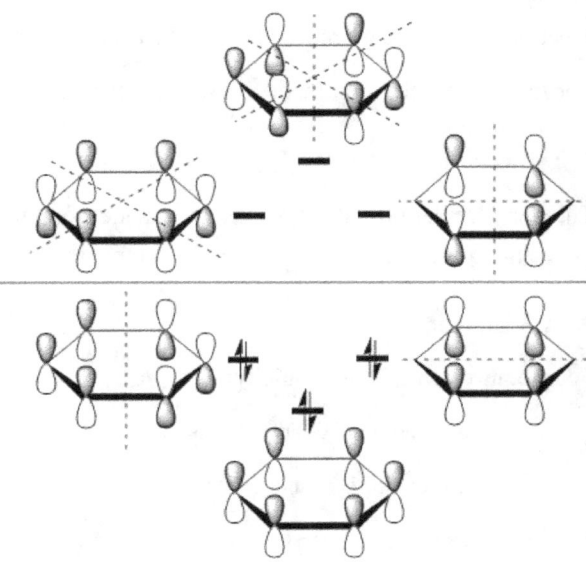

4. D is correct.

Because the nitrogen lone pair of the aniline is protonated, the group cannot donate electron density through the *pi* system of the aromatic ring.

Furthermore, the nitrogen atom has a formal charge of +1, and this group can function as a deactivating group because of its inductive effect.

5. A is correct.

Methyl substituents on benzene (toluene) are an ortho- / para-director and an activator (the electrophilic aromatic substitution rate is faster than benzene).

Friedel-Crafts alkylation reactions work best with electron-rich arenes.

Friedel-Crafts alkylation of toluene

6. D is correct.

I: The *ester* is the least activating moiety shown because the electron-withdrawing inductive effect of the electronegative oxygen on the carbonyl (ester group) reduces the electron density in the ring (i.e., deactivation).

II: *Methyl group* attached to the ring donates electron density *via* hyperconjugation and activates the ring slightly.

III: The lone pair of electrons on the ether resonates into the ring, which increases the electron density for EAS (i.e., activation).

7. C is correct.

Compared to benzene, electron-donating groups (such as ~OH) activate the ring towards EAS reactions.

Electron-withdrawing groups (such as the acetyl group) deactivate the ring, slowing EAS reactions.

Benzene undergoes bromination; however, its rate of bromination is less than for the phenol bromination.

8. D is correct.

The *pi* molecular orbitals of benzene are made from the overlap of six *p* orbitals. Because the number of molecular orbitals equals the number of atomic orbitals involved in the overlap, there are six molecular orbitals.

9. A is correct.

Four requirements for aromaticity:

    1) molecule is cyclic

    2) molecular is planar (flat)

    3) each atom is sp^2 hybridized (i.e., conjugated)

    4) the number of *pi* electrons satisfies Hückel's rule ($4n + 2$ *pi* electrons)

Aromatic rings can be positively or negatively charged if the 4 criteria are satisfied.

B, C, and D are antiaromatic with $4n$ *pi* electrons.

10. D is correct.

Phenol (i.e., hydroxybenzene) has an OH group on the benzene ring. The hydroxyl oxygen has two nonbonded pairs of electrons, which can be donated (*via* resonance) to the aromatic ring after the addition of an electrophile.

Electron-donating groups stabilize the cations formed upon adding a substituent to the *ortho* and *para* positions and therefore are *ortho/para*-directing activators.

The ~OH group is not notably bulky; there is little steric hindrance reducing substitution at the *ortho* position.

11. C is correct.

C: contains 4 π electrons and does *not* satisfy Hückel's rule for aromaticity ($4n + 2$ π electrons).

Additionally, the silicon atom is sp^3 hybridized and would not participate in aromatic conjugation.

12. A is correct.

Four requirements for aromaticity:

    1) molecule is cyclic

    2) molecular is planar (flat)

    3) each atom is sp^2 hybridized (i.e., conjugated)

    4) the number of *pi* electrons satisfies Hückel's rule ($4n + 2$ *pi* electrons)

Aromatic rings can be positively or negatively charged if the four criteria are satisfied.

Aromatic compounds have no saturated carbon atoms present in the aromatic ring.

The hybridization state of aromatic carbon atoms is sp^2, and the number of electrons in the aromatic system is consistent with Hückel's rule ($4n + 2$ π electrons).

13. D is correct.

The methyl group is an electron-donating substituent and therefore is an *ortho / para* director.

Due to steric hindrance at the *ortho* position, the *para* product is the major product.

14. C is correct.

The unknown cyclic hydrocarbon does not react with bromine in dichloromethane, carbon tetrachloride, or water. This means it cannot contain non-conjugated double or triple bonds; otherwise, the bromine would have added across the double bonds.

Since the unknown compound reacts when FeBr$_3$ is added, it must be benzene. The FeBr$_3$ acts as a *Lewis acid* and catalyzes aromatic electrophilic addition reactions.

Bromine is not a strong enough electrophile to disrupt the conjugated double bonding in the benzene ring.

However, adding the iron (III) bromide catalyst converts the bromine into a strong enough electrophile to add to the benzene ring.

15. D is correct.

The activating, deactivating, and directing (*ortho-* / *para-* and *meta*-directing) properties of aromatic substituents in electrophilic aromatic substitutions (EAS) and nucleophilic aromatic substitutions (NAS) are based on resonance stabilization or destabilization.

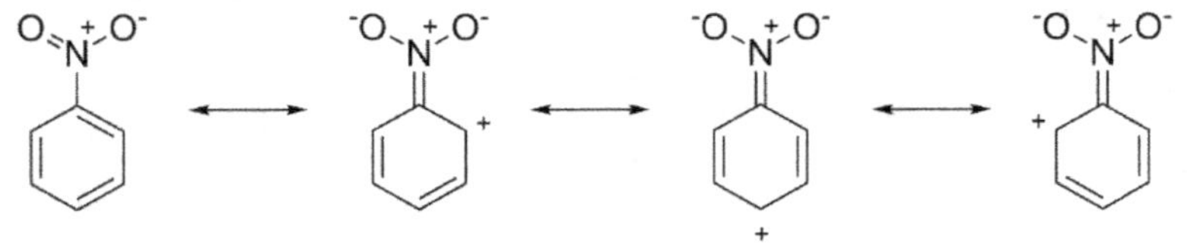

Resonance forms of nitrobenzene during EAS. Through resonance, electron-withdrawing groups introduce a partial positive charge at *ortho* and *para*. The *pi* bonds undergo delocalization for two hybrid structures:

In EAS, the aromatic ring acts as a nucleophile, and the partially positive sites are less nucleophilic, making the electron-withdrawing group *meta*-directing.

In NAS, the aromatic ring acts as an electrophile, and the partially positive sites are more electrophilic, making the electron-withdrawing group an *ortho-* / *para*-director.

16. E is correct.

The two arenes activated by nitrogen heteroatoms are more nucleophilic.

The amide does not donate as strongly as the amine, so ring 1 is less nucleophilic than ring 3.

Ring 2 is only activated by the alkyl group and is the least reactive of the three.

17. A is correct.

The reaction is a Friedel-Crafts acylation using acetic anhydride $(CH_3CO)_2O$ as an acylating agent. The product is methyl ketone.

All halogens are deactivators but direct *ortho* / *para* because the halogen (like *ortho* / *para* directors) has lone pairs of electrons (on the atom attached to the ring). The lone pair helps stabilize the positive charge on the ring present in the resonance hybrid intermediates.

In general, deactivators (except for the halogens) are *meta* directors.

18. D is correct.

In EAS, the aromatic ring acts as a nucleophile attacking an electrophile.

Due to its aromaticity, however, the aromatic ring is relatively unreactive and is a poor nucleophile.

Extremely reactive electrophiles (and the addition of a Lewis base) are used to overcome this limitation.

19. C is correct.

Halides are electron-withdrawing (i.e., deactivating) and are *ortho/para*-directing.

Halogens are the exception to the general rule, which states that electron-withdrawing species are deactivating and *meta*-directing.

The halogens (despite being electron-withdrawing due to their high electronegativity) are *ortho/para*-directing with lone pairs of electrons localized, *via* resonance, to the electron density at *o/p* positions.

Resonance hybrids show the anion at both ortho positions

Note that the resonance structures include an anion at the *para* position.

Therefore, like atoms with lone pairs of electrons attached to the ring, halogens are *ortho-para* directors.

The resonance structure forms a double bond between the halide bearing a formal positive charge and the phenyl ring; therefore, it is not a significant resonance structure and cannot overcome the deactivating (due to electronegativity) effect caused by induction (along the *sigma* bond).

20. E is correct.

When 1,3-cyclopentadiene reacts with sodium metal, it is converted from nonaromatic to aromatic.

According to Hückel's rule, a planar cyclic compound is aromatic if conjugated (adjacent sp^2 hybridized carbon atoms) and has $4n + 2$ *pi* electrons (where n is any whole number).

21. B is correct.

Carbonyl compounds are *meta*-directing in EAS reactions.

Since the nitro and carbonyl groups are deactivating, disubstitution is slow, and single *meta* substitution predominates.

22. D is correct.

Halogens are electron-withdrawing and deactivate a benzene ring toward electrophilic aromatic substitution, so bromobenzene undergoes nitration slower than benzene.

23. C is correct.

In electrophilic aromatic substitution, the aromatic ring acts as a nucleophile, attacking an electrophile that has been treated with a Lewis acid (e.g., $FeBr_3$, $AlCL_3$).

Deprotonation of the aromatic ring at the site of the attack reforms the double bond (an elimination reaction) and restores aromaticity.

The overall reaction substitutes an electrophile (e.g., Br, CH_3, RCO, HSO_3) for hydrogen on the aromatic ring.

24. D is correct.

The halogens are deactivating due to their high electronegativity. However, like *ortho / para*-directors, the halogens have a lone pair of electrons on the atom attached to the ring.

Alkyl chains do not have lone pairs of electrons on the C attached to the ring but are *ortho / para* directors due to hyperconjugation.

25. C is correct.

The reactivity of aromatic molecules toward *electrophilic aromatic substitution* (EAS) depends on the substituents on benzene.

Electron-donating substituents increase the electron density of the benzene ring and therefore activate benzene towards EAS.

Electron-withdrawing substituents deactivate the ring, making it less susceptible to EAS. The benzene ring is deactivated by the electron-withdrawing effects of the Cl and NH_3^+ substituents, and the ring is deactivated (compared to benzene) to EAS.

A: *p*-H_3CCH_2O–C_6H_4–O–CH_2CH_3 has two electron-donating ethoxy substituents and is highly reactive to EAS.

B: *p*-O_2N–C_6H_4–NH–CH_3 contains strong electron-withdrawing effects from the nitro (NO_2). The N of the NO_2 group has a formal charge of +, while the single-bonded O has a formal charge of –.

The NO_2 group offsets the strong electron-donating amino group (lone pair of electrons on N), so the molecule is only slightly reactive to EAS.

D: *p*-CH_3CH_2–C_6H_4–CH_2CH_3 contains two electron-donating (i.e., activating) ethyl substituents and is more reactive towards EAS.

E: benzene is the reference molecule to determine if substituents are activating (relative to benzene) or deactivating (relative to benzene).

26. C is correct.

The carbon atoms of benzene are sp^2 hybridized.

Non-planar molecules cannot be aromatic because the *pi* system must be planar (i.e., flat).

A *Kekule structure* is a Lewis structure with covalently bonded electron pairs drawn as lines. The Kekule structures illustrate the two most significant resonance contributors of benzene.

The alternating single and double drawn for benzene exist as a hybrid resonance structure from the delocalization.

Benzene with alternating double and single bonds – hybrid structure as bottom structure.

27. A is correct.

Addition reactions are typically not observed for aromatic compounds because the aromaticity is restored during their substitution reactions. When aromatic functional groups react, they may temporarily lose their aromaticity (i.e., high-energy resonance hybrids are the intermediates), and its restoration greatly increases the stability of the molecule.

Electrophilic aromatic substitution (EAS) reactions are favored over nucleophilic aromatic addition (NAS) reactions for aromatic compounds.

28. D is correct.

The double bonds in benzene are less reactive than in a non-aromatic alkene because addition (e.g., hydrogenation) disrupts the aromaticity (i.e., delocalization) of the ring, making it less stable.

Applying heat and high pressure with the Rh catalyst permits benzene to overcome the energy of activation necessary to transform the highly stable benzene molecule into a non-aromatic product.

29. D is correct.

$CH_3C_6H_5 + H_2$, Rh / C is a reduction reaction with a powerful reducing agent capable of disrupting the stability of the aromatic ring.

The regents reduce the benzene ring catalytically *via* hydrogenation to form cyclohexane.

Therefore, this is not an electrophilic aromatic substitution.

In general, an aromatic ring is especially susceptible to electrophilic aromatic substitution (EAS) with the Lewis acid (e.g., $FeBr_3$, $AlCl_3$ or H_2SO_4).

A: $CH_3C_6H_5 + C_6H_5CH_2CH_2Cl$ / $AlCl_3$ is an example of Friedel-Crafts alkylation (EAS), whereby toluene (benzene with a methyl substituent) reacts with an alkyl chloride with the Lewis acid aluminum trichloride ($AlCl_3$).

The Lewis acid removes chloride from the alkyl halide, forming a carbocation which is then attacked by the benzene ring.

B: $CH_3C_6H_5 + Br_2$ / $FeBr_3$ is an example of electrophilic aromatic substitution (EAS). $FeBr_3$ (similar to $AlCl_3$) is a Lewis acid.

Toluene is activating, and the Br substitutes in the *ortho / para* position.

C: $CH_3C_6H_5 + CH_3CH_2CH_2COCl$ / $AlCl_3$ is an example of Friedel-Crafts acylation (EAS), whereby toluene reacts with an acyl chloride in the presence of the Lewis acid aluminum trichloride ($AlCl_3$).

The Lewis acid removes chloride from the acyl halide, forming a carbocation which is then attacked by the benzene ring. Then a proton is removed to restore the aromaticity of the original ring structure.

E: $C_6H_6 + HSO_3$ / H_2SO_4 is an example of the EAS reaction for sulfonation of a benzene ring.

30. D is correct.

A: of the two substituents, the chloro group is *para*-directing, so it should be substituted first.

Additionally, Na / NH$_3$ results in the single *trans* hydrogenation of alkenes but does not substitute a nitro group in an EAS reaction.

B: of the two substituents, the chloro group is *para*-directing, so it should be substituted first.

C: while HCl / H$_2$O adds H and Cl across the double bonds of alkenes, these conditions do not substitute Cl in EAS reactions.

E: Cl$_2$ / CCl$_4$ are the conditions to dechlorinate an alkene.

Benzene does not undergo addition reactions due to aromaticity.

31. A is correct.

Benzene is aromatic and undergoes electrophilic aromatic substitution (EAS) with the addition of a Lewis acid (e.g., AlCl$_3$, FeBr$_3$ or H$_2$SO$_4$).

The reagents of SO$_3$ and concentrated H$_2$SO$_4$ are used to sulfonate aromatic compounds, whereby a SO$_3$H group is substituted onto the benzene ring.

Halogens are *ortho* and *para* directing deactivators.

Therefore, the product is a mixture of *ortho*- and *para*-bromobenzenesulfonic acid.

B: the bromine is not displaced from the aromatic benzene ring and replaced with hydrogen to form benzene.

C: the SO$_3$H group does not substitute for the bromine of bromobenzene.

D: *meta*- is not formed because halogens are *ortho* and *para* directing deactivators.

E: toluene is benzene with a methyl group attached.

32. A is correct.

Substituents on an aromatic ring affect the rate at which electrophilic aromatic substitution reactions occur.

Since the ring acts as a nucleophile in these reactions, electron-donating substituents increase the rate of reaction, while electron-withdrawing substituents decrease the rate of reaction.

The bromine substituent (III) is slightly deactivating (due to its electronegativity), making it less reactive than benzene (I).

The nitro group is extremely deactivating (due to resonance), making nitrobenzene (II) less reactive than bromobenzene (III).

33. B is correct.

Benzene is a cyclic aromatic hydrocarbon that has 4 degrees of unsaturation.

The *pi* bonds in the molecule are delocalized and impart aromaticity to the molecule.

34. C is correct.

Each molecule has nitro (~NO_2) and hydroxyl (~OH) functional groups to hydrogen-bond.

Proximity is needed for either intramolecular (within the molecule) or intermolecular (between molecules) bonding.

The alignment of the functional groups permitting hydrogen bonding depends on the shape of the molecule.

The melting points (transition from packed molecules in a solid to liquid phase), boiling points (transition from associated molecules in a liquid to independent molecules in the gas phase), and water solubility of polar molecules are related to the presence and quantity of intermolecular hydrogen bonding.

The *meta*- and *para*-nitrophenol are more water-soluble and have higher melting points than *ortho*-nitrophenol because *ortho*-nitrophenol tends to form intramolecular hydrogen bonds instead of intermolecular hydrogen bonds.

Intramolecular hydrogen bonds for *ortho*-nitrophenol make the molecule independent so that it takes less energy to disrupt the lattice structure of the molecules (melting).

Likewise, intramolecular hydrogen bonding reduces the molecule's ability to form hydrogen bonds with water, and therefore the molecule is less water-soluble.

A: *meta*- and *para*-nitrophenol form strong intermolecular hydrogen bonds, leading to a higher melting point.

The nitro and hydroxyl groups form hydrogen bonds with water, and the molecules are more water-soluble.

B: *ortho*-nitrophenol forms some intermolecular hydrogen bonds, but less than *meta*- and *para*- nitrophenol.

Meta-nitrophenol forms weak intramolecular bonds due to the large distance between the functional groups.

D: *para*-nitrophenol has the nitro and hydroxyl substituents at opposite ends of the flat molecule and cannot form intramolecular hydrogen bonds.

35. A is correct.

Hückel's rule predicts that for a monocyclic compound to be aromatic, there must be a fully conjugated *pi* (sp^2 hybridization at each atom in the ring) containing $(4n + 2)$ *pi* electrons.

Two *pi* electrons are contributed by each of the double bonds.

The lone pair of electrons on a double-bonded N is perpendicular to the *pi* cloud and does not count as the number of *pi* electrons for Hückel's rule.

Note: when N is in a single bond, the unshared electrons parallel the *pi* system count toward aromaticity.

Benzimidazoline is not fully conjugated because there is a ~CH_2~ (sp^3) group in the ring.

Cyclic conjugation is necessary for a molecule to be aromatic.

B: thiophene has an sp^2 hybridized sulfur (like oxygen) and has a lone pair of electrons counted as in the ring. The number of *pi* electrons is 6.

continued…

C: quinoline has 10 *pi* electrons and therefore is aromatic.

The lone pair of electrons on a double-bonded N is perpendicular to the *pi* cloud and does not count as the number of *pi* electrons for Hückel's rule.

D: thiazole has 6 *pi* electrons and therefore is aromatic.

The lone pair of electrons on a double-bonded N is perpendicular to the *pi* cloud and does not count as *pi* electrons for Hückel's rule.

Sulfur (like oxygen) is sp^2 hybridized and has a lone pair of electrons counted as in the ring.

E: imidazole has 6 *pi* electrons and therefore is aromatic.

The lone pair of electrons on a double-bonded N is perpendicular to the *pi* cloud and does not count as *pi* electrons for Hückel's rule.

The lone pair of electrons on a single bonded N is part of the *pi* cloud and counts as *pi* electrons for Hückel's rule.

36. D is correct.

Degrees of unsaturation:

        double bonds = 1 degree of unsaturation

        triple bonds = 2 degrees of unsaturation

        rings = 1 degree of unsaturation

Using the degree of unsaturation calculation:

        benzene (1 ring and 3 double bonds) = 4 degrees of unsaturation

Notes for active learning

Notes for active learning

Alcohols – Detailed Explanations

1. B is correct.

Alcohols follow the same trend for boiling points as alkanes, with longer chain molecules having higher boiling points. Hexanol is the longest chain and therefore has the highest boiling point.

2. E is correct.

When phenols are deprotonated (i.e., acting as an acid), a phenoxide ion (i.e., phenolate ion) is produced, as shown. The pK_a of phenols is much lower than it is for alcohols.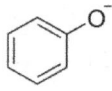

Methanol has a pK_a of 16, while phenols have pK_a values ≈ 9.9.

3. C is correct.

The formation of an inorganic ester by adding alcohol and phosphoric acid is an example of dehydration synthesis because an equivalent of water is lost during the process.

4. E is correct.

No carbon-oxygen *pi* bonds exist in alcohols (hydroxyl) functional groups.

Carbonyl (C=O) groups (i.e., carbon double bonded to oxygen) indicate ketones, aldehydes, or carboxylic acid derivatives (acyl halide, anhydride, ester, and amide).

5. A is correct.

Draw the structure of known compounds or the product of the reaction.

Ethanoate is an ester that has an *n*-propoxy substituent. Therefore, the alcohol that should be used is 1-propanol, and the remaining component is ethanoic acid.

6. C is correct.

The five-carbon molecule of *n*-pentanol has the largest alkyl portion (i.e., the greatest number of London forces). It possesses hydroxyl, which allows it to hydrogen bond (increases boiling point).

An overall molecular dipole exists for this molecule that contributes to its high boiling point.

7. B is correct.

The ~OH (hydroxyl) group is characteristic of alcohol and carboxylic acids (includes a carbonyl group).

8. D is correct.

Sodium dichromate ($Na_2Cr_2O_7$) is a strong oxidizing agent which converts primary alcohols to carboxylic acids.

Sodium dichromate converts secondary alcohols to ketones.

9. B is correct.

Thionyl chloride (SOCl$_2$) is a reagent that converts primary and secondary alcohols to alkyl chlorides.

Phosphorous tribromide (PBr$_3$) is a reagent that converts primary and secondary alcohols to alkyl bromides.

10. A is correct.

The molecule has an alkene on the righ, alcohol near the lower right portion, and a fused ether functional group.

11. B is correct.

Molecules containing rings or multibonds are molecules with degrees of unsaturation.

A completely saturated compound only consists of single bonds (*sigma* bonds).

12. C is correct.

Primary alcohols have an ~OH group attached to a carbon that is connected to another carbon atom.

Secondary alcohols possess an ~OH group bonded to a carbon attached to two other carbon atoms.

Tertiary alcohols have ~OH groups bonded to carbon atoms attached to three other carbon atoms.

13. C is correct. Carboxylic acids (~COOH) contain an ~OH group bonded to carbonyls (C=O).

Oximes (C=N–OH) are functional groups containing an ~OH group; they are used for reductive amination.

14. B is correct

The reaction between (*S*)-2-heptanol and SOCl$_2$ (thionyl chloride) proceeds *via* an S$_N$2 reaction mechanism.

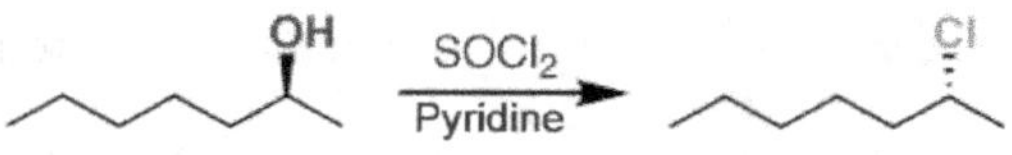

An addition-elimination sequence occurs at the S=O double bond, then substitution by Cl⁻ of the alcohol gives the inverted (*R*)-2-chloroheptane product.

15. A is correct.

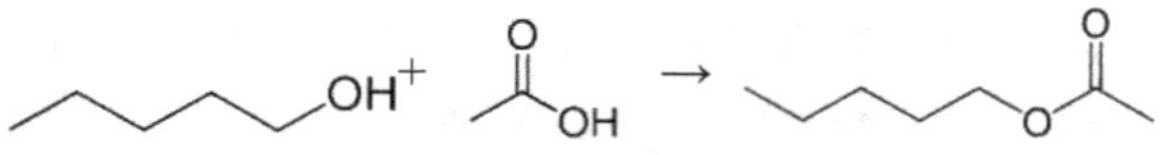

The reaction of 1-pentanol + acetic acid

The alkoxy group has five carbon atoms, and an acyl portion is an acetyl group.

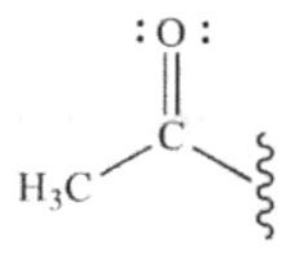

Acetyl group

Acetate is the root of this molecule.

16. B is correct.

Ethers follow the same trend as alkanes, so dihexyl ether has the highest boiling point because it has the greatest molecular weight and has the largest surface area.

17. E is correct.

The most notable intermolecular force for water is hydrogen bonding.

Alcohols donate and receive hydrogen bonds.

Thiols (S–H bonds) have much smaller dipoles than O–H, so they do not participate in hydrogen bonding.

18. B is correct.

The tosyl group adds with retention of stereochemistry because the C–O bond is not broken.

Tosylation of secondary alcohol makes it a better leaving group.

The chloride anion is a good nucleophile, displacing the tosylate to form the secondary alkyl chloride.

D: shows a product that has retained stereospecificity (no inversion).

This retention of stereochemistry occurs when an S_N2 reaction (i.e., inversion) is followed by a second S_N2.

19. D is correct.

For a given class of compounds, the lower the molecular weight, the lower the boiling point.

Alcohols follow the same trend for the boiling point as alkanes.

Ethanol is a two-carbon chain (i.e., lowest molecular weight) with hydrogen bonding (e.g., alcohols and carboxylic acids). It has the lowest molecular weight of the choices and therefore has the lowest boiling point.

20. D is correct.

HBr is a strong acid that protonates the oxygen atom of the alcohol. Then, either the bromide displaces the water to give 1-bromo-2-methylpropane, or a hydride shift occurs to form the tertiary carbocation, which can be trapped by the bromide resulting in 2-bromo-2-methylpropane.

21. D is correct.

The reaction between propanol and PBr_3 (phosphorous tribromide) proceeds *via* an S_N2 reaction mechanism.

An addition-elimination sequence occurs to the P–Br bond, then substitution by Br⁻ of the alcohol gives the inverted (*R/S*) product (i.e., bromopropane).

22. D is correct.

Primary alcohols are oxidized to carboxylic acids with an oxidizing agent (e.g., CrO_3 in HCl).

Oxidation:

primary alcohol → aldehyde → carboxylic acid

secondary alcohol → ketone

tertiary alcohol → no reaction

23. D is correct.

The *boiling points* of compounds are determined by two general factors: molecular weight and intermolecular interactions.

The higher the *molecular weight*, the harder it is to "push" it into the gas phase; the higher the boiling point.

Similarly, the stronger the *intermolecular interactions*, the more energy is required to disrupt them and separate the molecules in the gas phase, hence the higher the boiling point.

Alcohols participate in hydrogen bonding due to the hydroxyl (~OH) group.

The alkane, alkene, ether, and alkyl halide only participate in dipole-dipole interactions and London forces.

24. C is correct.

2-hexanol is secondary alcohol and can be oxidized to a ketone with an oxidizing agent (e.g., CrO_3 in HCl).

The carbon atom of the alcohol is only bonded to one other hydrogen atom; therefore, the highest oxidation state it can acquire is the ketone oxidation state (i.e., +2).

Oxidation: primary alcohol → aldehyde → carboxylic acid

secondary alcohol → ketone

tertiary alcohol → no reaction

25. D is correct.

Alcohols of the same chain length as alkanes, alkenes, and alkynes have higher boiling points due to hydrogen bonding of the ~OH group.

Alkanes, alkenes, and alkynes are not able to form hydrogen bonds.

26. B is correct.

Carbonyl (C=O) groups indicate ketones, aldehydes, or carboxylic acid derivatives (acyl halides, anhydrides, esters, and amides).

Ethers are noted as R–O–R and do not contain carbonyl groups.

27. A is correct.

For alkenes, allylic refers to an atom attached to a carbon adjacent (β) to the double bond.

For carbonyl (C=O) compounds (aldehydes, ketones, acyl halides, anhydrides, carboxylic acids, esters, and amides), the position adjacent to the C of the C=O is the α position.

The hydroxyl (~OH) group of allylic alcohols is one carbon-carbon *sigma* bond away from the double bond.

Vinyl refers to an atom attached to carbon on a double bond.

28. A is correct.

The *electronegativity* of the oxygen atom in alcohol helps stabilize the negative charge of the conjugate base (i.e., negative oxygen) of the alcohol.

A more stable conjugate base results in a stronger acid (more readily dissociating its proton).

Secondary alcohols are more acidic than tertiary alcohols.

29. A is correct.

Pyridinium chlorochromate (PCC) is a gentle oxidizing agent which converts primary alcohols to aldehydes.

PCC is used to convert secondary alcohols to ketones.

B: carboxylic acids require a more powerful oxidizing agent (e.g., Jones reagent; CrO_3, H_2SO_4 and acetone)

C: alkenes proceed *via* E_1 when the alcohol is subjected to mineral acid (e.g., H_2SO_4)

D: terminal alkyl halides are produced in two steps.

First, alcohol becomes an alkene (E_1 when the alcohol is subjected to a mineral acid, H_2SO_4).

The alkene is halogenated in *anti*-Markovnikov regiochemistry when peroxides (H_2O_2) are in the reaction.

E: an alkyne requires an initial alkene, followed by Br_2 (or Cl_2) for the dibromo compound, followed by elimination with two equivalents of a base to yield the alkyne.

30. C is correct.

Tertiary alcohol undergoes E_1 reactions faster (i.e., due to the stability of the carbocation intermediate) than secondary alcohols, which can dehydrate at a faster rate than primary alcohols.

31. D is correct.

A *tosyl group* (Tos) is $CH_3C_6H_4SO_2$ (derived from $CH_3C_6H_4SO_2Cl$) and forms esters and amides of tosylic acid.

Tosylates are used to increase the efficiency of the original hydroxyl as a leaving group.

Unlike PBr_3 or $SOCl_2$ (both *via* S_N2), the reaction mechanism preserves the bond between the carbon and the O of the hydroxyl. Therefore, no inversion of stereochemistry occurs (first step in this example) using a tosylate.

The first step with the tosylate results in retention of the chiral center, and the second step (i.e., Cl^- as a nucleophile) produces an inverted product.

32. D is correct.

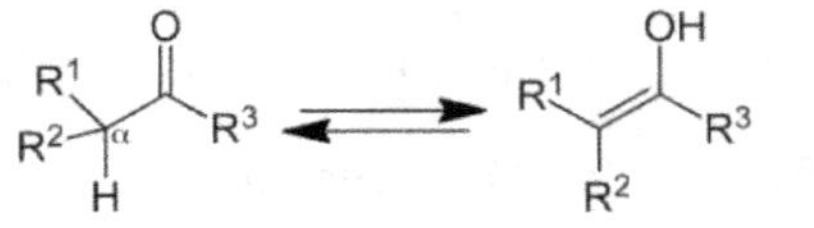

Aldehydes and ketones undergo tautomerization to exist in equilibrium between the keto and enol forms.

Most molecules (~99%) exist predominantly in the keto form because the carbon-oxygen (carbonyl) double bond is more stable than the hydroxyl on the double bond of the enol.

Phenols are a few aldehydes/ketones existing predominantly as enols form in keto-enol tautomer equilibrium.

The conjugated benzene ring system of phenol provides stability for the enol form.

The keto form of phenol lacks conjugation because the carbon in the ring is sp^3 hybridized.

When the phenol molecule assumes the keto form, aromaticity is lost, and the molecule becomes less stable.

Therefore, the keto form is non-aromatic and thus less stable.

Aromatic molecules are cyclic, planar, have conjugated double bonds (i.e., sp^2 at each atom), and satisfy Hückel's number of *pi* electrons (4n + 2, where n is an integer).

Anti-aromatic compounds are cyclic, planar, have conjugated double bonds, but have 4n (e.g., 4, 8, 12 and so on) *pi* electrons and therefore are unstable.

Nonaromatic compounds do not meet the four criteria needed for aromatic compounds: cyclic, planar, conjugated double bonds, and Hückel's number of *pi* electrons.

33. C is correct.

Treatment of *salicylic acid* (i.e., aspirin) with methanol (CH_3OH) and dry acid are conditions for synthesizing an ester in a mechanism is *Fischer esterification*:

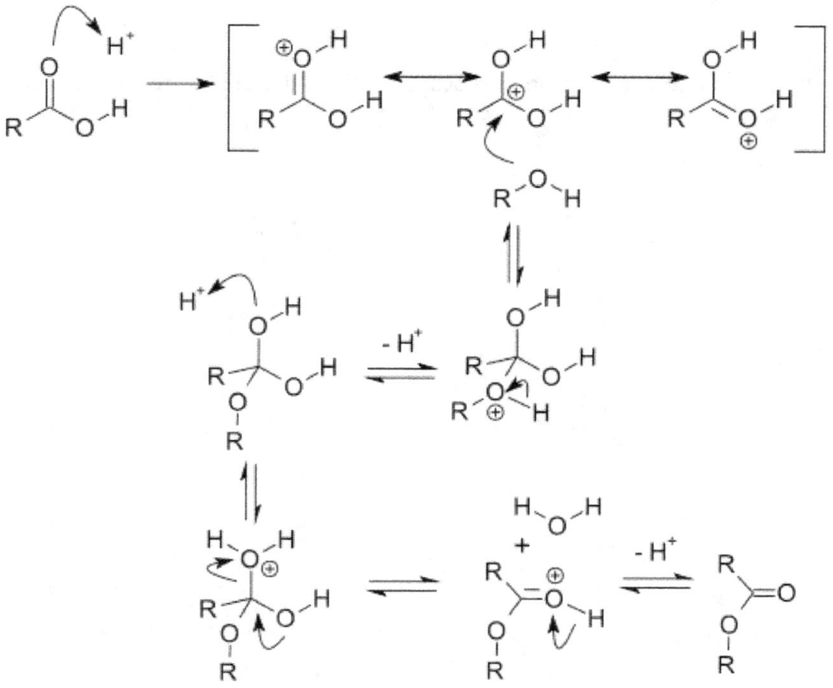

34. B is correct.

The reaction between (*R*)-2-hexanol and PBr_3 (phosphorous tribromide) proceeds *via* an S_N2 mechanism.

An addition-elimination sequence occurs in the P–Br bond, then substitution by Br⁻ of the alcohol gives the inverted (*S*)-2-bromohexane product.

35. B is correct.

Esterification occurs when a carboxylic acid reacts with alcohol under catalytic acidic conditions to form an ester + water.

A: C_6H_5OH and CH_3CH_2Br form an ether *via* Williamson (S_N2) ether synthesis when the alcohol reacts with the alkyl halide.

C: $CH_3COOH + SOCl_2$ is the S_N2 reaction of a carboxylic acid with thionyl chloride, and an acyl halide is formed.

D: $2CH_3OH + H_2SO_4$, form dimethyl ether from methanol and catalytic acid (e.g., sulfuric acid).

E: $CH_3CH_2Br + CH_3CH_2O^-Na^+$ forms an ether *via* Williamson (S_N2) ether synthesis when alcohol reacts with an alkyl halide.

The requirement is that the alkyl halide is less substituted because the reaction proceeds *via* S_N2.

36. B is correct.

The hydrobromic acid protonates the secondary alcohol that dissociates as water.

The secondary carbocation, formed as an intermediate, is repositioned to the tertiary position through an alkyl (i.e., methide) shift.

After the methide shift, this more stable tertiary carbocation is then attacked by the bromide ion.

Notes for active learning

Notes for active learning

Aldehydes and Ketones – Detailed Explanations

1. E is correct.

Ketones are functional groups that can no longer be oxidized by Cr^{VI} reagents but do not undergo further reactions because there are no C–H bonds to the carbonyl carbon of the ketone.

Aldehydes, however, may have subsequent oxidations to generate carboxylic acids.

2. A is correct.

Aldehydes can convert to the enol form, although the equilibrium constant favors the aldehyde.

Conversion to the enol form destroys the stereocenter because this isomer is achiral.

When the enol form converts back to the ketone, protonation may occur on either face of the group, which interconverts the configuration (i.e., producing R and S isomers).

The process is catalyzed by an acid or base.

3. D is correct. If a stronger reducing agent, such as lithium aluminum hydride ($LiAlH_4$), is used, the aldehyde is reduced to the primary alcohol.

Ester is reduced to an aldehyde with the mild reducing agent of DIBAL

4. B is correct.

For a ketone to be converted to an enolate, a strong base completely removes a proton from the alpha position.

If hydroxides or alkoxides are used instead of lithium amide bases, an equilibrium exists between conjugate forms of the ketone.

Methyllithium and diethylamine may add to the ketone instead of deprotonating it. They are too weak to deprotonate the ketone.

5. C is correct.

Ketones are molecules with alkyl or aryl (i.e., ring) substituents on either side of the carbonyl (C=O) functional group. Τηε ΙΥΠΑΧ ναμε φορ τηε μολεχυλε ισ βυταν–2–ονε ορ 2–βυτανονε.

6. D is correct.

1,3-dithiane serves as a nucleophilic carbonyl equivalent. Treatment of 1,3-dithiane with n-BuLi generates a carbanion between sulfur atoms of the molecule, and this anion can add to electrophiles such as alkyl halides.

Both C–H bonds of the 1,3-dithiane can be substituted with alkyl groups.

The hydrolysis of the dithiane to the carbonyl group requires mercury salts in an acidic solution.

7. A is correct.

The compounds ending in ~one suggest that the highest priority functional group is the ketone.

Compounds that end in ~ol may possess alcohol as the highest priority group.

8. D is correct.

These reaction conditions are the Clemmensen reduction reaction conditions.

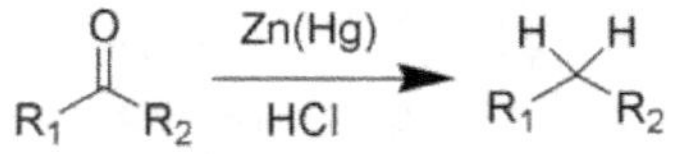

The Clemmensen reduction mechanism involves the protonation of the ketone oxygen atom.

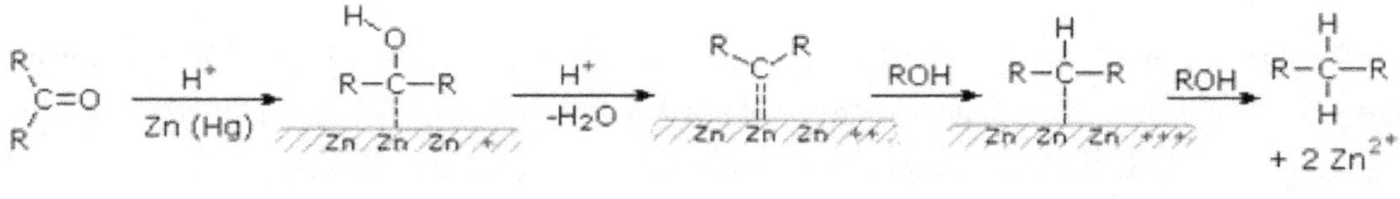

Clemmensen reduction reaction

9. D is correct.

When Grignard (R–MgBr or R–MgCl) or Gilman (R$_2$CuLi) reagents are combined with aldehydes, the product is a secondary alcohol.

This alcohol may or may not be chiral, depending on which Grignard reagent is used for alkylation.

The alkylation of ketones yields tertiary alcohols (see below).

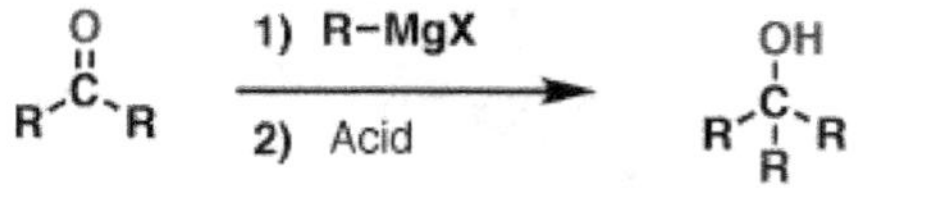

The alkylation of epoxides results in alcohol with an extended carbon chain.

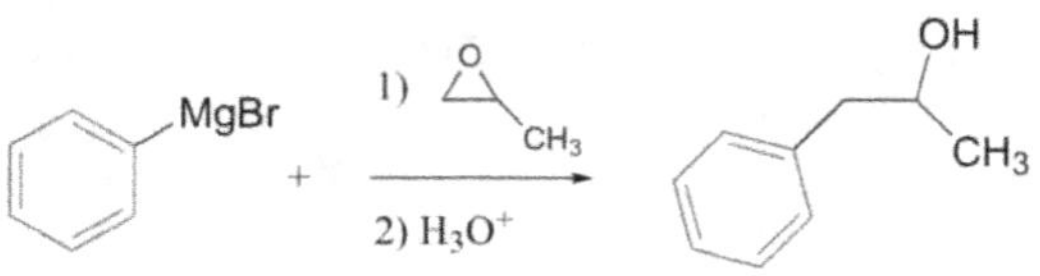

The Grignard reagent (i.e., carbanion in basic conditions) proceeds with the nucleophile (Grignard) attaching the less substituted carbon of the epoxide.

continued…

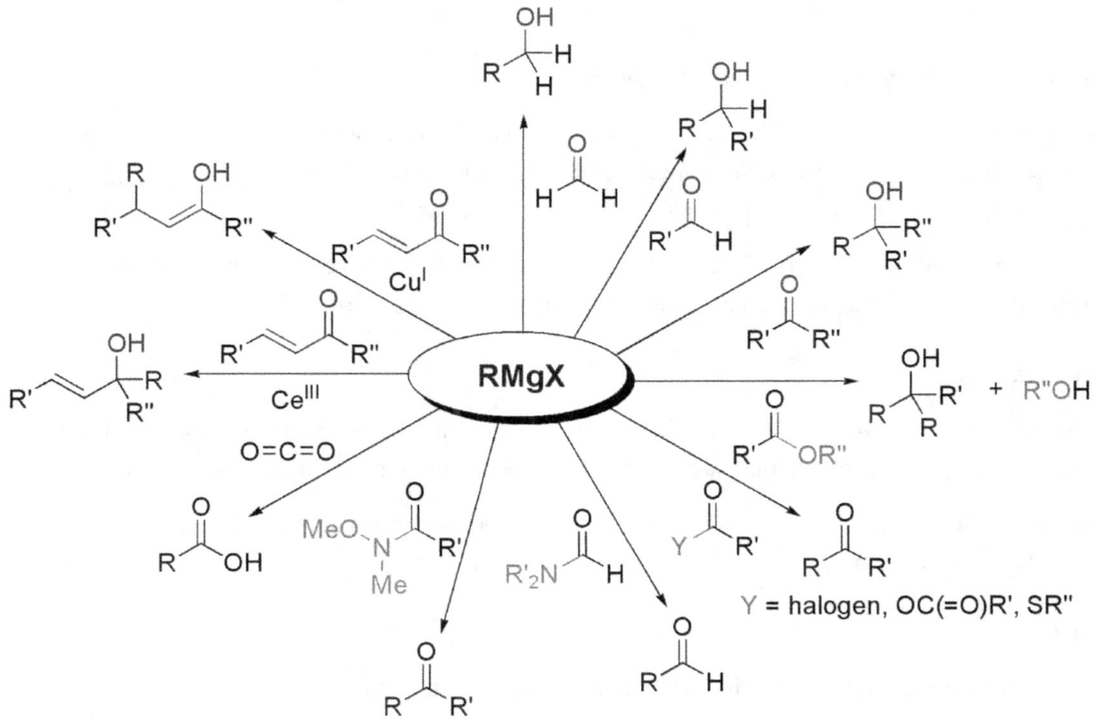

Grignard reaction (RMgX) involving several electrophiles and the resulting products

10. C is correct.

Benedict's test is used to detect reducing sugars.

An oxidized copper reagent is reduced by a sugar's aldehyde, and the aldehyde is oxidized to a carboxylic acid in the process.

Monosaccharides are reducing sugars, while some disaccharides (e.g., lactose and maltose), oligosaccharides, and polysaccharides are reducing sugars.

Reducing sugars (and alpha hydroxyl ketones) give a positive Benedict's test: a red-brown precipitate forms.

Fehling's solution gives a positive test for reducing sugars by changing from blue to clear and forming a red-brown precipitate.

Tollens' reagent forms silver ions (mirror) as a positive test for reducing sugars.

The Tollens test for aldehydes reduces silver cations [Ag^+] to reduced silver; the metal precipitates out of solution and coats the inner surface of the reaction flask.

Benedict's test and Tollens' test oxides an aldehyde to form a carboxylic acid.

Sugars able to do this have the aldehyde or hemiacetal functional groups; they are reducing sugars.

11. E is correct.

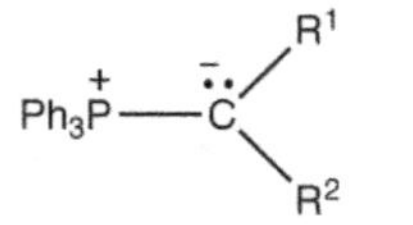

A common ylide is the phosphonium ylide as a Wittig reagent.

The double-charged (+ and –) resonance structure is the more dominant resonance contributor because the orbitals of phosphorus are larger than the valence orbitals of carbon, thus making the overlap of carbon and phosphorus p orbitals less favorable.

Phosphonium ylide

The minor resonance contributor for this functional group is the charge-neutral pentavalent phosphorus-containing structure formed from the overlap of the carbon and phosphorus p orbitals.

12. E is correct.

Hydrocarbons tend to have lower boiling points than molecules containing heteroatoms because hydrocarbons lack large molecular dipoles. These dipoles strengthen the intermolecular forces and raise boiling points.

Furthermore, molecules that can hydrogen bond have the highest boiling points because the hydrogen bond is a strong type of dipole interaction.

13. A is correct.

The treatment of a terminal alkyne with a dialkyl borane forms a vinyl borane.

The vinyl borane is oxidized to the enol through exposure to hydrogen peroxide and hydroxide base. This group tautomerizes to form an aldehyde.

14. B is correct.

The iodoform reaction (see mechanisms shown below) involves trihalogenating the methyl into a good leaving group. After tribromination of the methyl group has occurred, the group is expelled from the molecule by adding hydroxide to the ketone.

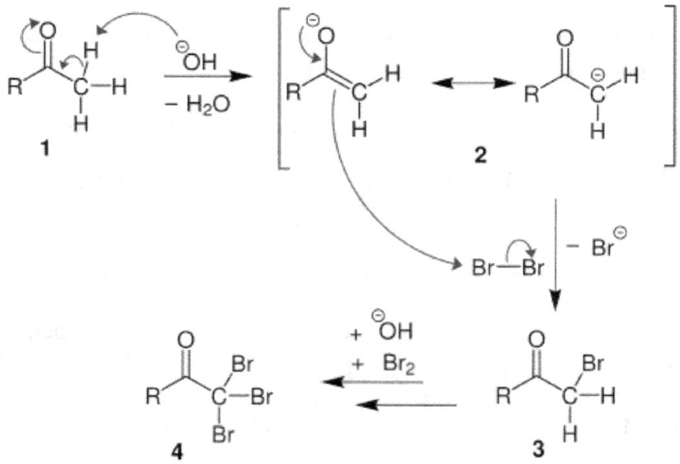

Molecule 4 undergoes tribromination before being replaced by the ⁻OH

Acidic conditions only provide the monobrominated compound from the starting ketone; therefore, basic conditions need to be used during the reaction.

15. A is correct.

Hemiacetals are intermediates in acetal formation reactions of aldehydes with two equivalents of alcohols.

$$\left[\begin{array}{c} R^1 \\ \diagdownOH \\ \diagupOR^2 \\ H \end{array}\right] \underset{H^+}{\overset{R^2OH}{\rightleftharpoons}} \begin{array}{c} R^1 \\ \diagdownOR^2 \\ \diagupOR^2 \\ H \end{array} + H_2O$$

hemiacetal → acetal

Hemiacetals are not stable groups and typically cannot be isolated.

The hemiacetal O–H bond collapses to expel alcohol and regenerate the carbonyl of the aldehyde. This typically occurs because carbonyl formation increases the entropy of the system (starting from one molecule and converting it to two molecules), and the carbonyl is almost as thermodynamically stable as the hemiacetal form.

16. D is correct.

Benedict's test (or Tollens' reagent) is used to detect reducing sugars. The aldehyde is oxidized to a carboxylic acid in the process. Monosaccharides are reducing sugars, while some disaccharides, oligosaccharides, and polysaccharides are reducing sugars.

Reducing sugars (and alpha hydroxyl ketones) give a positive Benedict's test: a brown precipitate forms (as does for Fehling's solution).

Tollens reagent forms silver ions (mirror) as a positive test for reducing sugars.

The Tollens test for aldehydes involves the reduction of silver cations [Ag^+] to reduced silver; the metal precipitates out of solution and coats the inner surface of the reaction flask.

The ethyl formate has a terminal (HC=O) group, but this group does not oxidize in the Tollens test conditions because the group is an ester and not an aldehyde.

The C–H of the formate ester cannot be oxidized to an O–H bond.

The mechanism for the oxidation of carbonyl C–H bonds typically proceeds through either the hydrate or nucleophilic attack on the carbonyl carbon atom, which is not easily done because of the ester alkoxyl group.

The ester is much less electrophilic than the aldehyde, so it is less reactive.

17. A is correct.

The first step in the reaction is the coordination of the carbonyl oxygen atom with the positively charged counter ion of the reducing reagent and the introduction of the hydride to the group.

The second step is the protonation of the alkoxide (i.e., negative oxygen atom) intermediate with a weak acid.

18. D is correct.

A *Michael acceptor* is an α, β unsaturated carbonyl – the double bond is between the first and second carbons from the carbonyl carbon.

The best Michael acceptor is the electrophile that best stabilizes the resulting negative charge at the alpha position when a nucleophilic attack (i.e., by Michael donor) occurs at the beta position.

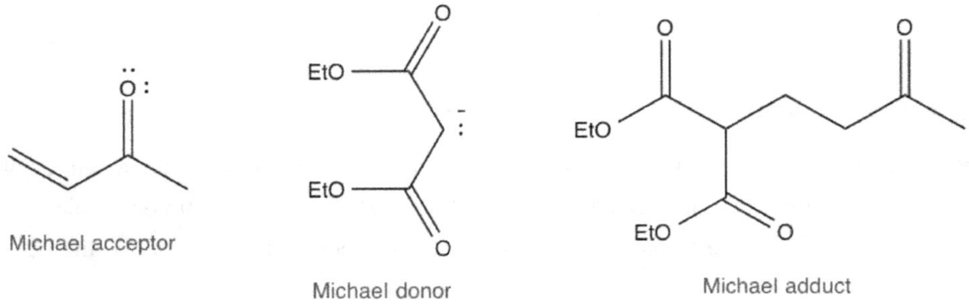

The mechanism for the Michael reaction is shown below:

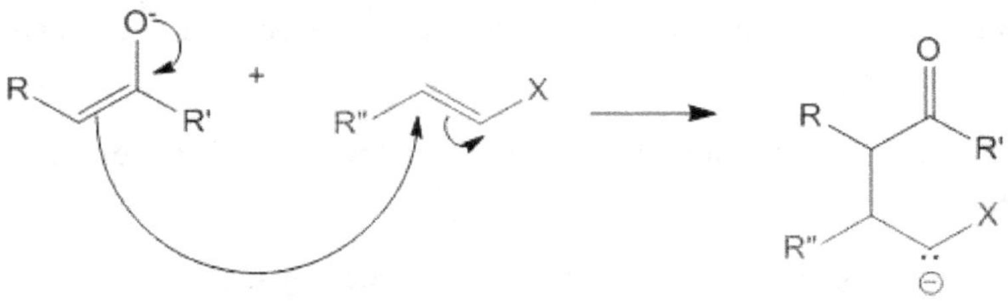

The Michael donor participates in a nucleophilic attack on the Michael acceptor

19. D is correct.

Ketones and aldehydes have similar chemical properties because both contain a carbonyl group not in conjugation with a heteroatom. They undergo similar reduction and alkylation reactions, although the products of these reactions may be slightly different.

Conversely, aldehydes can undergo an additional oxidation reaction to form carboxylic acids, unlike ketones, which cannot be oxidized further.

20. E is correct.

Tollens' reagent or Benedict's test detect reducing sugars (i.e., aldehydes). The aldehyde is oxidized to a carboxylic acid during the reaction.

Monosaccharides are reducing sugars, while some disaccharides, oligosaccharides, and polysaccharides are reducing sugars.

The Tollens' test for aldehydes involves the reduction of silver cations [Ag^+] to reduced silver; the metal precipitates out of solution and coats the inner surface of the reaction flask (i.e., mirror) as a positive test for reducing sugars. The test does not give a positive test with ketones because the ketone is already fully oxidized and lacks a C–H bond to be further oxidized and be converted to C–O bonds (e.g., aldehydes to COOH).

21. E is correct.

The hydrocyanation (i.e., adding ⁻C≡N) of ketones and aldehydes is essentially an alkylation reaction.

This alkylation reaction is sensitive to the steric environment of the carbonyl electrophile.

Formaldehyde (below) contains two hydrogen substituents and is the most reactive.

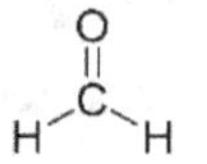

Aldehydes (below) are the second most reactive.

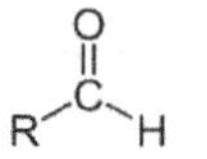

Unhindered (below) ketones are the next most reactive.

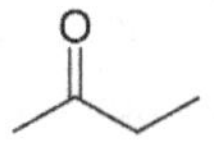

Hindered (below) ketones are the least reactive.

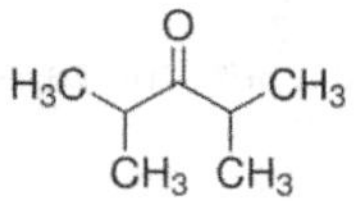

These alkyl groups block access to the *pi** orbital of the carbonyl, and the more sterically-hindered substrates react more slowly because the nucleophile (⁻C≡N) is impeded in its approach to the delta plus of the carbonyl carbon.

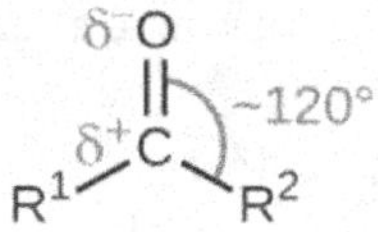

Carbonyl carbon with delta plus carbon (electrophile) indicated

22. D is correct.

The substituents on the carbonyl carbon can be alkyl, alkenyl, or aryl groups. The R groups do not need to be the same group (i.e., R and R').

Asymmetric ketones form chiral secondary alcohols if reduced or alkylated by a carbon nucleophile that is unlike the structure of the ketone alkyl groups.

Asymmetric ketones are "*prochiral*" electrophiles.

A: carboxylic acid functional group.

B: ester functional group.

C: aldehyde functional group.

E: anhydride functional group.

23. C is correct.

Reducing sugar can function as a reducing agent because it has a free aldehyde group or a free ketone group. Monosaccharides are reducing sugars, while some disaccharides, oligosaccharides, and polysaccharides are reducing sugars.

A reducing sugar becomes oxidized (e.g., aldehyde → carboxylic acid) from reducing another compound.

Benedict's test (or Tollens' reagent) is used to detect reducing sugars. An oxidized copper reagent is reduced by a sugar's aldehyde, and the aldehyde is oxidized to a carboxylic acid in the process.

Reducing sugars (and alpha hydroxyl ketones) give a positive Benedicts test: a red-brown precipitate forms. Fehling's solution gives a positive test for reducing sugars by changing from blue to clear and forming a red-brown precipitate.

The Tollens test for aldehydes involves the reduction of silver cations [Ag^+] to reduced silver; the metal precipitates out of solution and coats the inner surface of the reaction flask. Tollens' reagent forms silver ions (mirror) as a positive test for reducing sugars. The aldehyde is oxidized to the carboxylic acid when this occurs.

24. E is correct.

The compounds containing the carbonyl (C=O) group are ketones, aldehyde, carboxylic acids, and carboxylic acid derivatives (acyl halide, anhydride, ester, and amide).

Other groups may contain carbonyl groups, such as carbonate, carbamate, urea, etc., and these groups have the same oxidation state as carbon dioxide (O=C=O) with four C–X bonds.

25. E is correct.

Carbonyl groups contain lone pairs on the oxygen atom for hydrogen-accepting capabilities.

These groups lack a polarized *sigma* bond to hydrogen atoms (assuming the ketone is in the keto tautomer).

Since these molecules cannot donate hydrogen bonds, they cannot form hydrogen bonds with each other.

26. A is correct.

The larger the alkyl portion of an organic molecule, the less likely it can dissolve in water.

When ketones and aldehydes are dissolved in water, they are in equilibrium with their hydrated forms.

The hydrated form enhances the solubility of the compound in water.

27. B is correct.

Grignard + nitrile → imine salt + H_3O → ketone

$$RMgX + R'-C\equiv N \longrightarrow \underset{\text{imine salt}}{\overset{NMgX}{\underset{\|}{R-C-R'}}} \xrightarrow[\text{work-up}]{H_3O^+} \underset{\text{ketone}}{\overset{O}{\underset{\|}{R-C-R'}}}$$

Unlike esters, the nitrile is not subject to over-alkylation because the negatively charged imine intermediate generated from the alkylation is less electrophilic than the starting nitrile.

28. D is correct.

The oxidation of aldehydes to carboxylic acid can be done by exposing aldehydes to chromic acid in water and acetone or to potassium permanganate.

29. A is correct.

Nucleophiles attack sterically hindered alkyl halides at a much slower rate. Smaller or less sterically hindered substrates undergo reactions with nucleophiles at the fastest rate.

Bromobenzene undergoes the addition reaction at a negligible rate because the ring lacks an electron-withdrawing group that can activate the ring towards nucleophilic aromatic substitution.

Furthermore, the bromide of bromobenzene does not undergo S_N2 displacement reactions because the ring blocks access to the carbon-bromine *sigma** orbital.

30. E is correct.

Benedict's test (or Tollens' reagent) is used to detect reducing sugars. An oxidized copper reagent is reduced by a sugar's aldehyde; the aldehyde is oxidized to a carboxylic acid in the process.

Reducing sugars (and alpha hydroxyl ketones) give a positive Benedict's test: a red-brown precipitate forms. Fehling's solution gives a positive test for reducing sugars by changing from blue to clear and forming a red-brown precipitate.

The copper complex is reduced to form a red copper product that is less soluble in aqueous solutions. The detection of this precipitate means the molecule was oxidized.

31. D is correct.

Since the aldol reaction involves deprotonation (abstraction of H^+) by a strong base, the preferred solvents are neither acidic nor electrophilic.

Dimethyl ether, unlike the other solvents listed, does not contain an acidic proton.

32. C is correct.

Oxidation of primary alcohols produces aldehydes (by PCC or oxidation in dry conditions). Oxidation of secondary alcohols commonly produces ketones (e.g., by PCC) or carboxylic acids (e.g., Jones oxidation by CrO_3 in H_2SO_4).

A: Benedict's test (or Tollens' reagent) is used to detect reducing sugars. An oxidized copper reagent is reduced by a sugar's aldehyde, and the aldehyde is oxidized to a carboxylic acid in the process.

Reducing sugars (and alpha hydroxyl ketones) gives a positive Benedict's test: a red-brown precipitate forms. Fehling's solution gives a positive test for reducing sugars by changing from blue to clear and forming a red-brown precipitate.

B: Tollens' reagent forms silver ions (i.e., shiny mirror surface) as a positive test for reducing sugars.

E: aldehydes possess a hydrogen atom bonded to the carbonyl carbon, so they can be oxidized to produce carboxylic acids.

33. B is correct.

When *Grignard nucleophiles* alkylate ketones and aldehydes, the number of carbon-oxygen bonds decreases by one, and the number of carbon-carbon bonds increases by one.

primary alcohol ⇌ [O]/[H] ⇌ aldehyde ⇌ [O]/[H] ⇌ carboxylic acid

secondary alcohol ⇌ [O]/[H] ⇌ ketone ↛

tertiary alcohol ↛

Oxidation proceeds towards the right, while reduction proceeds to the left.

34. E is correct.

The tautomer forms of the ketone have the carbon-oxygen bond directly bound to the alkene carbon atom.

35. A is correct.

Carboxylic acid derivatives are similar in structure to ketones and aldehydes; however, one of the H or R groups has been replaced with a heteroatom, such as oxygen or nitrogen.

36. B is correct.

A ketone is being converted to an alkene. The oxidation state of ketones is larger than the oxidation state of alkenes, and therefore the ketone needs to be reduced.

Reduction of a ketone forms secondary alcohol.

The carbon skeleton rearranges, and this requires the formation of a carbocation. Exposure to phosphoric acid produces secondary carbocations, an alkyl shift, and deprotonation to yield the alkene product.

Notes for active learning

Notes for active learning

Carboxylic Acids – Detailed Explanations

1. C is correct.

Carboxylic acids (~COOH) are acidic functional groups with a hydroxyl (~OH) group directly bonded to the carbonyl (C=O). Carboxylic acids are acidic because the conjugate base is relatively stable due to resonance stabilization of the negative charge on oxygen. The molecule contains three hydroxyl groups and an arene.

2. D is correct.

A chemical equilibrium exists for this reaction, and water is given off as a byproduct in the process.

benzoic acid + Na⊕OH⊖ → sodium benzoate + H₂O

benzoic acid
insoluble in water

sodium benzoate
soluble in water

3. C is correct. The products formed from benzoic acid with NaOH are sodium benzoate and water.

4. B is correct.

Carboxylic acids are molecules that contain a terminal carbon atom with three bonds to oxygen. It consists of a hydroxyl (~OH) group and a carbonyl (C=O) group.

5. A is correct.

The carboxylic acid and alcohol starting components for the synthesis of an ester are determined by cleaving the *sigma* bond between the oxygen atom of the alkoxyl (O of the ether) group and the carbonyl (C=O) group.

6. C is correct.

The initial reaction between the amine and carboxylic acid is neutralization (i.e., proton transfer). However, at high temperatures, this proton transfer is reversible, and the reaction has enough energy to promote the nucleophilic attack of the carboxylic acid instead, leading to amide formation.

7. E is correct.

The acids contain a carboxylic acid functional group, which largely contributes to the boiling point. The molecule with the largest hydrocarbon region has more intermolecular forces (most of which are London forces) and has the largest boiling point. Stearic acid is a saturated fatty acid and has the highest boiling point.

The alkyl region of stearic acid consists of 17 saturated carbon atoms

Among the answer choices, no other acid possesses an alkyl region this large, and this molecule has the largest boiling point.

8. D is correct.

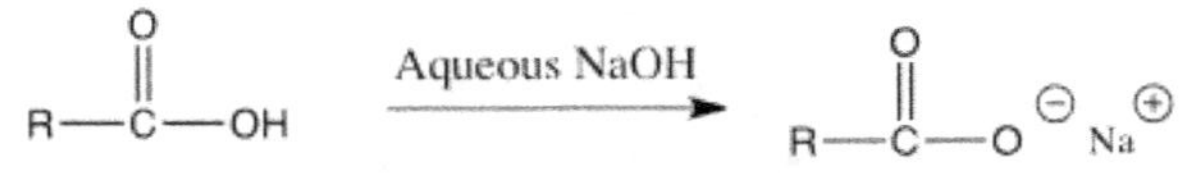

Carboxylic Acid

Carboxylate Salt
(More soluble in water)

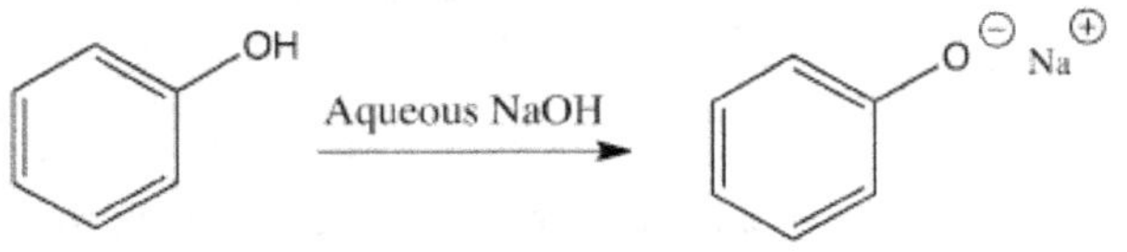

Phenol

Phenolate Salt
(More soluble in water)

The salt is a sodium carboxylate salt of the carboxylic acid with the byproduct of water

Because carboxylic acids are about 10 or more orders of magnitude more acidic than water, the conversion to the carboxylate form by exposing it to sodium hydroxide is irreversible.

Therefore, no equilibrium is associated with this reaction.

9. B is correct.

When carboxylic acids and alcohols are combined in acid catalysts (e.g., H_2SO_4), esters form as the product.

Water is produced as a byproduct in the reaction, and its removal from the reaction drives the reaction forward because an equilibrium exists.

10. E is correct.

The carboxylic acid (COOH) functional group accounts for at least two sites of hydrogen bonding.

Because the answers include carboxylic acids, the hydrocarbon regions of the molecules must be considered.

The molecule with the largest hydrocarbon region has more intermolecular forces (most of which are London forces) and has the largest boiling point.

Stearic acid is a saturated fatty acid and has the highest boiling point.

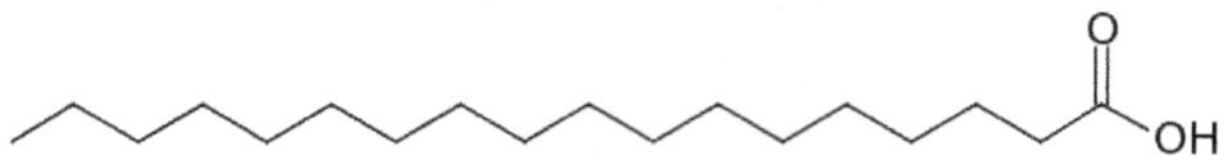

11. B is correct.

Acetic acid is a neutral compound, with the atoms in the molecule having no formal charge.

Carboxylic acids (COOH) are charged when protonated by other acids, as in Fischer esterification reactions or deprotonated by bases.

12. B is correct.

$$R-\underset{\underset{}{\overset{\overset{O}{\|}}{C}}}{}-OH + ROH \underset{}{\overset{H^+}{\rightleftharpoons}} R-\underset{\underset{}{\overset{\overset{O}{\|}}{C}}}{}-OR + H_2O$$

Esters from alcohols and carboxylic acids are condensations, so an equivalent of H_2O is lost as a byproduct.

13. C is correct.

The reaction used to combine alcohols and carboxylic acids is the Fischer esterification reaction.

$$CH_3-C\overset{O}{\underset{O-H}{\diagup\!\!\!\diagdown}} + CH_3CH_2OH \leftrightarrow CH_3-C\overset{O}{\underset{O-CH_2CH_3}{\diagup\!\!\!\diagdown}} + H_2O$$

Acid catalysis is needed to activate the carboxylic acid carbonyl group for the nucleophilic attack.

14. B is correct.

Carboxylic acids have some of the highest boiling points due to their hydrogen bonds forming dimers.

Alcohols contain the polar O–H bond and form hydrogen bonds, thus contributing to their high boiling point.

15. D is correct.

The negative charge of the carboxylate anion is stabilized by the electronegativity of the oxygen atom and by resonance stabilization. The negative charge delocalizes over the two oxygen atoms of the carboxylate functional group.

Resonance structures of the carboxylate anion

16. B is correct.

Molecules that possess the hydroxyl group are alcohols and carboxylic acids.

17. A is correct.

In addition to the polar functional groups, the molecular weight of a compound is an important influence on its boiling point. Because carboxylic acids are compared, the one with the highest molecular weight has the highest boiling point.

18. D is correct.

Carboxylic acids are more acidic than phenols; the pK_a of carboxylic acids is ≈5, and pK_a of phenols is ≈ 10.

The aryl (on the C=O) proton of benzaldehyde has a pK_a near the lower 40s, while the ketones (e.g., acetone) have pK_a values of ≈ 20.

19. A is correct.

Acid strength increases by the *inductive effect* of electron-withdrawing groups on neighboring carbon atoms.

[Structure of fluoroacetic acid: F-CH₂-C(=O)-OH]

Factors that affect the *acidity of a molecule*:

    1) the more electronegative the substituents, the greater the inductive effect;

    2) the closer the electronegative group is to the carbonyl carbon, the greater the effect;

    3) the greater the number of electronegative substituents, the greater the inductive effect.

Fluoroacetic acid has a $pK_a \approx 2.6$ and is more acidic than others because of the inductive influence of the electronegative fluorine atom.

B: *acetic acid* (i.e., vinegar) has a $pK_a \approx 5$ and is the least acidic since it has no electronegative substituents.

C: *bromoacetic acid* has a $pK_a \approx 3.6$.

D: *methoxyacetic acid* has a $pK_a \approx 3.6$

E: *phenol* has a $pK_a \approx 10.0$. The pK_a of phenol is much lower than cyclohexane due to resonance stabilization of the anion through the *pi* system of electrons in the aromatic ring.

20. A is correct.

When dissolved in the basic solution, the carboxylic acid is deprotonated to form the carboxylate salt.

The salt is charged, and this ion can engage in ion–dipole interactions with the solvent and is therefore soluble.

The aldehyde may react with the base to form charged species (either the geminal diol or the enolate), but these charged species are formed reversibly.

Therefore, the acid can dissolve, while the aldehyde is not.

21. E is correct.

Carboxylic acids exchange protons with bases.

When a neutral compound ionizes through the loss of a proton, the remaining charge of the larger fragment is negatively charged. The proton is positively charged, so the resulting ion is H^+.

The hydroxyl (~OH) group of a carboxylic acid (or of alcohol) can be replaced (i.e., substitution reaction) with special reagents (e.g., $SOCl_4$ or PBr_3) to form acyl halides (or alkyl halides for alcohol functional groups).

22. C is correct.

The high pH (basic conditions) of the aqueous solution causes the carboxylate to deprotonate and assume its conjugate base form.

This carboxylate form (i.e., the anion of the carboxylic acid) has a higher affinity for the aqueous layer than for the organic layer.

23. D is correct.

The carboxylic acid (~COOH) contains the most acidic functional group among the molecules listed.

Any proton has the potential to protonate a base, given that the base is sufficiently strong to remove the proton from an acid. This requires comparing the pK_a of the acid and base, whereby the base must have a higher pK_a.

Amines can deprotonate carboxylic acids; amide bases can deprotonate alcohols.

The use of strong organometallic bases (e.g., Grignard reagent) may be necessary for the deprotonation of neutral amines and hydrocarbons.

24. A is correct.

Citric acid is a chemical involved in the citric acid (TCA or Krebs) cycle.

Citric acid

The TCA cycle is a metabolic process for ATP production and commonly occurs in most aerobic organisms.

25. C is correct.

All the molecules contain the carboxylic acid functional group, which can hydrogen bond and ionize to increase its solubility in water.

Besides benzene, the choices differ only in the length of the carbon chain attached to the carboxylic acid.

Saturated carbon chains are hydrophobic, and therefore the shortest-chain carboxylic acid is most water-soluble.

26. B is correct.

The most polar molecule is the molecule that possesses the smallest alkyl portion.

Acidic functional groups enhance the intermolecular forces between molecules, with hydrogen bonding being the largest contributing factor.

The more polarized a hydrogen-heteroatom bond is, the more acidic it is and the stronger the intermolecular forces the molecule experiences.

Therefore, the carboxylic acid is the most polar molecule.

27. A is correct.

Nucleophilic acyl substitution (NAS) reactions are the most common reactions that carboxylic acid derivatives undergo.

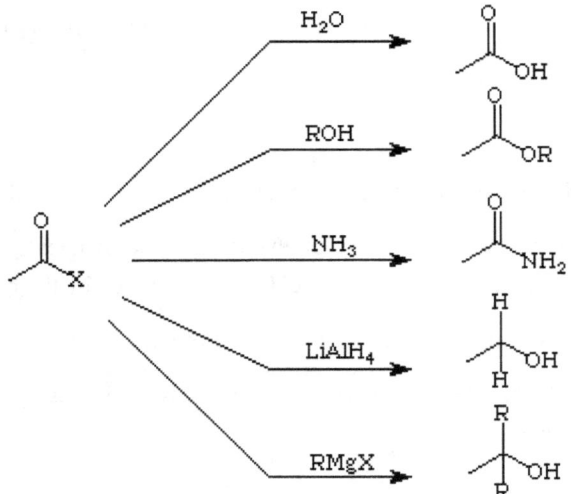

Example reactions involving acyl halides

28. A is correct.

Secondary alcohols cannot be oxidized further.

B: *primary alcohols* undergo oxidation by strong oxidizing agents (e.g., potassium permanganate, O_3 or Jones's reagent) to yield carboxylic acids.

C: *acidic or basic hydrolysis* of a nitrile yields carboxylic acids.

D: *Grignard reagents* reacting with CO_2 are a method for preparing carboxylic acids. The carbonation of a Grignard reagent (adding CO_2) forms the magnesium salt of a carboxylic acid. Subsequently, the magnesium salt is protonated and converted to a carboxylic acid when treated with mineral acid (e.g., H^+ from HCl).

E: carboxylic acids are formed by the oxidation of aldehydes.

29. B is correct.

Acids with smaller alkyl chains have lower boiling points.

Hydrogen bonding increases the boiling point. Increased molecular mass and hydrogen bonding are the factors that increase the boiling point.

Formic acid (below) has the molecular formula of CH_2O_2: it contains one carboxylic acid functional group and a hydrogen atom for the *R* group.

Formic acid

continued...

A: oxalic acid (below) has the molecular formula of CH_2O_4.

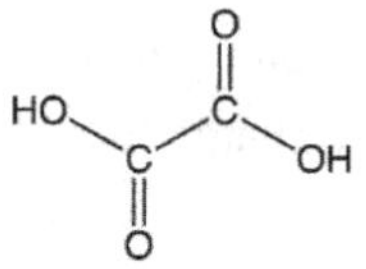

C: benzoic acid (below) has the molecular formula of $C_7H_6O_2$

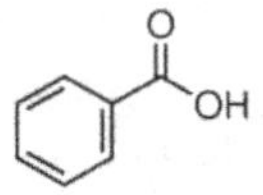

D: acetic acid (below) has the molecular formula of $C_2H_4O_2$.

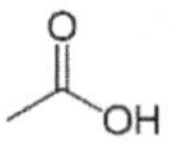

E: oleic acid (below) has the molecular formula of $C_{18}H_{34}O_2$.

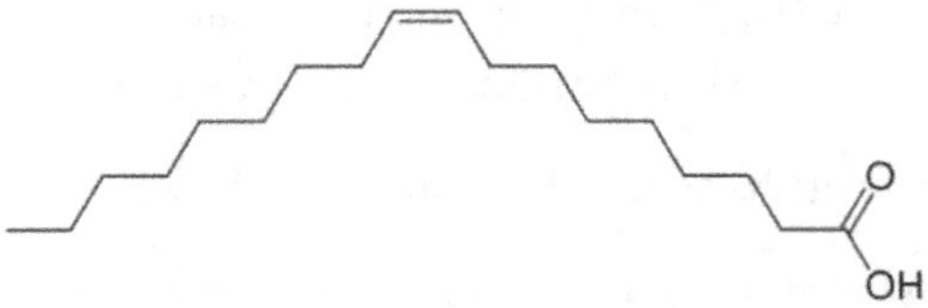

30. D is correct.

This reaction the *Fischer esterification reaction*.

Water is given off as the byproduct of this transformation.

To drive the reaction equilibrium forward, water should be removed from the reaction, or a large excess of one of the two components should be used.

31. C is correct.

Lithium aluminum hydride (LAH or $LiAlH_4$) is a powerful reducing agent and can reduce carboxylic acids and esters to form primary alcohols and reduce nitro groups to amines.

Sodium borohydride ($NaBH_4$) is a weak reducing agent and is only used for the reduction of aldehydes (to primary alcohols) and ketones (to secondary alcohols).

32. E is correct.

The chlorine substitution stabilizes the negative charge of the carboxylate through the inductive withdrawal of electron density through the *sigma* bonds.

The *ortho* substitution lowers the pK_a of the acid because its conjugation with the aryl ring decreases (*ortho* effect).

33. B is correct.

Exposing a carboxylic acid to sodium hydroxide produces water and the sodium carboxylate conjugate base. This ionic base has a higher affinity for the aqueous layer due to the negative charge, and this charge forms hydrogen bonds to the hydrogen atoms of water molecules.

34. A is correct.

In Brønsted-Lowry theory, an acid is a proton donor, and a base is a proton acceptor.

In solution, a strong acid dissociates its proton and exists predominantly in the deprotonated form as the acid's conjugate base. A strong acid forms a weak conjugate base because the acid's protons dissociate in solution.

The resulting anions of the deprotonated carboxylic acids are stable due to resonance. Resonance is a major contributor to stabilizing the anion by delocalizing the negative charge (using *pi* bonds) over several (in this example, two oxygen) atoms.

Acids that contain an electron-withdrawing substituent on the α-carbon (i.e., carbon adjacent to the carbonyl) tend to be stronger (donate proton more readily) because induction (*via sigma* bond) pulls the electron density.

Induction has a stabilizing effect on the carboxylate anion (the conjugate base of carboxylic acid).

Conversely, acids with electron-donating α substituents (i.e., methyl chains) tend to be weaker (less likely to dissociate the H^+).

The acid with two electron-withdrawing chlorine substituents forms the most stable carboxylate anion (when H^+ dissociates). Therefore, it is the strongest acid, which has the weakest (most stable) conjugate base.

B: $CH_3CH_2CH_2CO_2H$ is a carboxylic acid with only hydrogens and therefore lacks stability influences from electronegative atoms (F, N, O or Cl).

C: $(CH_3CH_2)_3CCO_2H$ is a carboxylic acid with a tertiary butyl substituent electron-donating *via* hyperconjugation (i.e., alkyl chains donate electrons).

D: $CH_3HNCH_2CH_2CH_2CO_2H$ is a carboxylic acid with an amino substituent that is strongly electron-withdrawing *via* electronegativity of the nitrogen, but the nitrogen is located at a large distance from the COO^- and therefore, its influence is minimal.

E: $CH_3Cl_2CCH_2CO_2H$ is a carboxylic acid that contains an electron-withdrawing substituent on the β-carbon that has less effect than at the α-carbon.

35. D is correct.

Hydrogen bonding a strong type of dipole-dipole interaction and raises the boiling point of organic compounds.

Dimer formed from hydrogen bonding of two carboxylic acids.

The carboxylic acid of acetic acid participates in hydrogen bonding, while the ester of methyl acetate does not form similar intermolecular hydrogen bonds.

36. A is correct.

The pK_a is the pH where half of the acid has dissociated to its conjugate base form.

The concentration of the acid and conjugate base is equal when the pK_a of the compound and the pH of the solution are the same.

Henderson-Hasselbalch equation:

$$pH = pK_a + \log[\text{conjugate base}] / [\text{acid}]$$

Notes for active learning

COOH Derivatives – Detailed Explanations

1. A is correct.

Amides are functional groups that possess an amino group bonded to a carbonyl moiety.

The reference molecule is a tertiary amine because it is bonded to three R groups.

2. D is correct.

The *hydrolysis of an ester* is essentially the reverse of Fischer esterification. To drive the reaction forward, excess water and an acid catalyst are used. The ester hydrolyzes to its two simpler components: the corresponding alcohol and carboxylic acid.

The *R* component of the alcohol product and the alkoxy group of the ester are the same.

3. A is correct.

The reaction best suited to occur under "typical conditions," rather than harsh conditions, is the aminolysis of the ester to form an amide. Amides are less electrophilic than esters, and basic conditions are needed to convert the amide back to the ester.

4. C is correct.

The hydrolysis of an ester is essentially the reverse of Fischer esterification. To drive the reaction forward, excess water and an acid catalyst are used.

The ester hydrolyzes to its two simpler components: the corresponding alcohol and carboxylic acid.

The *R* components of the alcohol product and the alkoxy group of the ester are the same.

5. A is correct.

Esters are functional groups belonging to carboxylic acid derivatives (along with acyl halides, anhydrides, and amides).

Esters have the same oxidation state as carboxylic acids but have different properties than carboxylic acid groups, such as enhanced electrophilic properties, decreased polarity, and higher pK_a values.

6. D is correct.

Anhydride functional groups have oxygen between two carbonyl carbons. The anhydride functional group does not exist in the reference molecule.

7. C is correct.

The reaction above is a hydrolysis reaction, and it is the reverse process for the Fischer esterification reaction. In excess water, the acid catalyst activates the ester carbonyl group to promote nucleophilic attack by water. This produces the corresponding alcohol and carboxylic acid (in the presence of H^+).

8. A is correct.

Sodium hydroxide can function as a base to reversibly deprotonate the N–H of the amide (due to resonance stabilization of the anion). However, the irreversible reaction pathway forms carboxylate and amine products.

9. E is correct.

Summary of some reactions involving carboxylic acids:

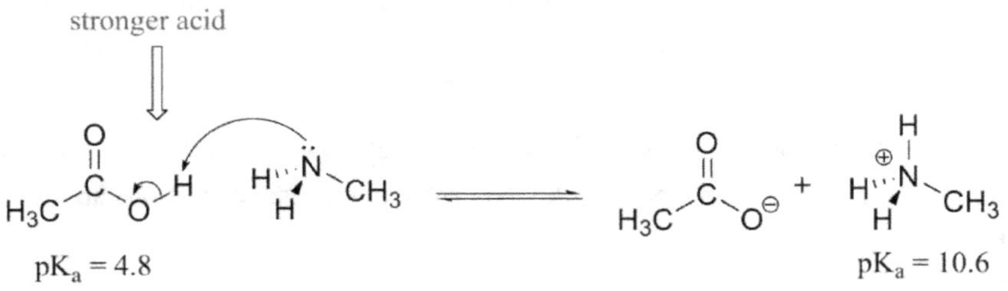

[reaction scheme showing acetic acid ($pK_a = 4.8$) with methylamine as stronger acid, in equilibrium with acetate and methylammonium ($pK_a = 10.6$)]

When forming an amide from a carboxylic acid and an amine, higher temperatures are typically needed to drive the reaction forward.

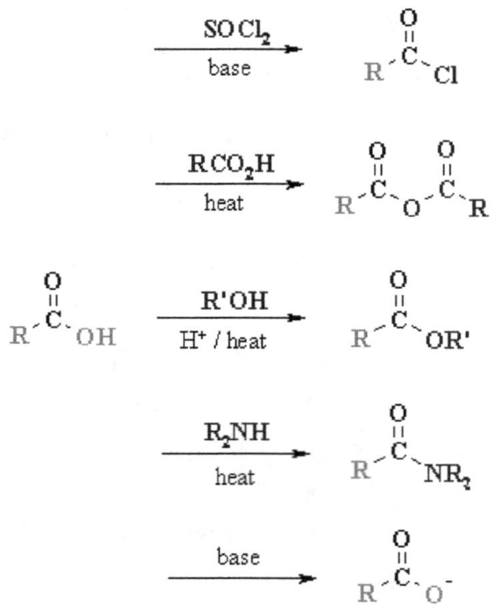

Reactions for a carboxylic acid with specific reagents:
- $SOCl_2$ / base → acyl chloride (R–C(O)–Cl)
- RCO_2H / heat → anhydride (R–C(O)–O–C(O)–R)
- $R'OH$ / H^+ / heat → ester (R–C(O)–OR')
- R_2NH / heat → amide (R–C(O)–NR₂)
- base → carboxylate (R–C(O)–O⁻)

Reactions for a carboxylic acid with specific reagents

The *reactivity of the carboxylic acid derivatives* is:

acyl chloride > anhydride > carboxylic acid ≈ ester > amide > carboxylate.

An equivalent of water is lost (i.e., dehydration) in the process, so this reaction is a condensation reaction.

10. E is correct.

Acid chlorides and anhydrides combine with alcohols to form esters and with amines to form amides.

Acid chlorides are more reactive than anhydrides; chloride is a better leaving group than the carboxylate anion.

These factors contribute to the enhanced reactivity of the acid chloride.

11. B is correct.

The synthesis (below) of amides from carboxylic acids and amines usually requires heating the reaction to high temperatures. This is necessary to create the amide bond and drive away the water byproduct, so the chemical equilibrium favors product formation.

$$CH_3-CH_2-\overset{\overset{O}{\|}}{C}-OH \ + \ H-\underset{\underset{H}{|}}{N}-CH_3 \ \rightarrow \ CH_3-CH_2-\overset{\overset{O}{\|}}{C}-\underset{\underset{H}{|}}{N}-CH_3 \ + \ H_2O$$

carboxylic acid   amine   amide   water

12. B is correct.

6-APA is the core of penicillin and is the chemical compound (+)-6-aminopenicillanic acid.

Amides are more stable than carboxylic acids, their hydrolysis is higher in energy.

Penicillin with 2 amide bonds

13. C is correct.

An *amide* is an amine that is directly bonded to a carbonyl group. The carbonyl group accepts the lone pair of electrons on the nitrogen atom through resonance action.

Amides ($pK_a \approx 20$–25) are less electrophilic than esters, and primary and secondary amides have acidic N–H protons ($pK_a \approx 35$).

14. D is correct.

The hydrolysis of esters using basic conditions (e.g., sodium hydroxide) is not a catalytic reaction. It is a base-promoted reaction because the hydroxide is consumed in the reaction and is not regenerated.

On the contrary, acid-catalyzed hydrolysis reactions of esters regenerate the acid catalyst.

15. B is correct.

Group 1 is a ketone; 2 is an amide, and 3 is an ester.

Ketones are more electrophilic functional groups than esters. Esters are more electrophilic than amides.

16. D is correct.

Nitriles (~C≡N) have the prefix *cyano* and are like alkynes in that both functional groups contain triple bonds.

However, the nitrile possesses an *sp* hybridized nitrogen atom, giving rise to its alternate chemical reactivity (pK_a = 25).

Nitriles are more electron deficient than alkynes and are more susceptible to nucleophilic attack by molecules such as organolithium compounds and Grignard reagents.

Alkynes undergo addition reactions with strong electrophiles, such as bromine.

17. A is correct.

Esters are similar in structure to carboxylic acids, but instead of having a hydroxyl group bonded to the carbonyl carbon atom, an alcohol-derived group is present.

These groups are less electrophilic than ketones and aldehydes, and they have the same oxidation state as carboxylic acids.

Important esters in the body include triacylglycerides or triglycerides.

18. E is correct.

Saponification is the base-promoted hydrolysis of esters. This type of hydrolysis is typically applied to form soap compounds.

For example, triacylglycerol is a lipid that can undergo saponification, the breakdown of fatty esters with bases, such as sodium hydroxide or potassium hydroxide.

carboxylate ester   sodium hydroxide   sodium carboxylate   alcohol

Base hydrolysis (saponification) of an ester to form a carboxylate salt and an alcohol

19. B is correct.

The reactant benzoyl chloride is an acyl halide that forms a ketone when reacted with an equimolar quantity of a Grignard reagent. Excess Grignard reagent converts the ketone to tertiary alcohol.

Acyl halide + Grignard reagent → ketone + Grignard reagent → tertiary alcohol

Sample addition reactions of *carbonyl compounds and excess Grignard*:

      Formaldehyde + excess Grignard → primary alcohol

      Aldehyde + excess Grignard → secondary alcohol

      Ketone + excess Grignard → tertiary alcohol

20. A is correct.

There are 4 alcohol functional groups present.

A cyclic ester is present. A cyclic ester is a *lactone*.

```
      H
      |
  R — C — O — R
      |
      OH
```

Hemiacetal: alcohol and ether on the same carbon

```
      OR"
      |
  R — C — H
      |
      OR'
```

Acetal: two ethers on the same carbon

21. A is correct.

In the addition reaction between water and benzoyl chloride, water donates electrons (acting as a nucleophile) to the electrophilic carbonyl of benzoyl chloride.

PhC(O)Cl + H₂O → PhC(O)OH

The chlorine atom, which is not present in the product, acts as a leaving group during the subsequent elimination step whereby the carbonyl carbon reforms.

22. C is correct.

PhCOOH + SOCl₂ → PhCOCl + NH₃ → PhC(O)NH₂

The first step in this reaction involves substituting the ~OH in benzoic acid with the ~Cl group from thionyl chloride (SOCl₂).

The resulting compound is benzoyl chloride – a highly reactive acyl halide that undergoes nucleophilic substitution. Treating this molecule with NH₃ (ammonia) results in substituting the ~Cl by the ~NH₂ group to form an amide (benzamide).

A and B: benzoyl chloride is highly susceptible to nucleophilic substitution, so chlorine is not part of the final product if another nucleophile (NH₃) is present.

D: *p*-aminobenzaldehyde is a benzene ring with an NH₂ substituent *para*- to the aldehyde group, replacing carboxylic acid. The carboxylic acid would not be reduced to the aldehyde. The condition for electrophilic aromatic substitution (EAS) requires an electrophile (e.g., Cl₂) and a Lewis acid (e.g., FeCl₃).

E: 3-chloro-4-aminobenzaldehyde is a benzene ring with an aldehyde replacing the carboxylic acid. The carboxylic acid would not be reduced to the aldehyde.

The molecule is named with chlorine in the *meta* position and an amine group at the *para* position.

23. D is correct.

The compound contains an aromatic ring is called *benzyl*.

The amide functional group is denoted by:

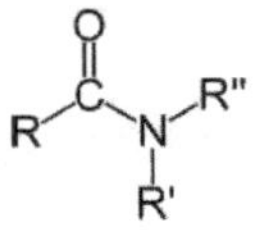

where R is an alkyl chain (or H).

An ether functional group is denoted by R–O–R'

Additionally, the molecule contains an aromatic ring, phenol group (i.e., hydroxyl attached directly to a benzene ring), and an alkene (i.e., double bond as a *trans*-alkene).

24. A is correct.

Fischer esterification involves a carboxylic acid and alcohol with an acid catalyst.

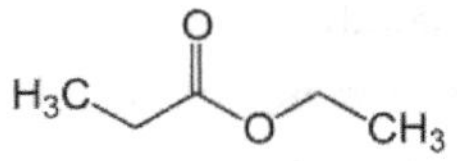

Ethyl propanoate

Esters have the general formula: R–COO–R'

The suffix for an ester is *–oate*. The prefix is the substituent attached to the oxygen adjacent to the carbonyl carbon, and the root is the substituent attached to the carbonyl oxygen.

Excess alcohol is used to drive the chemical equilibrium forward.

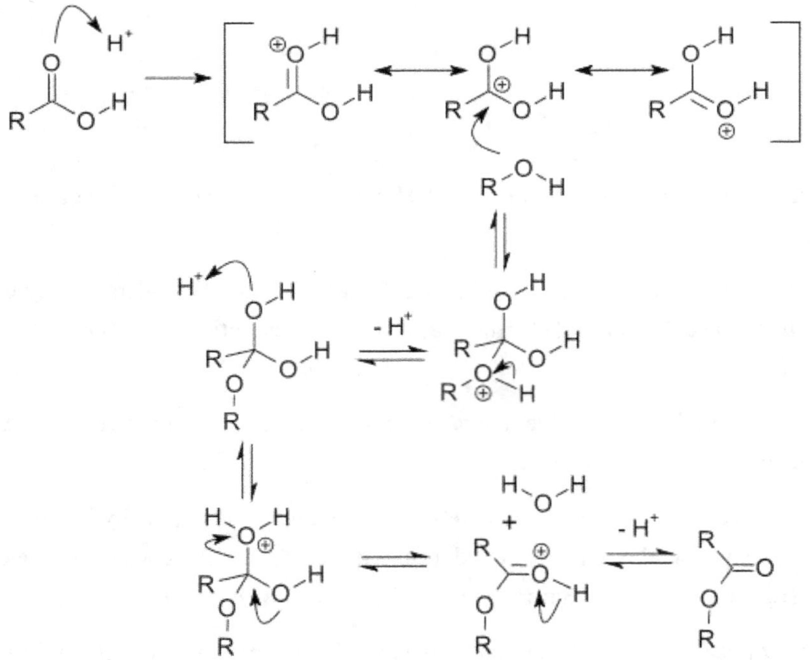

Fischer esterification mechanism

25. A is correct.

The *amide linkage* is present between individual amino acids, and these bonds are *peptide bonds*.

The amide bond is formed when the lone pair of electrons on the nitrogen of the amino group makes a nucleophilic attack on the carbonyl of the other amino acid. This process is classified as condensation (*via* dehydration) and results in the loss of water as the peptide bond forms.

26. C is correct.

The benzene ring (aromatic group) is bonded to the carbonyl of an amide functional group.

27. C is correct.

Amides are carboxylic acid derivatives made of amines, which are carboxylic acid components.

The *hydrolysis of carboxylic acid derivatives* forms the corresponding carboxylic acid and heteroatom-containing component (e.g., an amine or alcohol).

28. A is correct.

Consider the susceptibility of different compounds to nucleophilic attack. All the molecules undergo nucleophilic attack, but the molecule that undergoes attack the easiest is propionyl bromide.

Acyl halides are the most reactive towards nucleophiles of the carboxylic acid derivatives (acyl halides, anhydrides, esters, carboxylic acids, and amides) electron-withdrawing effects of oxygen and the stability of the halide anion as a leaving group.

B: *benzyl bromide* is susceptible to nucleophilic attack due to the electronegative substituents, but not as much as propionyl bromide. Acid halides are more electrophilic than alkyl halides, aldehydes, or ketones.

C: *propanal* is susceptible to nucleophilic attack since it has a carbonyl carbon but no electronegative substituents, as does propionyl bromide have.

D: *butanoic acid* has a carbonyl carbon that is slightly susceptible to nucleophilic attack because the double-bonded oxygen has an electron-withdrawing effect. However, comparing butanoic acid and its functional derivative propionyl bromide, propionyl bromide has withdrawing effects from the oxygen and the bromide.

29. C is correct.

Acyl halides and alcohols form esters.

This addition-elimination reaction begins with the nucleophilic attack of the hydroxyl group to the electrophilic carbonyl of the acid halide in an addition reaction.

This is followed by the elimination of chloride to generate the corresponding ester.

Pyridine is a common basic solvent used in organic chemistry:

HCl is generated as a byproduct in this reaction; therefore, a base (e.g., pyridine) is needed.

30. C is correct.

This reduction requires two equivalents of lithium aluminum hydride (i.e., powerful reducing agent).

The reactive intermediate involved after the first addition of hydride is a hemiacetal. This hemiacetal may reversibly open to form the aldehyde, which can be reduced to form the second primary alcohol.

Reduction:

    carboxylic acid / ester → aldehyde → primary alcohol

    ketone → secondary alcohol

Reduction of a carboxylic acid or ester requires $LiAlH_4$.

Reduction of an aldehyde or ketone can proceed with either $LiAlH_4$ or the milder reducing agent $NaBH_4$.

31. B is correct.

Treatment of the ester with potassium hydroxide (KOH) and heat results in the nucleophilic attack of hydroxide (⁻OH) to the carbonyl in an addition-elimination reaction.

The ring product exists as a negatively charged species (cyclohexanol as the alkoxide) because of a base (KOH) until the second step of the H^+ workup. After acidic workup, the alkoxide is protonated to form cyclohexanol.

32. B is correct.

The longest continuing chain in the product is four carbon atoms long.

Therefore, the root name is *butan,* and the suffix is ~*amide* because it is an amide (i.e., R–CONR₂) functional group.

For this example, the nitrogen contains a methyl group.

N-methylbutanamide

33. B is correct.

As the acidity of a group increases, its basic properties decrease.

The carbonyl (C=O) group of the (R-CONR₂) amide withdraws electron density from the nitrogen atom inductively and through conjugation. This causes the amide's N–H bond to be much less basic (pK_a of an amine ≈ 10 and pK_a of an amide ≈ 35).

2-butenoic acid   +   *isobutanol*

34. A is correct.

The reaction is the *acid hydrolysis of an ester*.

When esters are hydrolyzed, they yield carboxylic acids and alcohols.

From the hydrolysis of this ester, 2-butenoic acid and isobutanol are formed.

B: $CH_3CH(CH_3)_2$ is an alkane, and neither type of compound can be obtained by hydrolyzing an ester.

C: $HOOCCH_2CH(CH_3)_2$ may look like the carboxylic acid produced from the hydrolysis of the ester, but the carbonyl double bond is absent.

D: $CH_3CH=CHCHO$ is an unsaturated aldehyde.

E: $HOCH_2CH_2CH(CH_3)_2$ is alcohol with an additional carbon in the chain.

35. B is correct.

Acid bromides have the best leaving group (i.e., most stable anion), and therefore are the most reactive.

Acyl halides (RCOCl or RCOBr) are so reactive that they are not typically found in nature due to the moisture in the atmosphere, which converts them to carboxylic acids.

The *reaction rate* depends on factors such as the *leaving group*, the *steric environment*, and *substituents bonded* to the carbonyl carbon.

The *more stable the leaving group* (as an anion), the *faster* the reaction is.

Also, a more electrophilic carbonyl (i.e., induction from neighboring groups) undergoes nucleophile attack more rapidly.

Additionally, steric hindrance (i.e., bulky groups) at the reaction center leads to slower reaction rates.

36. B is correct.

The *hydrolysis of an ester* is the reverse process of condensation because water is introduced to the ester to produce carboxylic acid and alcohol.

A chemical equilibrium exists for the acid-catalyzed process, and the reaction is driven forward using a large excess of water.

A chemical equilibrium does not exist for the base-promoted process because the alcohol product cannot add into the carboxylate to reform the ester.

Notes for active learning

Amines – Detailed Explanations

1. B is correct.

Primary amines have one alkyl group bonded to the nitrogen atom, secondary amines have two alkyl groups bonded to the nitrogen atom, and tertiary amines contain nitrogen atoms bonded to three alkyl groups.

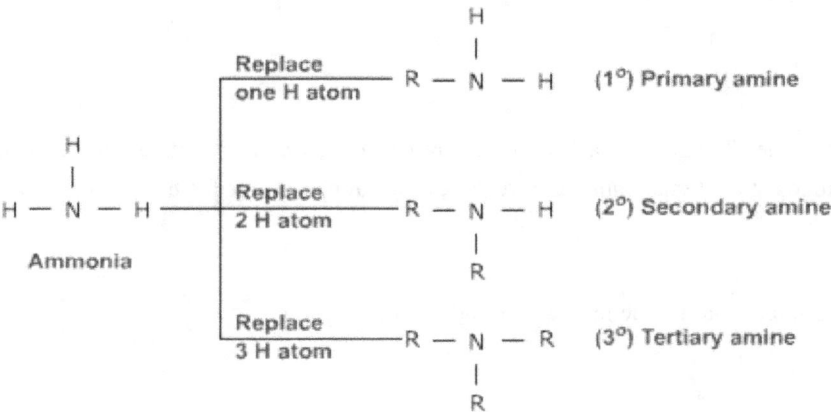

2. C is correct.

Amines possess nitrogen atoms, with lone pairs being more basic and nucleophilic than oxygen atoms and halogens such as fluorine. This is because the nitrogen atom is a less electronegative heteroatom; therefore, the nonbonding valence electrons of nitrogen have more energy and are more reactive.

3. A is correct.

Nitrogen bonded to alkyl and two hydrogens is a primary amine. Nitrogen bonded to alkyl and three hydrogens results in a positive charge on the nitrogen use the original lone pair on the nitrogen to bond to an H^+. The positive charge on the molecule allows it to be associated with a molecule with an anion (e.g., Cl^-).

Quaternary ammonium salts can be formed from the *N*-alkylation of tertiary amines.

When tertiary amines are converted to ammonium salts, the inductive effects of the nitrogen atom are larger, and the magnitude of the carbon–nitrogen dipoles increases.

4. E is correct.

Arenediazonium salts synthesized from primary aromatic amines are compounds with an N_2^+ group attached to the aromatic ring. They are useful for synthesizing many compounds because they can easily undergo replacement reactions in which molecular nitrogen is released, and a nucleophilic substituent attaches to the aromatic ring in its place.

For example, in the presence of cuprous halides, diazonium salts release molecular nitrogen and form halogen-substituted arenes – the Sandmeyer reaction. In cuprous cyanide, nitrogen is released and replaced by the cyanide ion, thus forming aromatic nitriles.

In the same way, in cold aqueous solutions, the hydroxyl group replaces the nitrogen to form phenol. Therefore, compounds I, II, and III can be obtained by replacing nitrogen in arenediazonium salts.

5. B is correct.

Ethylamine is the product since nitrile reduction adds carbon to the alkyl chain.

Cyanide ($^-$CN) is nucleophilic and can add to methyl bromide, which results in the expulsion of the bromide leaving group. The product of this reaction is cyanomethane.

The nitrile group can be reduced by lithium aluminum hydride ($LiAlH_4$ or LAH) or hydrogenation to form ethylamine.

6. D is correct.

Primary amines have one alkyl group attached to the nitrogen atom; secondary amines have two alkyl groups attached to the nitrogen atom; tertiary amines have three alkyl groups attached to the nitrogen atom.

7. B is correct.

The polar amino group of *p*-toluidine makes it slightly soluble in water:

p-toluidine is soluble in acidic water due to the formation of ammonium salts.

8. A is correct.

Amines are derivatives of ammonia. These compounds contain carbon-nitrogen single bonds and are known to be sufficient bases for strong acids.

Carbonyl-containing (C=O) compounds are aldehyde, ketone, acid chloride, anhydride, carboxylic acid, ester, and amide.

9. A is correct.

Although primary and secondary amines can accept and donate hydrogen bonds, the hydrogen bonding forces are greater for primary amines because these molecules possess two N–H bonds.

Quaternary ammonium salts cannot form hydrogen bonds because the N has four bonds, and no lone pairs of electrons remain on the nitrogen.

10. E is correct.

Oxygen and nitrogen atoms can act as proton acceptors if not conjugated with electron-withdrawing groups.

The amine is the most basic and nucleophilic functional group.

Detailed Explanations: Amines

11. B is correct.

Amines are compounds that contain basic nitrogen atoms that can function as nucleophiles if steric interactions do not prevent them from doing so.

The nitrogen lone pairs of amides are tied up in conjugation with the carbonyl group and are therefore not basic or nucleophilic.

12. A is correct.

When amines are added to water at neutral pH, the solution becomes basic, and hydroxide ions ($^-$OH) are produced.

Alternatively, when added to water at neutral pH, carboxylic acids generate hydronium ions (H_3O^+).

13. A is correct.

The boiling points of compounds are determined by two general factors: molecular weight and intermolecular interactions.

The higher the molecular weight, the harder it is to "push" it into the gas phase, and the higher the boiling point.

Similarly, the stronger the intermolecular interactions, the more energy is required to disrupt them and separate the molecules in the gas phase, hence the higher the boiling point.

Dimethylamine (below) has one N–H bond (donor) and one lone pair on the nitrogen (acceptor):

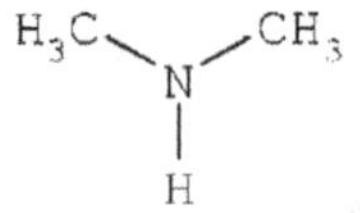

$$H_3C-\underset{\underset{H}{|}}{N}-CH_3$$

The other answer choices can accept and donate hydrogen bonds, which increases their boiling points.

E: ethanolamine has the following structure:

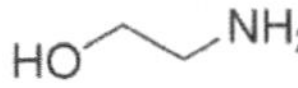

$$HO-CH_2-CH_2-NH_2$$

Ethanolamine can participate in hydrogen bonding: three with the hydroxyl (two acceptors and one donor) and three with the amino (two donors and one acceptor).

Hydrogen bonding increases the boiling point of molecules.

14. C is correct.

Tertiary amines possess no N–H bonds, while secondary amines have one N–H bond, and primary amines contain two N–H bonds.

15. E is correct.

The nitrogen atom of amines is basic because it possesses a lone pair of electrons that add to the acidic protons of acids, resulting in its protonation.

$$H-\overset{\overset{H}{|}}{\underset{\underset{H}{|}}{N^{\oplus}}}-H$$

Ammonium cation

The *conjugate acid of amines* is the ammonium cation, and the nitrogen atom is covalently bonded to four other atoms. Because the nitrogen atom is bonded to four other atoms, it develops a positive formal charge.

Although the nitrogen atom has a formal charge, this positive charge character is distributed over the less electronegative substituent groups.

16. A is correct.

Carbonyl (C=O) groups indicate aldehydes, ketones, carboxylic acid, or the four carboxylic acid derivatives (i.e., acyl halide, anhydride, ester, and amide).

Amines do not contain carbonyl groups.

17. B is correct.

A *quaternary ammonium salt* has no lone pairs of electrons on the nitrogen atom and therefore cannot function as a nucleophile.

1°, 2° or 3° amines can attack alkyl halides by donating their lone pairs of valence electrons.

Ammonium salts have no available valence electrons and can function as acids (proton donors), but not a nucleophile.

18. D is correct.

Ammonia is a basic compound with a low boiling point and exists as a gas at room temperature.

It is basic, colorless, has a pungent smell, and can be toxic if ingested.

19. C is correct.

Amine salts are soluble in water because nitrogen has a positive charge and can form ion–dipole interactions with water.

A: nitrogen donates its lone pair of electrons when forming a salt; positive charge forms on the nitrogen, not a negative charge.

B: the amine salt has a higher molecular weight than the amine, but it is not why the amine salt is more soluble.

D: some amines are soluble in water. Amines follow a similar solubility pattern to carboxylic acids: up to 6 carbons, they are relatively soluble.

Detailed Explanations: Amines

20. C is correct.

The most notable intermolecular force for water is hydrogen bonding.

$CH_3–CH_2–NH_2$ interact with water through hydrogen bonding. Hydrogens, bonded directly to F, O or N, participate in hydrogen bonds. The hydrogen is partial positive (i.e., delta plus or $\partial+$) due to the bond to these electronegative atoms. The lone pair of electrons on the F, O or N interacts with the $\partial+$ hydrogen to form a hydrogen bond.

$CH_3–CH_2–NH_2$ is ethylamine. Amines (i.e., nitrogen) can form hydrogen bonds, but thiols (i.e., sulfur) cannot.

21. D is correct.

Salts form as solids with tightly compact repeating unit structures, requiring more heat to boil than other substances, typically liquids.

22. A is correct.

Amines are typically bases due to the lone pair of electrons on the nitrogen.

The name suggests that three methyl groups are bonded to the central nitrogen atom.

23. C is correct.

Amine salts contain a positively charged, tetravalent nitrogen atom and an anionic counterion.

1°, 2° and 3° amines are neutral species.

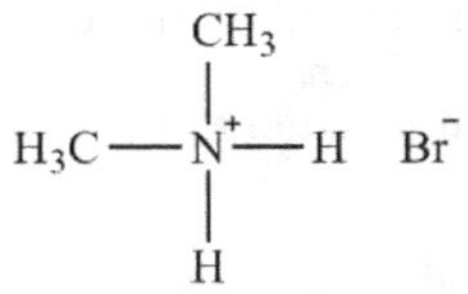

dimethylammonium bromide

A: sulfanilamide

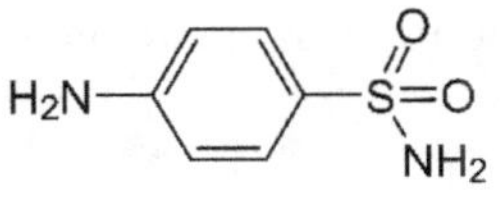

B: thioacetamide

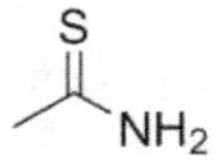

D: histamine

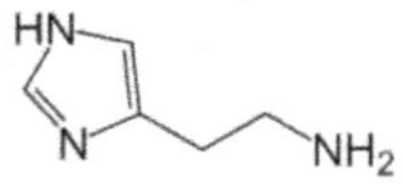

E: pyridoxine

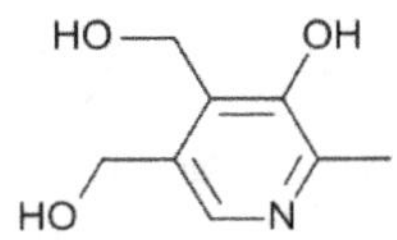

24. A is correct.

Amines tend to be basic and nucleophilic if they are not sterically bulky.

Attaching acyl groups to amines causes the nitrogen lone pair to delocalize into the carbonyl π* orbital, thus significantly reducing the nitrogen's nucleophilic and basic properties.

The *amide* ($RCONR_2$) group reacts with electrophiles on the carbonyl oxygen atom because a more stable cationic intermediate is generated.

25. C is correct.

The conjugate base of an acid is essentially the deprotonated form. The most acidic protons of the molecule are the N–H protons, with a pK_a in the mid-30s.

26. B is correct.

Amines are one of several functional groups that contain nitrogen atoms.

The other answer choices only contain carbon, hydrogen, or oxygen atoms.

27. D is correct.

Ammonia groups enhance the water solubility of compounds because of the hydrogen bonding and molecular dipoles that amines possess.

The greater the number of N–H bonds (i.e., primary > secondary > tertiary > quaternary) an amine possesses, the more hydrogen bonds the amine can form and the more soluble it is.

Furthermore, acid can be added to aqueous solutions of amines to enhance the solubility of the molecule.

28. C is correct.

The molecule has a basic site on the nitrogen atom.

The lone pair of electrons on the nitrogen atom can be protonated to form an ammonium cation.

29. A is correct.

Hydrogen bonding increases the boiling point of a molecule.

Amines with N–H bonds typically have higher boiling points compared to tertiary amines.

Tertiary amines can only accept a hydrogen bond, while primary and secondary amines can donate and accept hydrogen bonds.

30. B is correct.

The most basic site in the molecule is the *N*-methyl tertiary amine because the lone pair of this nitrogen atom is not in conjugation with an electron-withdrawing group or part of an aromatic ring.

31. C is correct.

Amines are Brønsted-Lowry bases and Lewis bases because nitrogen's lone pair of electrons can bind a proton.

It is favorable for the amine to abstract a proton from the acid.

Electron-donating groups attached to the nitrogen in the amine make the amine more basic because by donating electron density, they stabilize the positive ion formed.

Therefore, electron-donating groups (alkyl chains *via* hyperconjugation) destabilize the lone pair of electrons on the amine and make it more reactive.

Conversely, substituents (e.g., electronegative atoms) are electron-withdrawing, and the amine is less basic.

Compared to hydrogen atoms, alkyl groups are electron-donating, and therefore basicity decreases in the order trimethylamine > methylamine > ammonia in the gas phase.

The gas phase is specified because, in aqueous solutions, hydrogen bonding plays a role in stabilizing the salt and may result in a different order.

In addition reactions, electronegative atoms, such as the fluorine of $(CF_3)_3N$, strongly reduce the basicity of the amine because of the inductive electron-withdrawing effect of the electronegative atom.

In this example, it is unfavorable for the nitrogen to acquire a positive charge in forming a salt.

32. A is correct.

Amines can be protonated with Brønsted acids to produce ammonium salts.

Amides can be formed from primary and secondary amines by acylation of the nitrogen of primary and secondary amines.

33. E is correct.

Hydrogens, bonded directly to F, O or N, participate in hydrogen bonds.

The hydrogen is partial positive (i.e., delta plus or ∂+) due to the bond to these electronegative atoms.

The lone pair of electrons on the F, O or N interacts with the ∂+ hydrogen to form a hydrogen bond.

None of the other molecules can form hydrogen bonds because hydrogen is not attached directly to F, O or N.

34. B is correct.

Exposing *bulky amines* to an acid such as HCl allows it to become more soluble in water.

The charged ammonium cation has stronger dipole interactions with water compared to the neutral amine form.

35. E is correct.

A tertiary amine (R_3N) has three alkyl groups bonded to the nitrogen atom.

A secondary amine (R_2NH) has two alkyl groups bonded to the nitrogen atom.

A primary amine (RNH_2) has one alkyl group bonded to the nitrogen atom.

A, C, and D: are primary amines.

B: is a secondary amine.

36. A is correct.

Structure 1: *cyclic secondary enamine*

Structure 2: *secondary amine*

Structure 3: *cyclic quaternary amine*

The *secondary amine* (2) contains one N–H bond, and therefore, this amine can form hydrogen bonds.

The *cyclic amine* (3) and the *enamine* (1) do not contain N–H bonds, so their boiling points are lower.

Aside from hydrogen bonding considerations, molecules with formal charges (molecule 3) interact *via* electrostatic forces, which increase their boiling point.

Notes for active learning

Notes for active learning

Amino Acids, Peptides, Proteins – Detailed Explanations

1. A is correct.

In this reaction, an amino acid is prepared from a carboxylic acid.

The first step in this reaction is converting the carboxylic acid to the *alphabromo acid bromide intermediate*. This intermediate is then hydrolyzed in water to regenerate the carboxylic acid. The alpha bromide is displaced upon exposure to ammonia and heat to form alanine.

Both enantiomers of the compound are expected for this reaction.

2. D is correct.

Threonine is an amino acid that contains secondary alcohol as part of the side chain.

The alcohol has a permanent dipole and is capable of hydrogen bonding.

Therefore, the type of intermolecular interactions expected is dipole-dipole interactions (hydrogen bonding is a subset of dipole-dipole interactions).

3. E is correct.

Glycine is an achiral molecule that lacks a substituent at the alpha position. The unsubstituted alpha position is methylene, and a methylene group has a symmetrical center (i.e., bonded to two hydrogens).

4. E is correct.

Tyrosine contains a phenol *R* group, phenylalanine has a phenyl group, tryptophan has an indole group, and histidine has an imidazole group.

5. A is correct.

The primary structure of a protein is the linear sequence of amino acids.

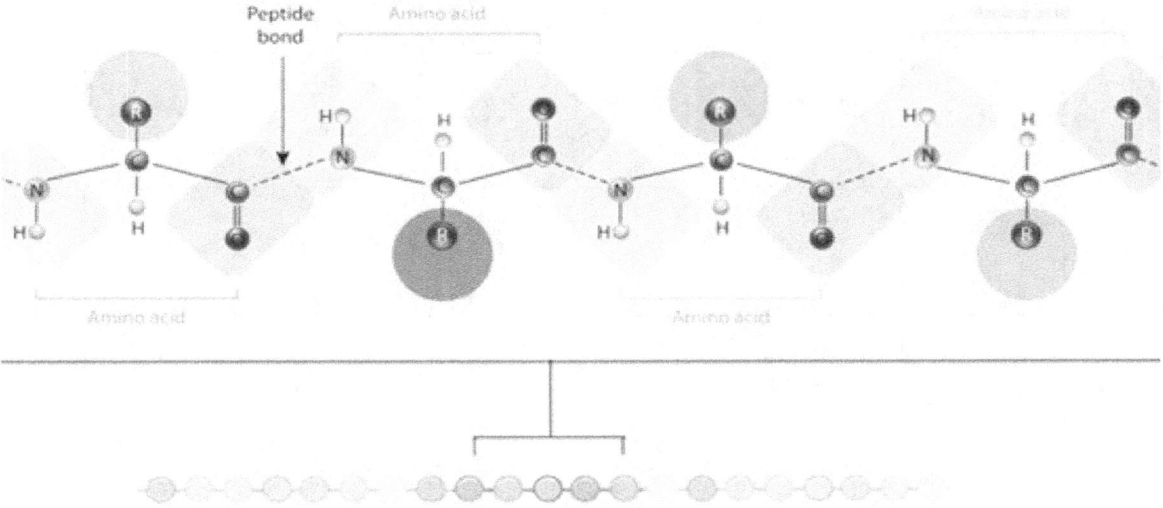

6. D is correct.

Hydrophilic amino acids are found near the external regions of proteins. They line the walls of protein ion channels because these molecules form more favorable interactions with the aqueous environment than hydrophilic amino acids.

7. C is correct.

Nonpolar amino acids have side chains composed of hydrogen and carbon atoms. Incorporating electronegative heteroatoms, such as oxygen and nitrogen atoms, can make the side groups polar, acidic, or basic.

8. B is correct.

All the molecules in the body are divided into four categories. These biomolecular categories are carbohydrates, lipids, nucleic acids, and proteins.

9. A is correct.

Amino acids are linked in peptides by amide bonds. The peptide forms in a condensation reaction *via* dehydration (loss of water) when the lone pair on the nitrogen of an amino group of one amino acid makes a nucleophilic attack on a carbonyl carbon.

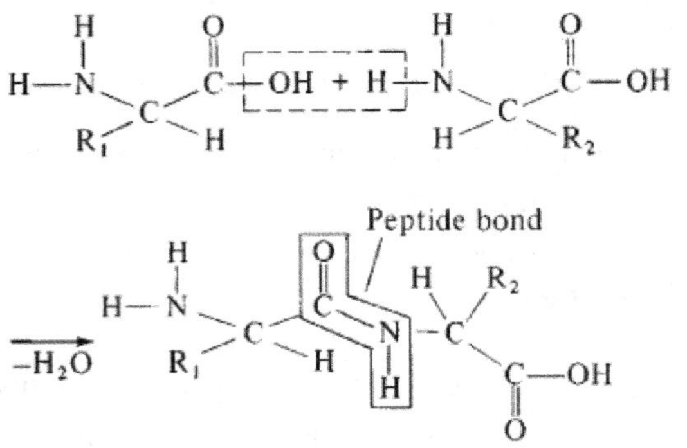

Peptide bonds form from the condensation reaction of 2 amino acids

10. C is correct.

Valine is the only amino acid listed with a hydrophobic (carbon and hydrogen only) side chain.

Group	Characteristics	Name	Example (-Rx)
non-polar	hydrophobic	Ala, Val, Leu, Ile, Pro, Phe Trp, Met	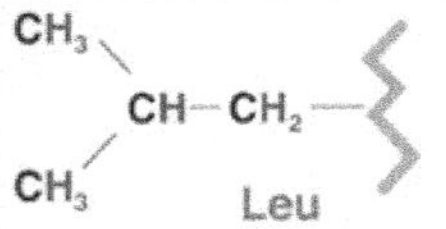 Leu
polar	hydrophilic (non-charged)	Gly, Ser, Thr, Cys, Tyr, Asn Gln	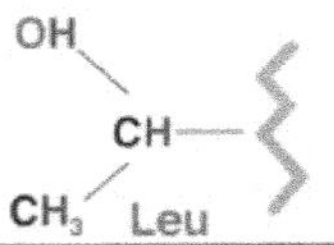 Leu
acidic	negatively charged	Asp, Glu	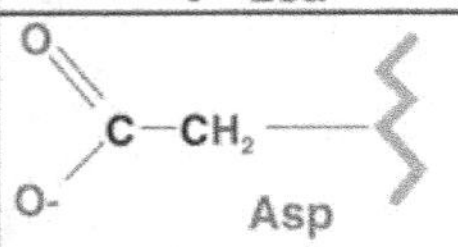 Asp
basic	positively charged	Lys, Arg, His	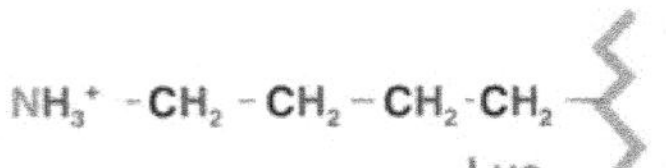 Lys

11. C is correct.

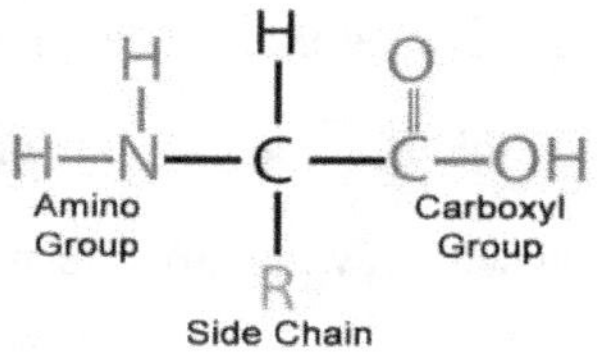

Amino acid – glycine

Amino acid with the amino group, carboxyl group, and α-carbon attached to the side chain (R group)

Glycine has no alkyl substituents at the *alpha* position (R = hydrogen) and is the smallest known amino acid.

12. D is correct.

The *isoelectric point* of an amino acid is determined by calculating the average pK_a values of the carboxylic acid and the ammonium cation.

When more than one ionizable group is present, the two pK_a numerically closer are used to calculate.

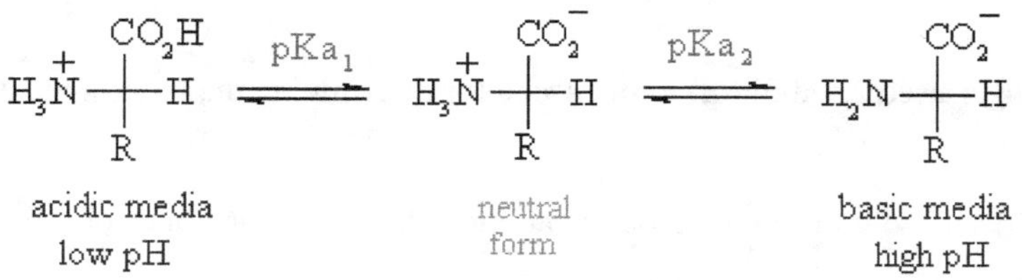

acidic media — low pH ⇌ (pKa_1) neutral form ⇌ (pKa_2) basic media — high pH

For acidic amino acids, the isoelectric point can be calculated by averaging the two lower pK_a values.

For basic amino acids, the two highest pK_a values are averaged to determine the isoelectric point.

Amino acid	α-CO₂H pK_a^1	α-NH₃ pK_a^2	Side chain pK_a^3	pI
Arginine	2.1	9.0	12.5	10.8
Aspartic Acid	2.1	9.8	3.9	3.0
Cysteine	1.7	10.4	8.3	5.0
Glutamic Acid	2.2	9.7	4.3	3.2
Histidine	1.8	9.2	6.0	7.6
Lysine	2.2	9.0	10.5	9.8
Tyrosine	2.2	9.1	10.1	5.7

pK_a values for the 7 amino acids with ionizable side chains

13. A is correct.

The amine N–H bond is acidic because the nitrogen atom is in conjugation with the carbonyl group.

The N–H bond of saccharine is much more acidic because the functional group is an amide and a sulfonamide.

The sulfone group (~SO_2) acts as an electron-withdrawing group to stabilize the negative charge generated from deprotonation.

The pK_a of saccharin is 1.6.

14. C is correct.

2,4-dinitrofluorobenzene is used to identify the *N*-terminus of a peptide. It reacts with the *N*-terminal amino acid and remains attached even after complete acid hydrolysis, which allows for isolation of the *N*-terminal amino acid from the rest of the polypeptide.

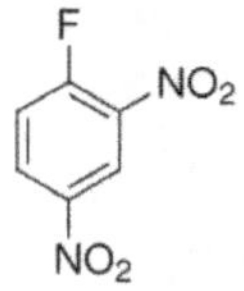

In 1945, Frederick Sanger used 2,4-dinitrofluorobenzene for determining the N-terminal amino acid in polypeptide chains.

Dinitrofluorobenzene reacts with the amine group in amino acids to produce dinitrophenyl-amino acids.

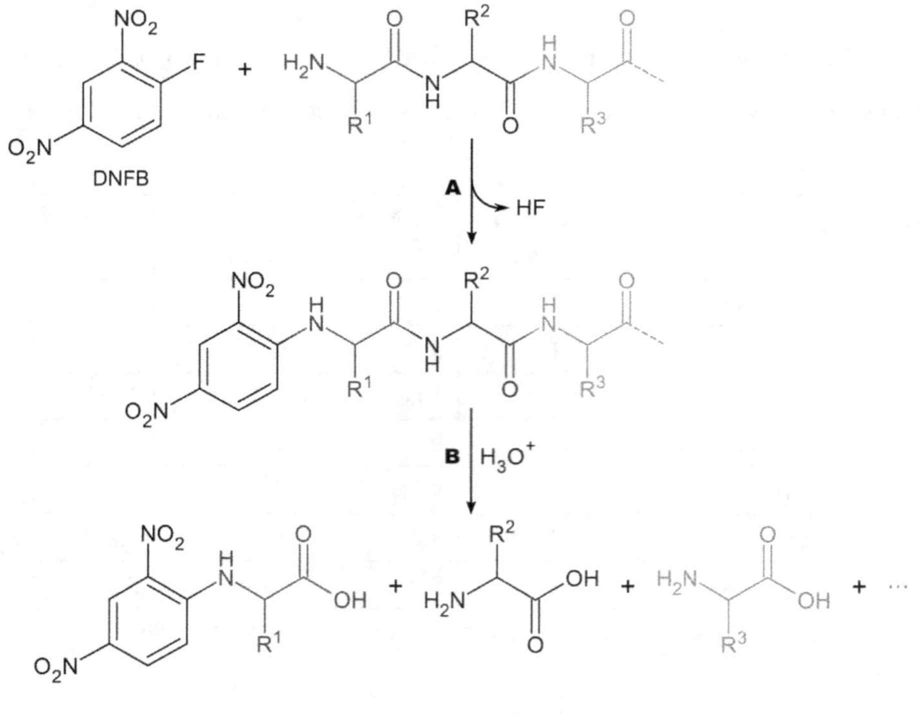

15. E is correct.

Hormones are substances secreted by a gland and released into the blood to affect a target tissue/organ.

Insulin is a hormone composed of amino acids (i.e., peptide hormones) and is a protein molecule. Insulin is composed of two peptide chains (A chain and B chain). *Two disulfide bonds* link the chains, and an additional disulfide is formed within the A chain. The A chain consists of 21 amino acids and the B chain – of 30 amino acids in most species.

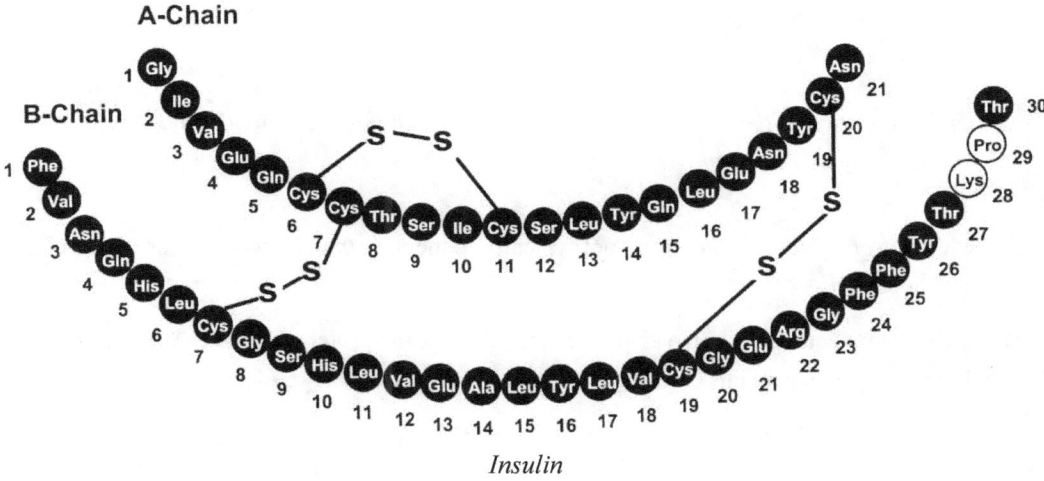

Insulin

Hormones can be lipid molecules, such as steroid derivatives (e.g., testosterone, progesterone, estrogen).

16. B is correct.

The *alpha helix structure* is one of the two common types of secondary protein structure. The alpha helix is held together by hydrogen bonds between every N–H (amino group) and the oxygen of the C=O (carbonyl) in the next turn of the helix; four amino acids along the chain.

The typical alpha helix is about 11 amino acids long.

The other type of secondary structure is the *beta-pleated sheet*.

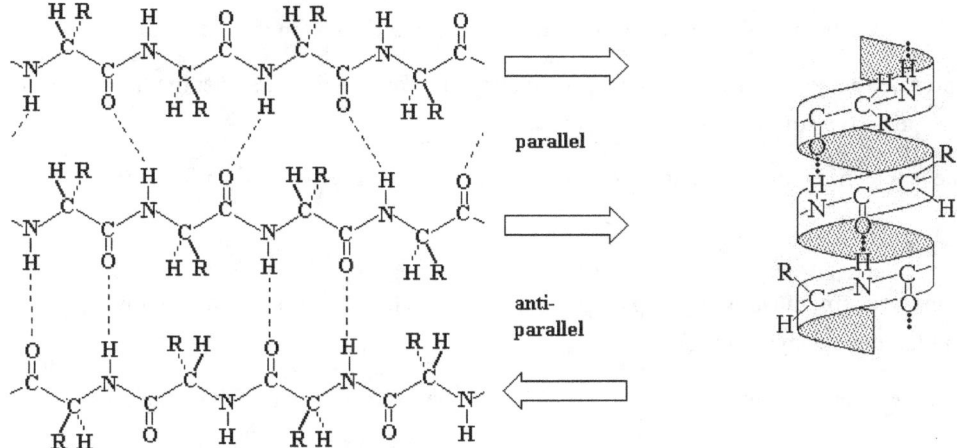

Beta pleated sheets are parallel or anti-parallel (a reference to the amino terminus).

17. C is correct.

A pH of 8 indicates a basic solution whereby the carboxylic acid is deprotonated, and the amino group remains protonated (until the pH reaches the pK_a of the amine at ≈ 9.8).

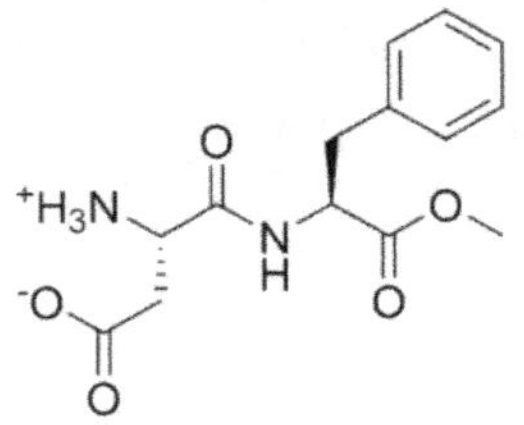

Zwitterion: carboxyl is deprotonated (negative), and the amino is protonated (positive)

18. D is correct.

Amino acids with *hydrophobic side chains* are not typically found near the external regions of proteins but rather in the interior regions.

Hydrophobic side group containing amino acids are located along the protein interfaces with lipid layers, such as the external amino acids of proteins embedded in lipid membranes.

19. B is correct.

The *isoelectric point* is the pH at which the net charge of the amino acid is 0.

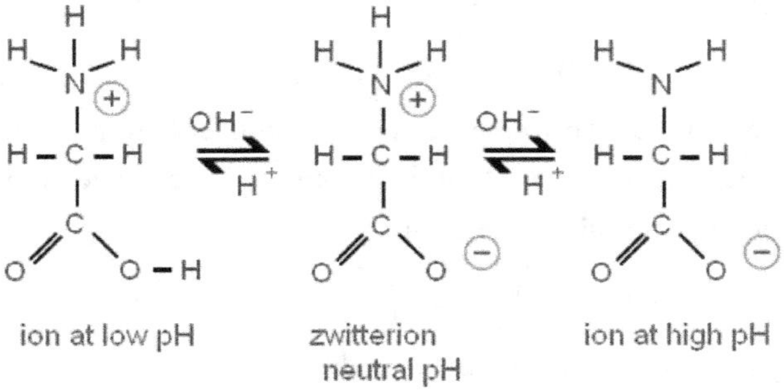

ion at low pH   zwitterion   ion at high pH
                       neutral pH

Aspartic acid is an amino acid that contains an additional carboxylic acid in the side chain.

For this amino acid to maintain an overall neutral polarity, the carboxylic acid must maintain its protonated form, which requires the solution to be acidic.

Therefore, the isoelectric point of aspartic acid is the lowest for the given list of amino acids.

20. D is correct.

In glutamate metabolism, the amino group of glutamic acid is removed through an oxidative deamination reaction with glutamate dehydrogenase.

Ammonia is released during this process and forms urea to be excreted as urine from the body.

21. C is correct.

The primary structure of a protein is the amino acid sequence, which is formed by covalent peptide linkages.

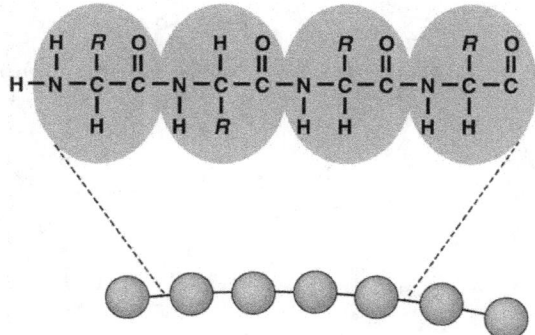

The amino acids (circles) are joined by covalent peptide bonds (lines)

A: only proteins containing more than one peptide subunit have a quaternary structure.

B: proteins are *denatured* by heating, and they lose their conformation above 35-40 °C, not retained.

D: many proteins contain more than one peptide chain (i.e., have a quaternary structure).

22. C is correct.

The reaction mechanism by which 2,4-dinitrofluorobenzene reacts with an amine is *nucleophilic aromatic substitution* (NAS).

The *nitro groups* are electron-withdrawing, and therefore, the benzene ring is electrophilic.

23. B is correct.

Amino acids are the building blocks of proteins.

Humans need 20 amino acids, some are made by the body (i.e., nonessential), and others must be obtained from the diet (i.e., essential).

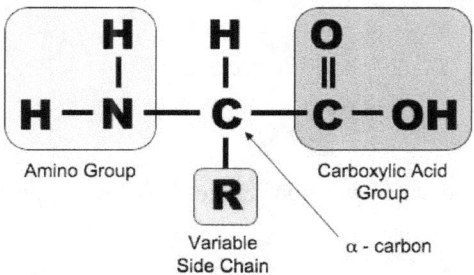

Amino acids contain an amine group, a carboxylic acid, an α-carbon, and an R group.

The following table is shown not for memorization but for identifying characteristics (e.g., polar, nonpolar) for the side chains.

continued…

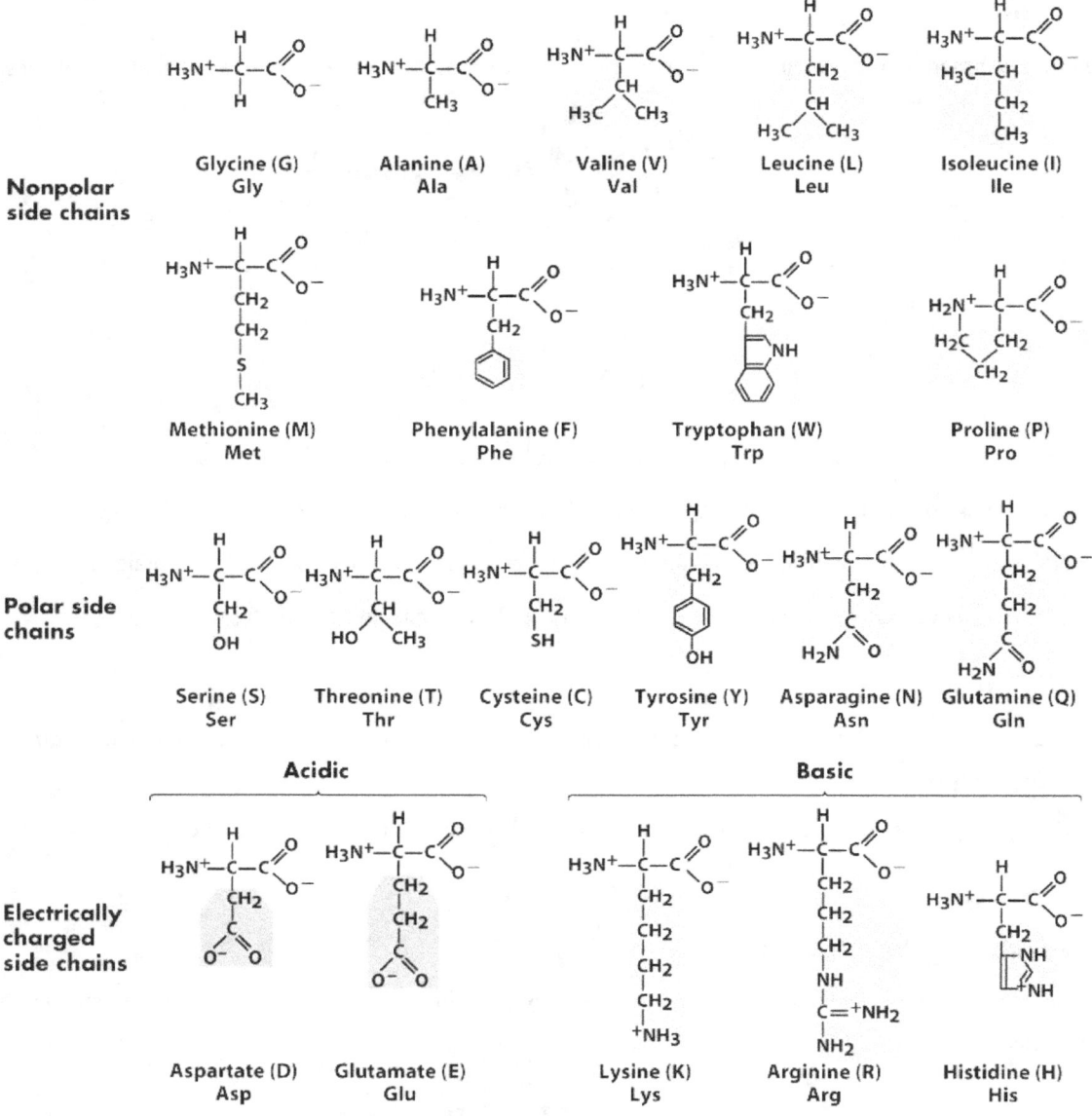

The 20 naturally occurring amino acids.

24. A is correct.

Human hair can be curly without reducing agents due to the cross-linking positions of the disulfide bonds.

The notation [O] represents oxidation (i.e., from reducing agents such as beta-mercaptoethanol), while [H] represents reduction (i.e., from an oxidation agent such as O_3).

The reducing agents cleave the disulfide bond linkages between peptides, enabling the hair to be restructured.

Detailed Explanations: Amino Acids, Peptides, Proteins

25. B is correct.

Peptide bonds link the primary sequence of amino acids in a protein. These bonds form between individual amino acids that then form a peptide chain.

Peptide bonds are not altered when a polypeptide bends or folds to form a secondary structure. New peptide bonds do not form when the polypeptide folds into a three-dimensional shape (i.e., tertiary structure).

A: interactions between charged groups (*electrostatic interactions*) can arise, especially in the tertiary structure.

C: *hydrogen bonds* participate in the secondary and tertiary structure of a protein. In the secondary structure, the polypeptide chain folds to allow the carbonyl oxygen and amine hydrogen to lie nearby. As a result, hydrogen bonding occurs to form sheets, helices, or turns. Likewise, hydrogen bonding may serve to stabilize the tertiary structure of a protein.

D: *hydrophobic interactions* are essential in the tertiary structure of a protein. For example, in an aqueous environment, the hydrophobic side chains of the amino acids may interact to arrange themselves towards the inside of the protein.

26. C is correct.

Amino acids are the basic building blocks for proteins.

Two amino acids (dimer) with peptide bonds indicated by arrows

The peptide bond is rigid due to the resonance hybrids involving the lone pair of electrons on nitrogen, forming a double bond to the carbonyl carbon (and oxygen develops a negative formal charge).

27. D is correct.

The *isoelectric point* is a pH when the amino and carboxyl end of the protein (or the side chains) results in a net neutral charge on the protein. Only charged species (i.e., anion and cations) migrate under the influence of an electric field in gel electrophoresis.

Ions are molecules that possess a certain charge, either positive (cation) or negative (anion).

The *zwitterion* (i.e., containing a positive and negative region within the same molecule) form is the dominant form at the isoelectric point, and it possesses two charges. However, the charges are opposite and cancel, forming an overall electrically neutral compound.

Neutral proteins do not migrate within the electric field during gel electrophoresis.

28. C is correct.

Protonation or deprotonation of an amino acid residue changes its ionization state: it may become positively or negatively charged or neutral. The process may lead to changes in the interactions among amino acid side chains, as some ionic bonds may be compromised from the lack of opposite charge pairing.

Certain hydrogen bonding interactions may be modulated if Lewis bases are protonated with Brønsted acids, impairing their ability to accept hydrogen bonds from nearby amino acid residues.

29. B is correct.

Amino acids contain amino (~NH$_2$) and carboxylic acid (~COOH) functional groups.

The α carbon is between the amino and carboxyl groups and is attached to the R (sidechain) group.

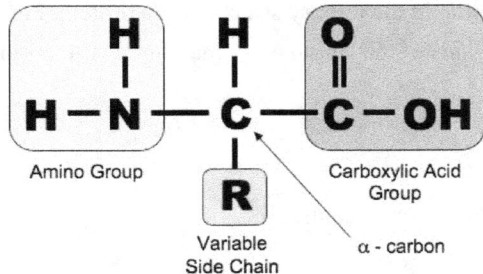

30. B is correct.

The π electron clouds of aromatic rings are not very basic because the aromatic molecule is stable.

Lone pairs on carbonyl oxygen (sp^2 hybridized) are more basic than π electrons but not as basic as the lone pair on nitrogen (sp^3 hybridized).

The nitrogen on an amine is more basic than the nitrogen on the amide because the electrons on the amide participate in resonance.

31. E is correct.

The amino acid *valine*:

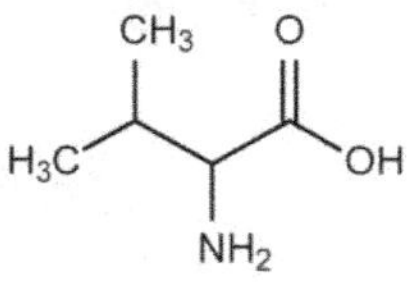

Valine has an R group (i.e., side chain) of isopropyl attached to the α carbon of the amino acid backbone.

The first step in the reaction sequence below is the Hell-Volhard-Zelinsky reaction (shown below).

continued…

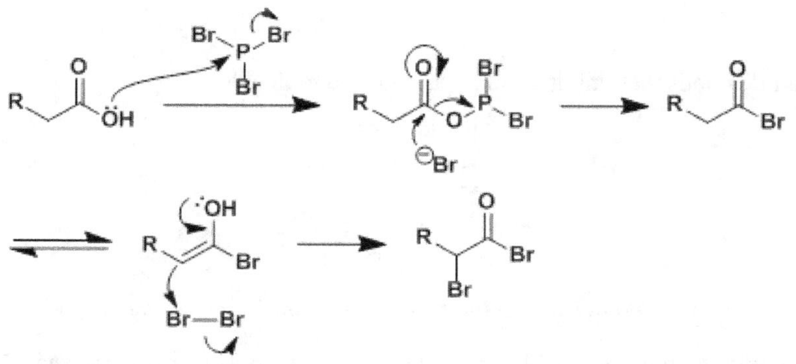

The mechanism for the Hell-Volhard-Zelinsky reaction

The phosphorus tribromide (PBr₃) is electrophilic and can be attacked by the carboxylic acid to generate the acid bromide.

The enol tautomer of this acid bromide can brominate at the alpha position to give the α-bromo acid bromide. The acid bromide can be subjected to water to hydrolyze the group to the carboxylic acid.

In excess ammonia, the *alpha* bromide is displaced to produce the valine amino acid product.

32. A is correct.

The primary structure of proteins consists of the linear sequence of amino acids (i.e., residues) in the polypeptide chain.

The secondary structure of proteins involves the regularly occurring motifs (alpha-helix and beta-pleated sheets) derived from intramolecular localized (within ten amino acids) hydrogen and disulfide bonding.

The tertiary structure involves amino acid R groups and interactions across different motifs (alpha helix, beta-pleated sheets, and loop structures).

The quaternary structure of proteins involves two or more polypeptide chains. The quaternary structure is maintained between the different polypeptide chains (e.g., 4 chains of 2 α and 2 β in hemoglobin) by hydrophobic interactions and disulfide bridges between cysteines.

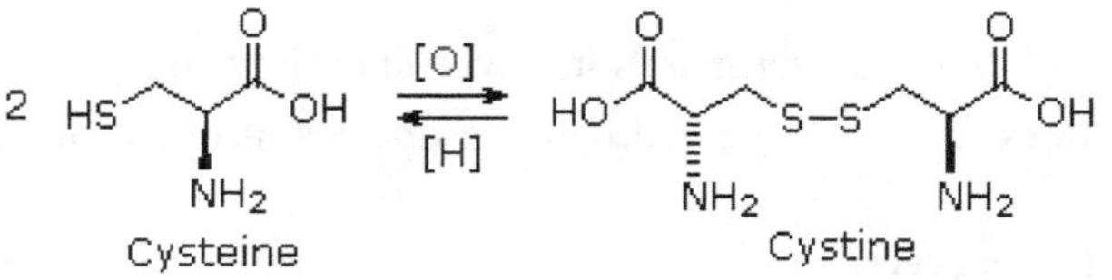

Two cysteine residues form a cystine covalent bond

33. A is correct.

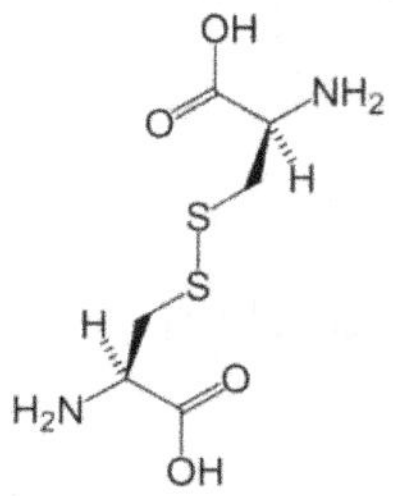

This covalent bond is a disulfide linkage that contributes to the protein's tertiary (3°) and quaternary (4°) structure (for multiple polypeptide chains).

Disulfide linkage of S–S covalent bond between two cysteine residues

34. A is correct.

Peptide bonds are amide bonds that link individual amino acid molecules.

Two amino acids with the peptide bond indicated by the arrow.

The peptide bond involves 4 atoms: C=O, N and H. The hydrogen must be antiperiplanar (180°) relative to the carbonyl oxygen to permit the lone pair on the nitrogen to participate in a resonance structure and confer rigidity on the peptide bond.

Resonance structure of peptide bond involving the lone pair of electrons on nitrogen.

The number of bonds between each amino acid (i.e., or monomers) equals n – 1 (where n is the number of monomers).

Therefore, 10 – 1 = 9 peptide bonds.

35. D is correct.

The primary structure of proteins refers to the linear sequence of amino acids.

Hydrogen bonding is important for the secondary (alpha-helix and beta-pleated sheet) and tertiary (i.e., overall, 3-dimensional shape) structures of proteins.

The *hydrophobic interactions* involved in tertiary and quaternary (i.e., two or more polypeptide chains) structures arise from the hydrophobic side chains of the amino acid residues.

36. E is correct.

The *isoelectric point* of an amino acid deals with the average pK_a of the acidic functional groups present in the molecule, including the ammonium and carboxylic acid functional groups.

The isoelectric point is a characteristic of the entire protein molecule (not just the side chain) where the net charge is zero.

37. E is correct.

All the described causes of protein denaturation induce changes in the intermolecular forces between the side chains of the residues or amino and carboxylic acid groups.

This denaturation is accomplished by breaking weak bonds or changing the polarity or charge character of key stabilizing groups.

38. D is correct.

None of the amino acids in the peptide chain contain R groups with charges.

The only charges in the molecule should be the ammonium cation ($^+NH_4$) and the carboxylate anion (COO^-).

39. C is correct.

The general structure of an amino acid (where R is the side chain):

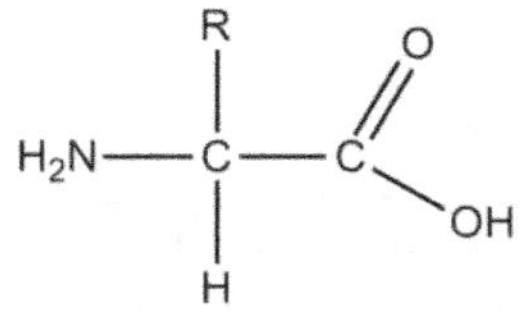

Amino acid – by convention, the amino terminus is drawn on the left.

The three amino acids with basic side chains are lysine (K), arginine (R), and histidine (H).

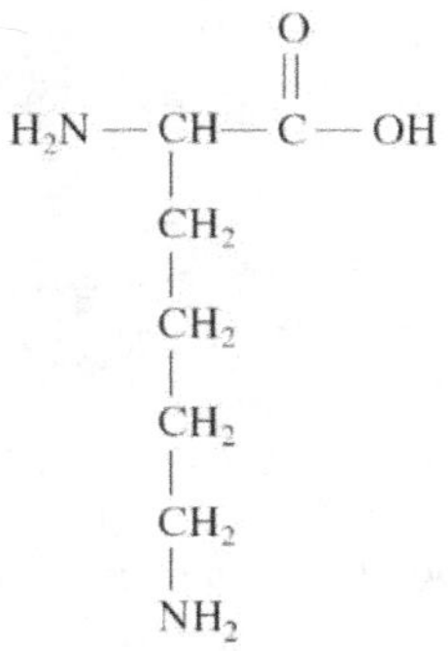

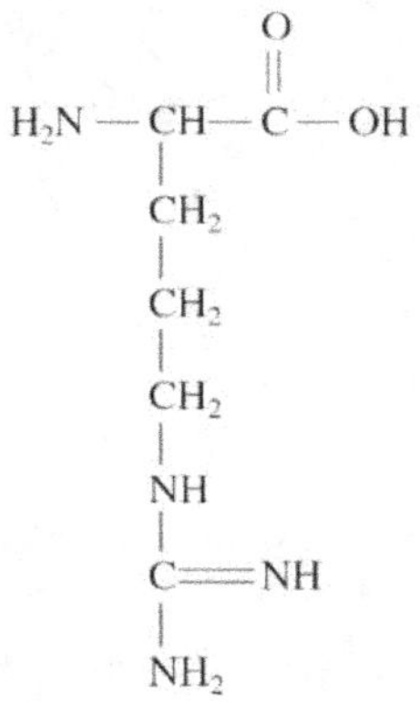

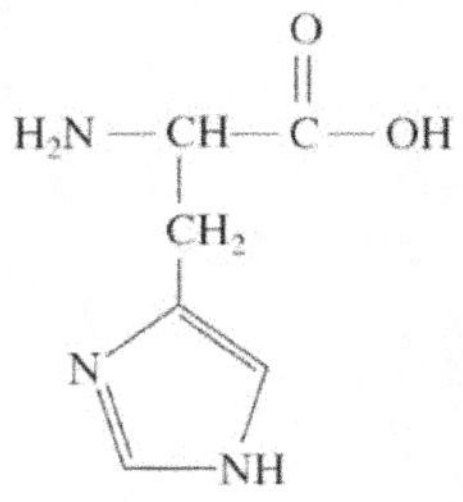

The side chain of threonine contains secondary alcohol.

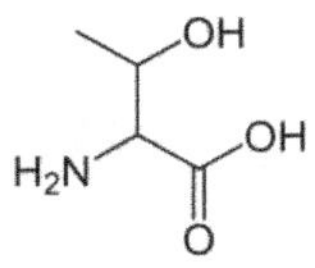

Threonine

Lysine

Arginine

Histidine

Alcohols can be protonated upon exposure to strong acids but are not basic at neutral pH.

40. A is correct.

A zwitterion is a neutral molecule with a positive and negative charge.

The isoelectric point of an amino acid is the average pK_a of the ammonium group and the carboxylic acid.

The isoelectric point is the pH where the carboxylate ion and ammonium cation are dominant in the solution.

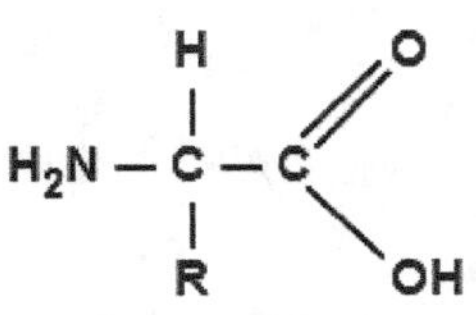

Amino acid

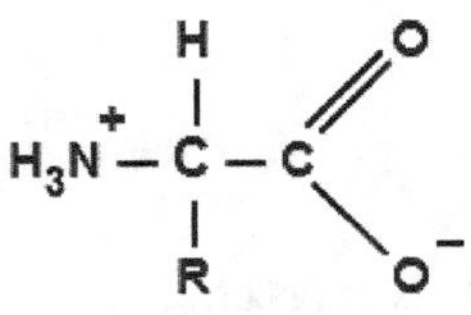

Zwitterion

41. D is correct.

Collagen is a protein that supports hair, nails, and skin.

Collagen is composed of a triple helix, and the most abundant amino acids in collagen include glycine, proline, alanine, and glutamic acid.

Much of the excess protein consumed in an animal's diet is used to synthesize collagen.

42. B is correct.

Secondary structure for proteins involves localized bonding.

The most important intermolecular interaction is hydrogen bonding, responsible for maintaining the alpha helix and beta pleated (parallel and antiparallel) sheet structures.

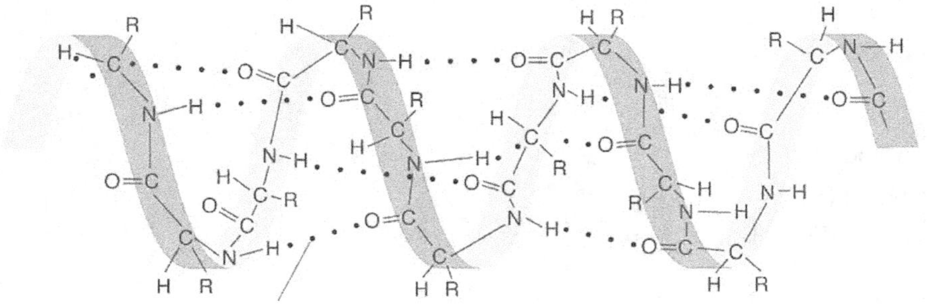

Alpha helix structure with hydrogen bonding shown as dotted lines

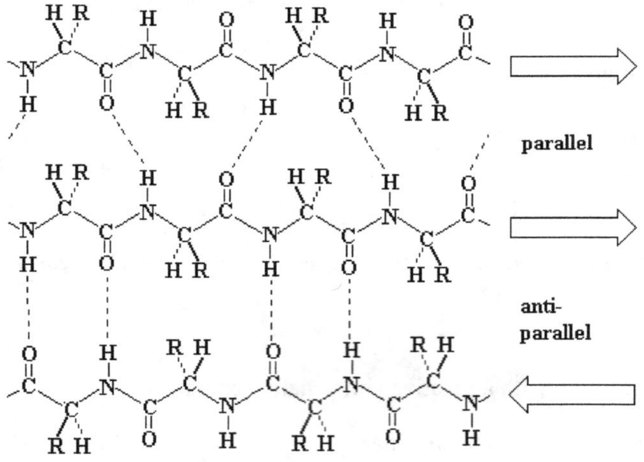

Beta pleated sheets (parallel and antiparallel) with hydrogen bonding shown

43. C is correct.

The standard conditions for breaking the covalent bonds during peptide hydrolysis are concentrated HCl and several hours of reflux.

The reaction time depends on partial or complete hydrolysis of the peptide.

44. B is correct.

Essential amino acids are obtained from the diet.

Nonessential amino acids can be synthesized by the body and do not need to be consumed.

Semi-essential (conditionally essential) amino acids can be synthesized within the body under special physiological conditions (e.g., in premature infants and under severe catabolic distress).

The nine essential amino acids for humans are histidine, isoleucine, leucine, lysine, methionine, phenylalanine, threonine, tryptophan, and valine.

The six conditionally essential amino acids for humans are arginine, cysteine, glycine, glutamine, proline, and tyrosine. The five nonessential amino acids are alanine, aspartic acid, asparagine, glutamic acid, and serine.

45. B is correct.

A *polypeptide chain* can undergo short-range bending and folding to form β sheets or α helices.

These structures arise as the peptide bonds can assume a partial double-bond character and adopt different conformations.

The arrangement of groups around the relatively rigid amide bond can cause *R* groups to alternate from side to side and interact with one another.

The carbonyl oxygen (C=O) in one region of the polypeptide chain could become hydrogen-bonded to the amide hydrogen in another region of the polypeptide chain. This interaction often results in forming beta-pleated sheets or an alpha helix.

Localized bending and folding of a polypeptide *do not* constitute a protein's primary structure.

A: *primary structure* of a protein is the amino acid sequence; individual amino acids are linked through peptide (i.e., amide) linkages.

C: *tertiary structure* is the 3-D shape that arises by further folding the polypeptide chain. Usually, these nonrandom folds give the protein a particular conformation and associated function.

D: *quaternary structure* is the spatial arrangement between two or more associated polypeptide chains (often linked by disulfide bridges between cysteine residues).

E: *zymogen* (or *proenzyme*) structure refers to an inactive enzyme that requires modification (i.e., hydrolysis to reveal the active site or change in configuration) to become an active enzyme.

46. B is correct.

Proteins are biological macromolecules composed of bonded amino acids with peptide (amide) bonds.

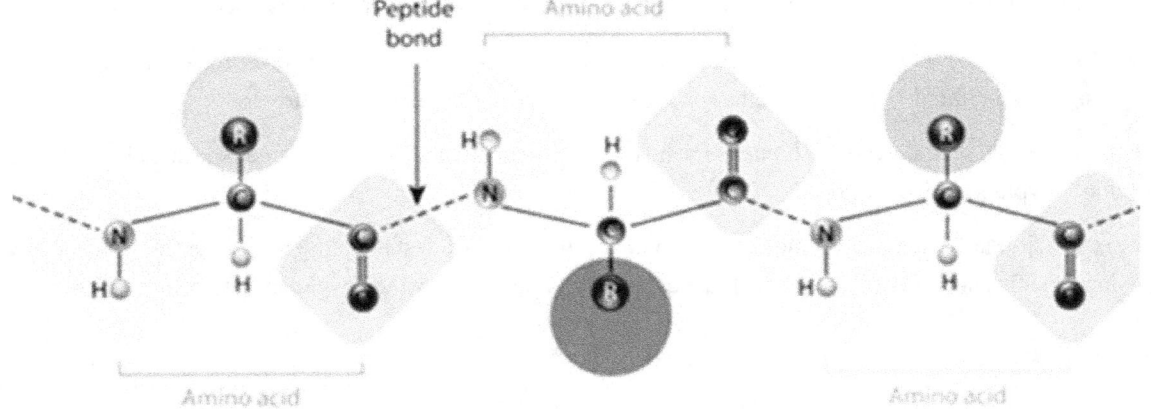

Three amino acids residues of nascent (i.e., growing) polypeptide

47. C is correct.

Sulfur-sulfur bonds are disulfide bonds. The bonds are formed from the dimerization of two cysteine residues.

Cysteine is the only molecule among the common amino acids that possess a thiol (or SH) group.

48. E is correct.

The amino group of the other common amino acids contains primary amine functional groups.

Proline contains a five-membered ring as part of its structure.

Notes for active learning

Notes for active learning

Lipids – Detailed Explanations

1. D is correct.

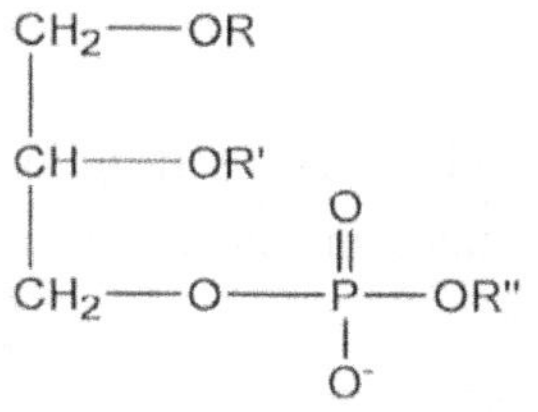

$$\begin{array}{l} CH_2-OR \\ | \\ CH-OR' \\ | \\ CH_2-O-\overset{\overset{\displaystyle O}{\|}}{\underset{\underset{\displaystyle O^-}{|}}{P}}-OR'' \end{array}$$

Glycerophospholipid

The ceramide lipid shown is a derivative of the triglyceride molecule. It is composed of glycerol, two fatty acid groups, and a phosphate group. The purpose of the phosphate group is to increase the polarity of the molecule so it can be used for cell membranes.

Sphingolipids (or glycosylceramides) are a class of lipids containing a backbone of sphingoid bases, a set of aliphatic amino alcohols that includes sphingosine. They are important for signal transmission and cell recognition.

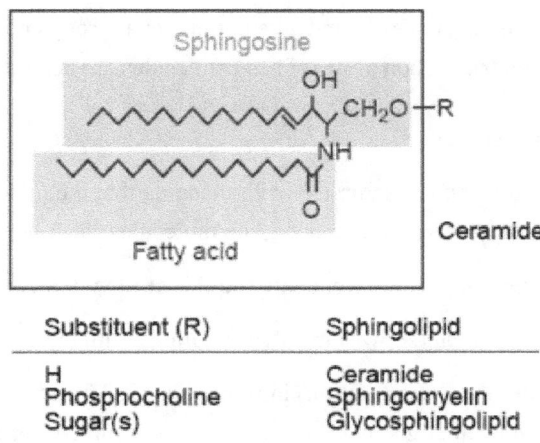

Sphingolipid (R group is: H, phosphocholine, sugars, ceramide, sphingomyelin, or glycosphingolipid)

Eicosanoids (e.g., prostaglandins and leukotrienes) are signaling molecules made by the oxidation of 20-carbon fatty acids. Eicosanoids are derived from ω-3 or ω-6 fatty acids.

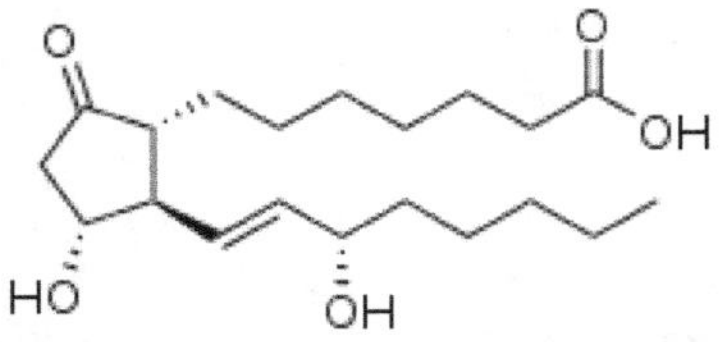

Prostaglandin E1 (with characteristic 5-membered ring)

Waxes are organic compounds that characteristically consist of long alkyl chains.

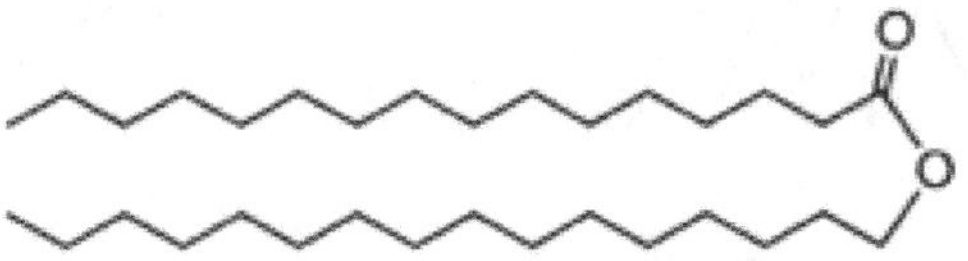

Wax

2. D is correct.

Fatty acids are biological molecules that contain long hydrocarbon groups.

These long hydrophobic chains can have alkene groups, but branching is not typically observed.

3. D is correct.

The plasma membrane is made of phospholipids (i.e., type of lipid). These molecules mostly possess nonpolar characteristics due to the long hydrocarbon chains, making the membrane permeable to nonpolar materials and semipermeable to polar or charged molecules.

4. A is correct. Fatty acids and triglycerides do not contain stereocenters.

In some fats, the ester and alkene functional groups possess a plane of symmetry and therefore do not possess asymmetric carbon atoms. These molecules are achiral and are not optically active.

5. D is correct.

Palmitic acid is a saturated acid, meaning that the molecule does not contain a double bond. Because this molecule lacks a double bond, the molecules stack better to form solids, and the fat has a higher melting point.

Unsaturated fats are frequently liquids at room temperature.

Linolenic acid (shown) is polyunsaturated with three *cis* double bonds.

Alkenes (i.e., unsaturation) introduce "kinks" in the chain that give the unsaturated fat an overall bent structure. This molecular geometry limits these fats from clustering closely to form solids.

Therefore, (relative to chain length), polyunsaturated molecules have the lowest melting point.

Position *omega* (ω) of the double bond(s) is the number of carbon atoms from the *terminal methyl group*.

Number of *carbon atoms* is noted from the carboxyl end

Palmitic Acid

Oleic Acid

Saturated and unsaturated fatty acids – note the omega (ω) position 9 of the oleic acid double bond

Linoleic Acid

Arachidonic Acid

Omega-6 polyunsaturated fatty acids. Linoleic is ω-6,9, and arachidonic acid is ω-6,9,12,15 fatty acids.

α-Linolenic Acid

CH_3 ⌇⌇⌇⌇⌇⌇⌇ COOH

Eicosapentaenoic Acid

CH_3 ⌇⌇⌇⌇⌇⌇⌇⌇ COOH

Docosahexaenoic Acid

CH_3 ⌇⌇⌇⌇⌇⌇⌇⌇⌇ COOH

Omega-6 polyunsaturated fatty acids. α-linoleic acid is ω-3,6,9, eicosapentaenoic acid is ω-3,6,9,12,15 and docosahexaenoic acid is ω-3,6,9,12,15,18.

Steroids (see below) are lipids used for cell signaling.

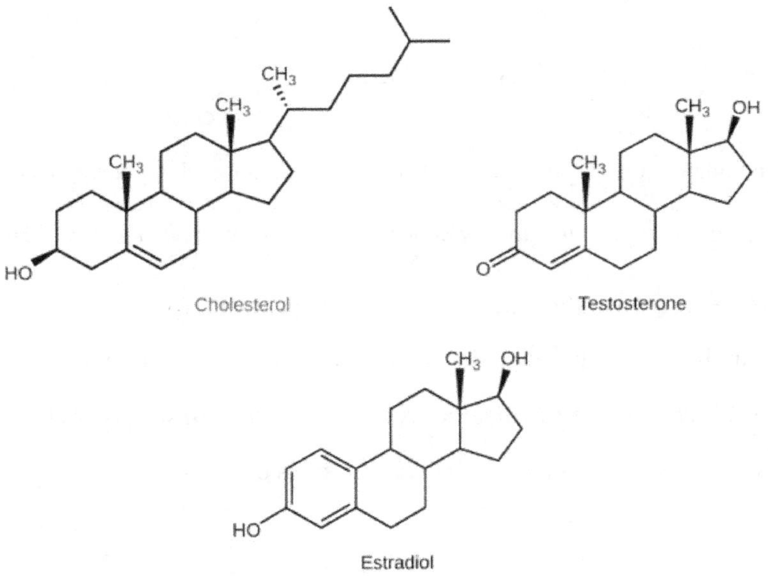

Cholesterol

Testosterone

Estradiol

Cholesterol is the precursor molecule for several steroid hormones such as testosterone, estradiol, progesterone, and aldosterone

6. D is correct.

As unsaturated fats become hydrogenated to form saturated fats, the melting point increases.

An equivalent of hydrogen is added across the double bonds of unsaturated fats during hydrogenation, increasing the molecular weight of the compound.

Furthermore, saturation helps to enhance the stacking ability of these compounds in the solid-state.

The more saturated the compound is, the higher its melting point.

7. B is correct.

The two ends of a fatty acid compound are the hydrophobic, nonpolar end composed of hydrogen and carbon atoms, and the hydrophilic, polar end of the molecule, which is composed of oxygen atoms and can hydrogen bond with water molecules.

8. B is correct.

Although cholesterol plays a pivotal role in the synthesis of other steroids and the integrity of cell membranes, too much cholesterol in the blood results in plaque deposits in blood vessels.

9. D is correct.

Steroid molecules are one of the two kinds of fat molecules, which are composed of fused rings.

Triacylglycerides are composed of long hydrocarbon chains with functionalized head groups.

10. C is correct.

Saponification is the base-promoted hydrolysis of esters. This type of hydrolysis (i.e., saponification) is typically used to form soap compounds.

$$\underset{\text{carboxylate ester}}{R-\overset{\overset{O}{\|}}{C}-O-R'} + \underset{\text{sodium hydroxide}}{NaOH} \rightarrow \underset{\text{sodium carboxylate}}{R-\overset{\overset{O}{\|}}{C}-O^-Na^+} + \underset{\text{alcohol}}{HO-R'}$$

Base hydrolysis (saponification) of an ester to form a carboxylate salt and an alcohol

Soap is hard or soft depending on the counter-ion of the carboxylate salt.

If the base used to hydrolyze the fat is sodium hydroxide, a hard soap is produced.

If potassium hydroxide is used for the base hydrolysis reaction, then a soft soap is produced.

Furthermore, other kinds of bases can give rise to these two kinds of soaps.

11. C is correct.

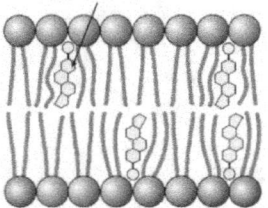

Cholesterol is a lipid molecule known as a steroid compound.

The fused ring structure of steroid molecules makes them rigid and has fewer degrees of motion due to the few conformations available for cyclic molecules *vs.* acyclic molecules.

Molecules such as phospholipids lack fused-ring structures, and they exist as straight-chained molecules.

Therefore, cholesterol in the cell membrane acts as a bidirectional regulator of membrane fluidity: at high temperatures, it stabilizes the membrane and raises its melting point, whereas, at low temperatures, it intercalates between the phospholipids and prevents them from clustering and stiffening.

12. B is correct.

Triglycerides are one of two common types of lipids in the body. This fat comprises four components, including one equivalent of *glycerol* (a triol) and *three fatty acid* molecules.

These components are combined to form the three ester groups of the triglyceride.

13. D is correct.

Saturated fats lack the alkene double bond. Unsaturated fats convert to saturated fats through a process of hydrogenation.

In this process, hydrogen gas is catalytically added to the alkene groups of the fatty acids to convert them to alkane groups.

14. D is correct.

The hydrogenation of unsaturated fats adds hydrogen across the double bonds of the fat. This causes the fats to be saturated and increases their melting points. Because of their ability to form solids, the consumption of hydrogenated fats should be limited for health.

15. B is correct.

Triglycerides (or triacylglycerides) are used for storage and exist in the adipose tissue of animals.

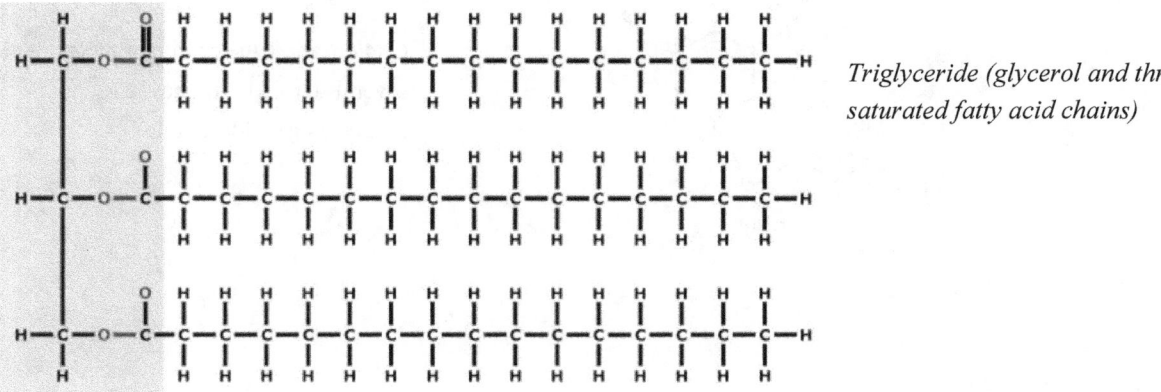

Triglyceride (glycerol and three saturated fatty acid chains)

Phospholipids (below) are the largest component of semi-permeable cell membranes

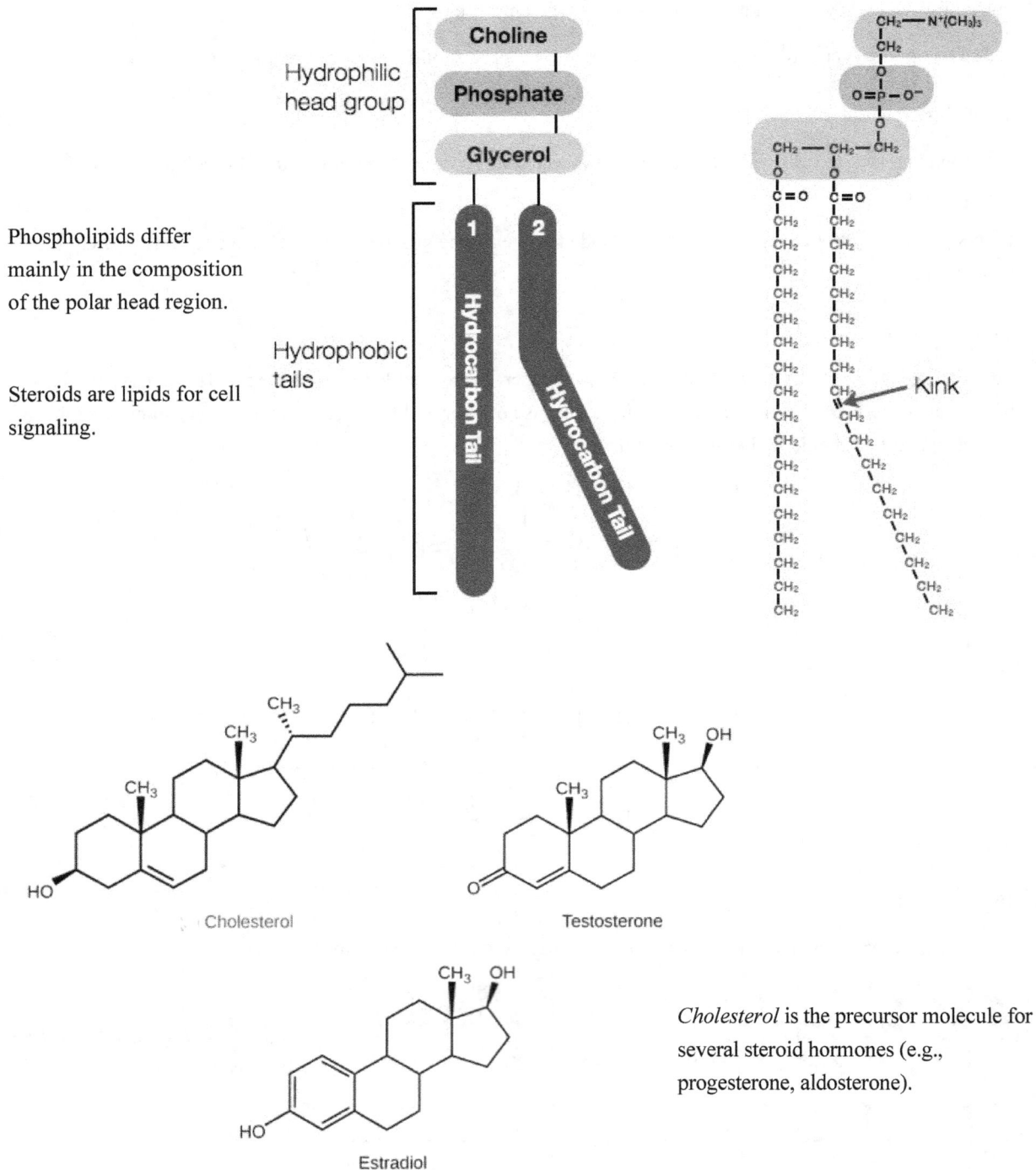

Phospholipids differ mainly in the composition of the polar head region.

Steroids are lipids for cell signaling.

Cholesterol is the precursor molecule for several steroid hormones (e.g., progesterone, aldosterone).

Detailed Explanations: Lipids

16. C is correct.

The straight-chain fatty acid that possesses the greatest number of alkenes in its structure is the fat that most likely is a liquid at room temperature.

Stearic acid is fully saturated, while linoleic acid has two double bonds.

The triacylglyceride of these fatty acids is a liquid (or oil) at room temperature.

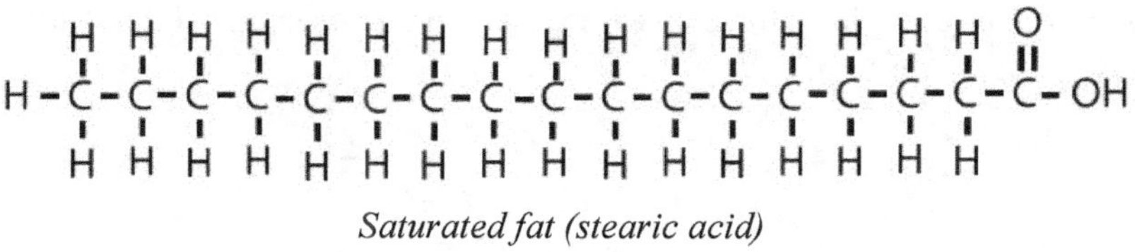

Saturated fat (stearic acid)

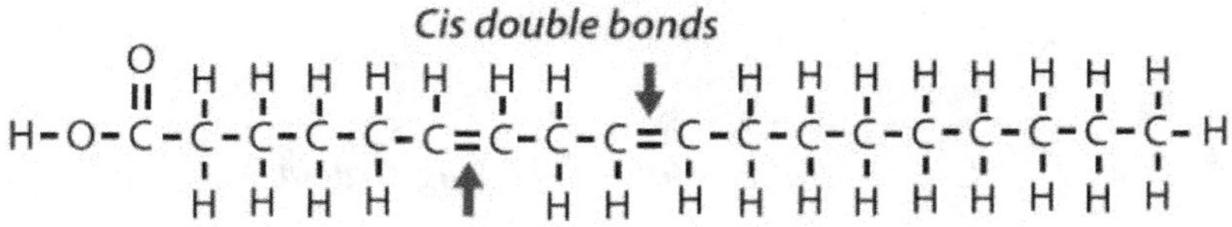

Unsaturated fat (linoleic acid)

17. B is correct.

An omega-3 is when the alkene (E/Z) double bond is three carbon atoms from the methyl end.

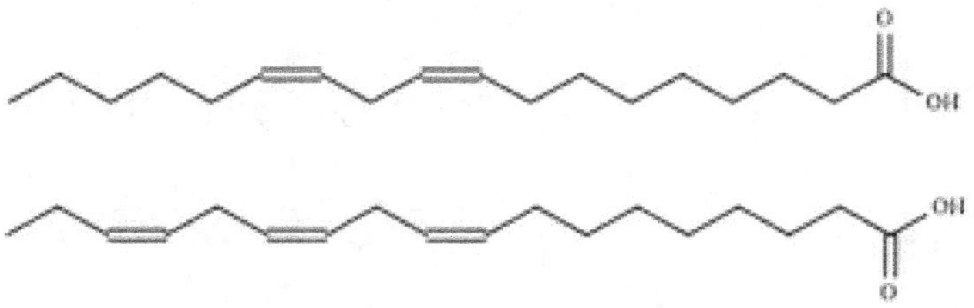

Linoleic acid (top) is ω-6,9 and linolenic acid (bottom) is ω-3,6,9. Note the positions of the double bonds and cis/trans (Z / E) relationships for double bonds

Linolenic acids are omega-3 fatty acids.

Linoleic acids are omega-6 fatty acids.

These molecules contain double bonds and are *unsaturated fats*.

18. D is correct.

Estradiol (shown) is a steroid hormone derived from cholesterol.

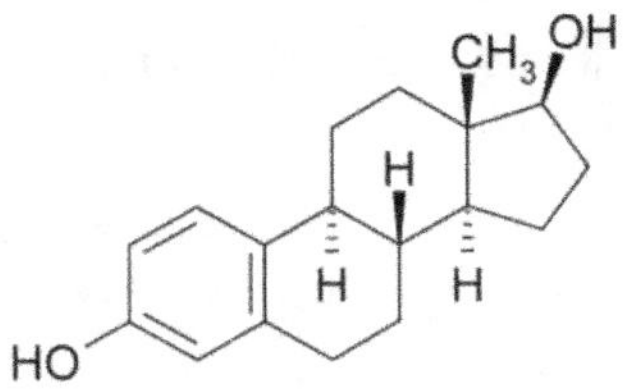

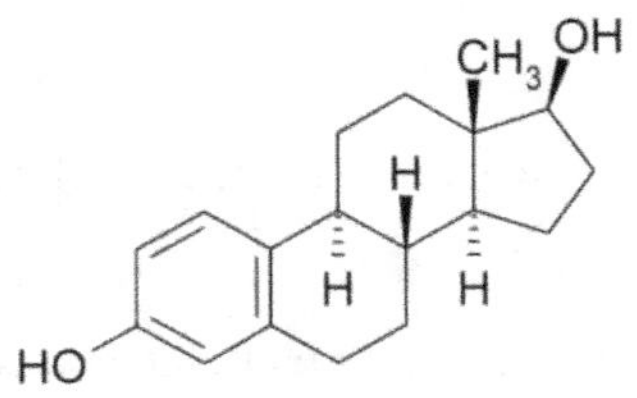

Estradiol is a derivative of cholesterol

Cholesterol is a lipid of four fused rings; three fused rings are six-membered, and the fourth is five-membered.

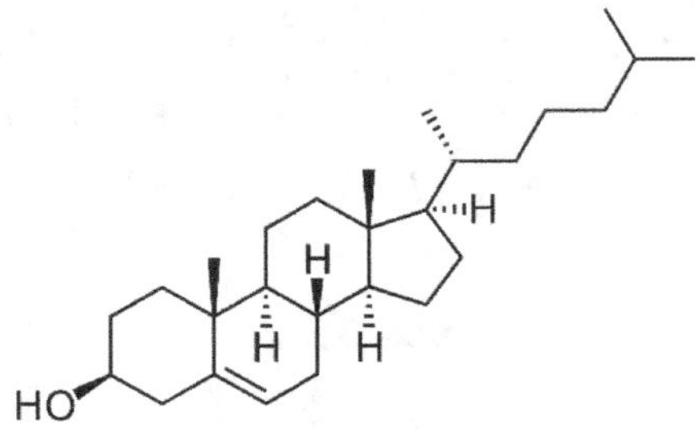

Cholesterol is a four-fused ring structure

Cholesterol is a steroid that makes up one of two types of lipid molecules.

Triglyceride (i.e., glycerol backbone with three fatty acid chains) is the other lipid (shown below).

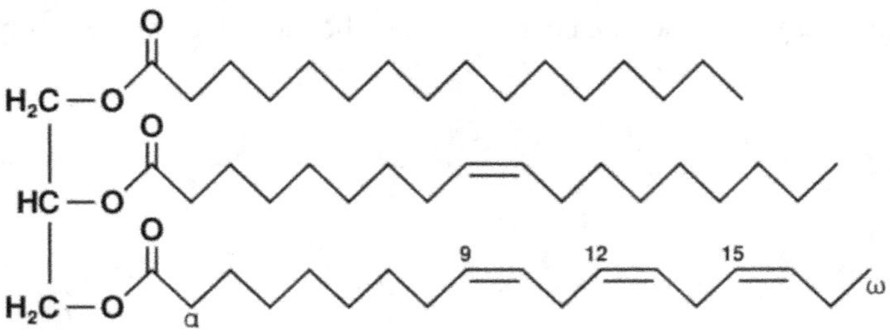

Triglyceride is glycerol backbone with three fatty acid chains

The carbon chain is numbered (example above) from the carboxyl end. Chemists number the double bonds as shown. Nutritionists specify the ω position from the terminal methyl group.

Position *omega* (ω) of the double bond(s) is the number of carbon atoms from the *terminal methyl group*.

19. C is correct.

When micelles form in aqueous environments, the hydrophobic tails cluster. The polar head regions of these molecules are exposed on the surface of the micelle and exposed to the aqueous environment.

20. D is correct.

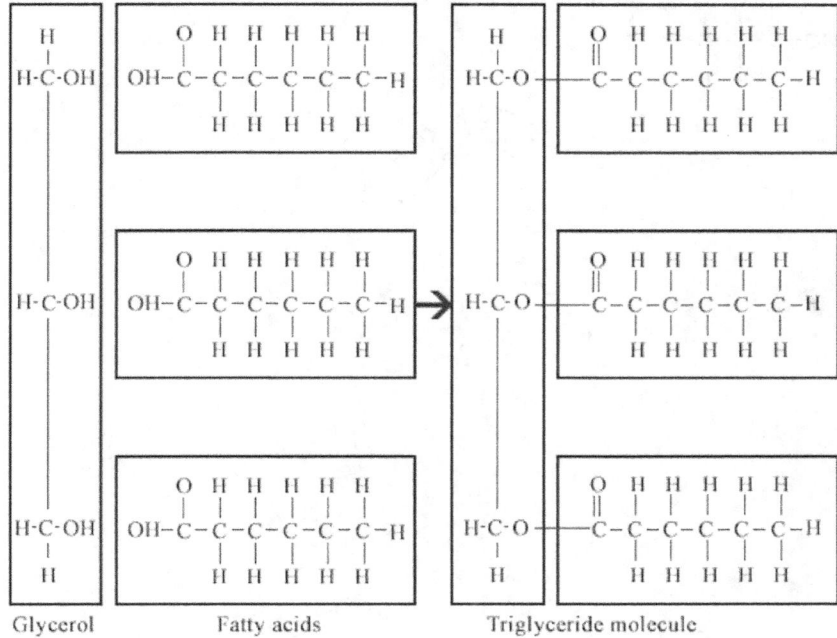

Triacylglycerides (or triglycerides) are made from one equivalent of glycerol and three equivalents of fatty acids.

The molecule contains a three-carbon chain that is flanked by oxygens. The three fatty acids (–COR chains) are attached *via* an ester linkage.

A triester is produced when condensation (*via* dehydration) occurs between the glycerol and three fatty acids.

21. B is correct.

Not all lipids are entirely hydrophobic. The ionic and polar heads of soaps and phospholipids, respectively, enable the molecules to interact with aqueous or polar environments.

The bulk of these molecules are hydrophobic because they largely consist of hydrocarbon chains or rings.

22. A is correct.

For unsaturated fats, the molecules are more likely to exist as oils (i.e., liquids) at room temperature.

Saturated fats tend to be solid at room temperature because the reduced forms of these molecules have better stacking properties, which allow them to form solid states.

23. B is correct.

The alkene molecules typically found in fatty acids tend to be *Z* alkenes.

Saturated Fatty Acid

The double bond of the alkene (i.e., unsaturated) prevents the fatty acid molecules from stacking closely. The result is lowering of the melting point for the fat molecule. This may influence the state of matter of the oil, as unsaturated fats tend to be liquid at room temperature, and saturated fats tend to be solids.

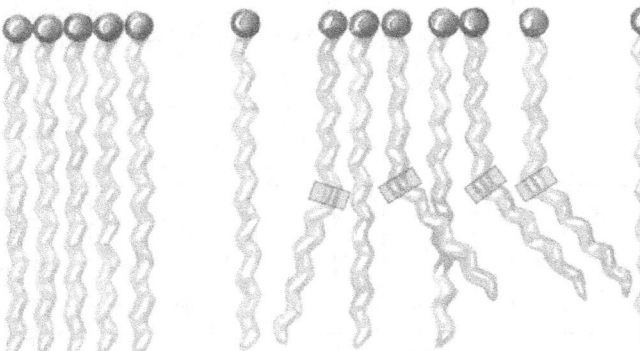

24. D is correct.

For omega acids, the number indicates the position of the first double bond from the non-carboxylic acid end (i.e., near the alkyl chain end) of the molecule (hence the "omega" part of the name).

Examples of omega-3 and omega-6 fatty acids.

25. C is correct.

Triacylglycerols are composed of a glycerol substructure and three fatty acids condensed to form a triester.

26. E is correct.

Triacylglycerol is a lipid that can undergo saponification, which is the breakdown of fatty esters with bases, such as sodium hydroxide or potassium hydroxide.

Saponification reactions are base promoted because sodium hydroxide is consumed in the reaction and not regenerated.

The amount of base needed to saponify 1 gram of fat is the *saponification number*.

27. A is correct.

In *esterification reactions*, the ester is made from the condensation of carboxylic acids and alcohols.

Fats are formed from the condensation (*via* dehydration) of fatty acids and glycerol.

For the reaction to occur, an acid catalyst (or enzyme) normally needs to be used. An acid catalyst is needed because carboxylic acids are not electrophilic, and the group needs to be activated by protonation.

Protonation of the carboxylic acid carbonyl oxygen atom causes the carbon-oxygen dipole to increase, allowing alcohol to add to the group.

Protonation of a geminal hydroxyl group and loss of the group as water leads to ester formation.

28. A is correct.

Oils are isolated from plant sources and may consist of several different fatty acids. These oils may contain saturated fat, but other fats present in the mixture are unsaturated fat molecules.

The saturated fats that may be present in oil are palmitic and stearic acid, and the unsaturated fats found in these oils include oleic, palmitoleic, and linoleic acids, as well as others.

29. B is correct.

There are two overall categories of lipids: long-chain lipids (e.g., triglycerides) and smaller, polycyclic lipids, such as steroids (e.g., cholesterol and its derivatives, such as estrogen and testosterone).

Lipid molecules are fat-soluble, and these molecules are largely soluble in organic/hydrophobic.

30. D is correct.

In hydrogenation reactions, *Z* and *E* alkenes are reduced to alkanes, and this process is catalyzed by transitions metals, such as nickel or palladium.

In this reaction, hydrogen (H_2) is added across the double bond of the alkene.

31. E is correct.

Many unsaturated fats are *omega fats* because they contain alkene groups.

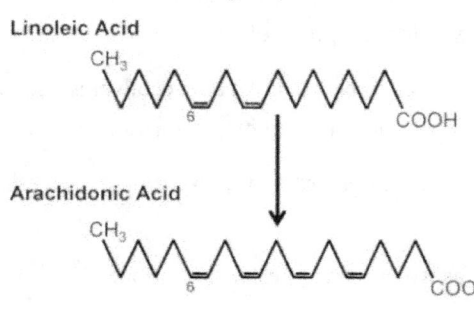

Sample fatty acids showing the positions of double bonds

Omega-3 fats have double bonds that appear three carbon atoms from the methyl end.

Omega-6 fatty acids contain double bonds six carbon atoms away from the methyl end. The double bonds tend to have Z (*cis*) geometry.

32. C is correct.

Long alkyl chains present on alkoxy portion and the carboxy backbone of the ester are characteristic features of waxes.

33. A is correct.

Glycerol can condense with fatty acids to expel water.

The new functional group produced is an ester joined by a glycosidic bond.

Condensation reactions (*via* dehydration) of glycerol and three fatty acids showing a glycosidic (ester) linkage.

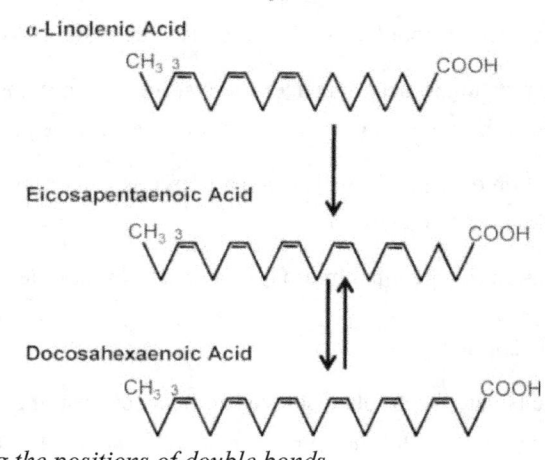

Detailed Explanations: Lipids

34. D is correct.

Amphipathic refers to molecules that possess hydrophobic and hydrophilic elements.

Biological molecules, such as fatty acids and some amino acids, have hydrophobic and hydrophilic regions.

A: *amphoteric* substance can function as an acid or a base, depending on the medium. Examples include metal oxides or hydroxides, which are amphoteric depending on the oxidation state of the element.

B: *enantiomeric* compounds are chiral molecules whose molecular structures have a non-superimposable mirror image relationship to each other.

C: *amphiprotic* molecules can donate or accept a proton (H^+), depending on the conditions (e.g., amino acids).

E: *allotropic* is the property of some elements to exist in two or more different forms, in the same physical state because the atoms are bonded differently (e.g., carbon exists as graphite, diamond, and fullerene).

35. B is correct.

Lipids (i.e., fats) are long-chain ester-linked molecules such as triglycerides. The term "lipid" is sometimes used interchangeably with the word "fat," however, lipids include cyclic biomolecules, such as steroids (e.g., cholesterol and its derivatives such as estrogen and testosterone).

Triacylglycerol is a lipid that can undergo saponification, breaking fatty esters with bases, such as sodium hydroxide or potassium hydroxide.

D: terpenes are small alkene-containing hydrocarbon building blocks that can be combined and cyclized to form steroids. Terpenes are simple lipids.

E: steroids do not contain ester groups, so they cannot be hydrolyzed to form soaps.

36. D is correct.

Fatty acids are long molecules with a hydrophobic chain and a hydrophilic region terminating in a carboxylic acid.

Myristic acid: an unsaturated 14 carbon fatty acid

Soaps are formed from the hydrolysis of fatty acids under basic conditions, known as saponification.

The positively charged counterion (e.g., Na^+, K^+) of the hydroxide base added to the reaction becomes the counter ion of the soap.

A: *emollient* is a topical agent designed to increase the hydration of the epidermis softer.

B: *ester* is a functional group in organic chemistry: R–O–R'.

37. B is correct.

Amphipathic refers to molecules with hydrophobic and hydrophilic elements (e.g., detergents, phospholipids of biological membranes).

Lipid molecules are common examples of amphipathic compounds because they possess hydrophobic tails and polar heads. The polar head group is often composed of electronegative heteroatoms, such as oxygen and nitrogen atoms.

38. D is correct.

Saturated fats tend to be solid at room temperature because they lack alkene groups. The alkene groups in fat molecules lower the melting point for these compounds.

For example, butter is a dairy product made from the fat of cow's milk, and it is solid at room temperature and is mainly composed of saturated fat molecules.

39. C is correct.

Although *waxes* are lipid molecules that contain esters as part of their structure, waxes only contain a single ester functional group.

A *monoalcohol* is used to form waxes, whereas glycerol is used to form triglycerides and phospholipids.

40. D is correct.

Fatty acids are used to make fat molecules known as triglycerides.

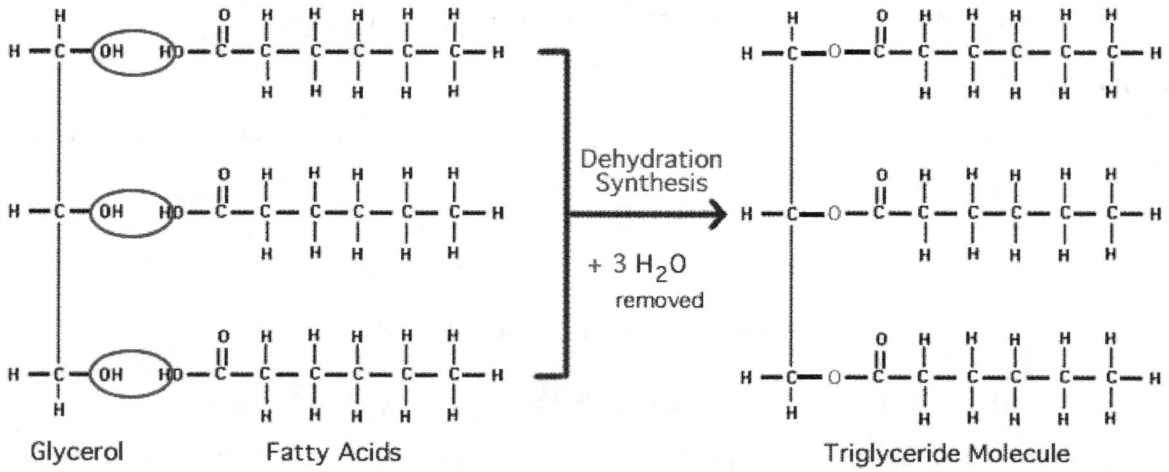

Formation, via dehydration (removal of H_2O), of triglyceride from glycerol and 3 fatty acids.

Fatty acids are made from one equivalent of a triol known as glycerol and three equivalents of acid-containing groups of fatty acids.

41. E is correct.

Dietary triglycerides are composed of glycerol and three fatty acids.

The hydrolysis of triglycerides yields glycerol and three fatty acid chains.

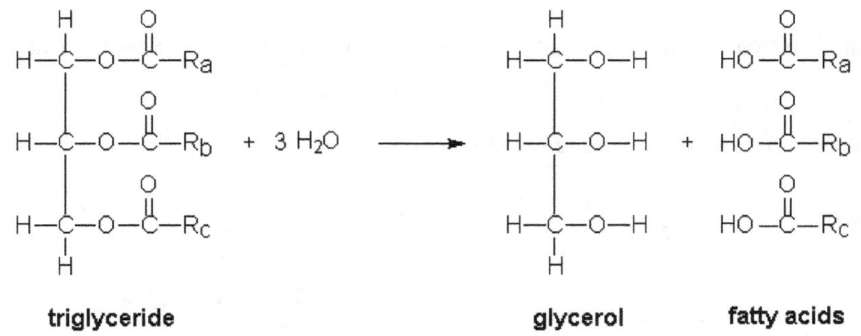

Hydrolysis of a triglyceride

42. B is correct.

The *cis* double bond of unsaturated fatty acids causes these molecules to stack less efficiently and less tightly.

The melting point for these compounds is lower than for saturated fats, and therefore, most unsaturated fatty acids are liquids at room temperature.

The double bonds indicate unsaturated fatty acids.

The unsaturated fatty acid contains a *cis* double bond at the 6th position.

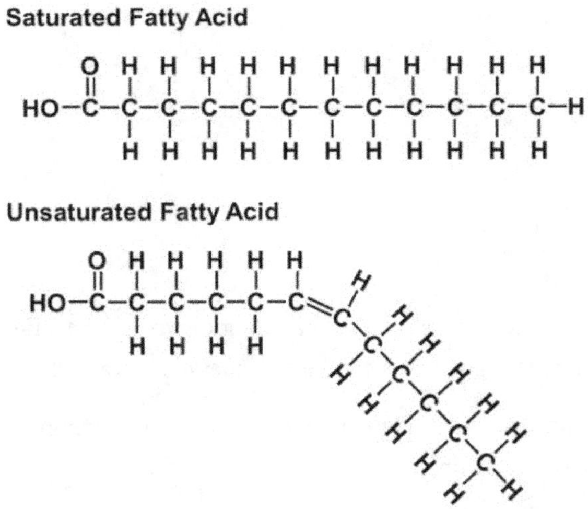

43. C is correct.

Glycerol is alcohol that possesses three hydroxyl groups; one hydroxyl group is bonded to each of the carbon atoms of glycerol. Glycerol is, therefore, a triol molecule, and this molecule is a critical component in the structure of triglyceride molecules.

44. C is correct.

Phospholipids are important lipids that make up the bilayer structure of the membranes of cells, organelles, and other enclosed cellular structures.

Phospholipids are two (same or different) fatty acid molecules, a phosphate group, and a glycerol backbone.

The phospholipid contains a hydrophobic (i.e., fatty acid tail) region and hydrophilic (polar head) region.

The hydrophobic regions point toward each other in the membrane bilayer, while the polar heads point towards the inside (i.e., cytosolic) or outside (i.e., extracellular) sides of the bilayer.

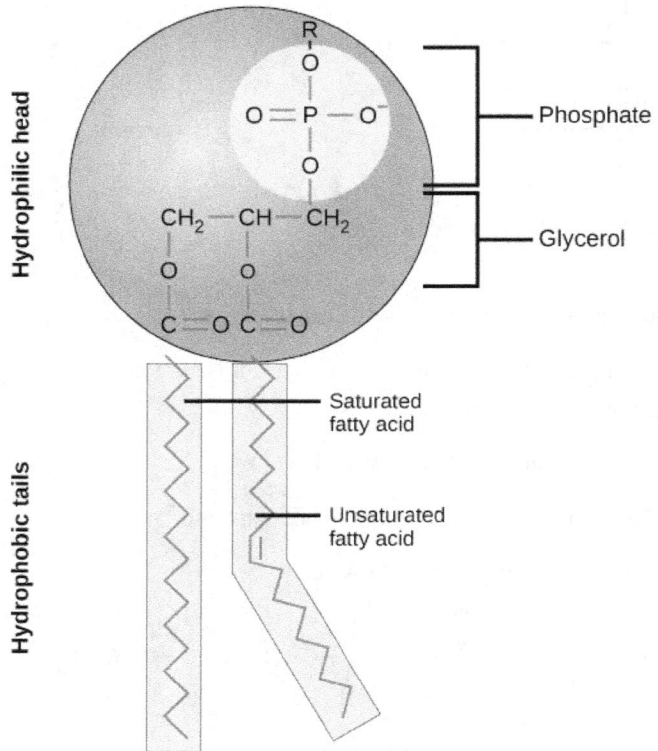

45. C is correct.

Fatty acid molecules are composed of long alkyl chains that do not involve branching. This is because each two-carbon segment of the fatty acid straight-chain can be enzymatically converted to an acetyl-CoA derivative.

Branching in the alkyl chain may impede the oxidative degradation of these compounds because a beta-ketone must be accessed before cleavage of the chain.

Secondary alkyl groups can be oxidized to alcohol but cannot be converted to ketones for this step.

46. D is correct.

Lipids (i.e., fats) is a term used to describe long-chain ester-linked molecules such as triglycerides. The term "lipid" is sometimes used interchangeably with "fat," however, lipids include cyclic biomolecules, such as steroids (e.g., cholesterol and its derivatives such as estrogen and testosterone).

Glycerol is a 3-carbon chain with three hydroxyl groups. Each hydroxyl undergoes a condensation reaction with the carboxylic acid end of a fatty acid chain to form a lipid.

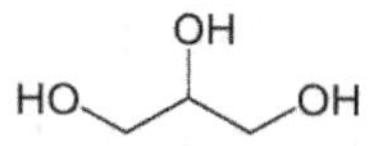

Glycerol

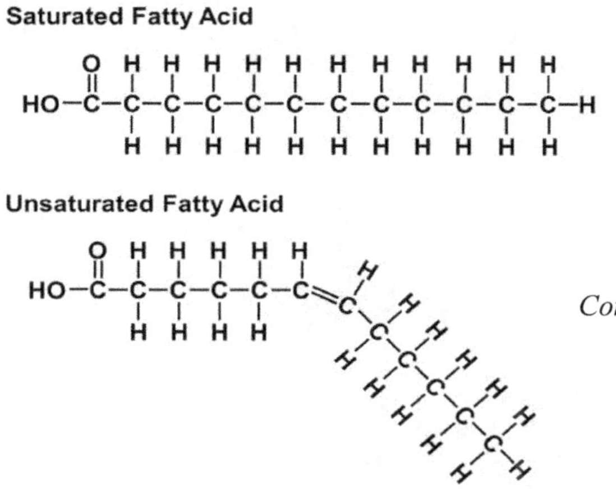

Saturated Fatty Acid

Unsaturated Fatty Acid

Comparison of saturated and unsaturated fatty acids

The unsaturated fatty acid contains double bonds, and the unsaturated fatty acid is a *cis* alkene in the example.

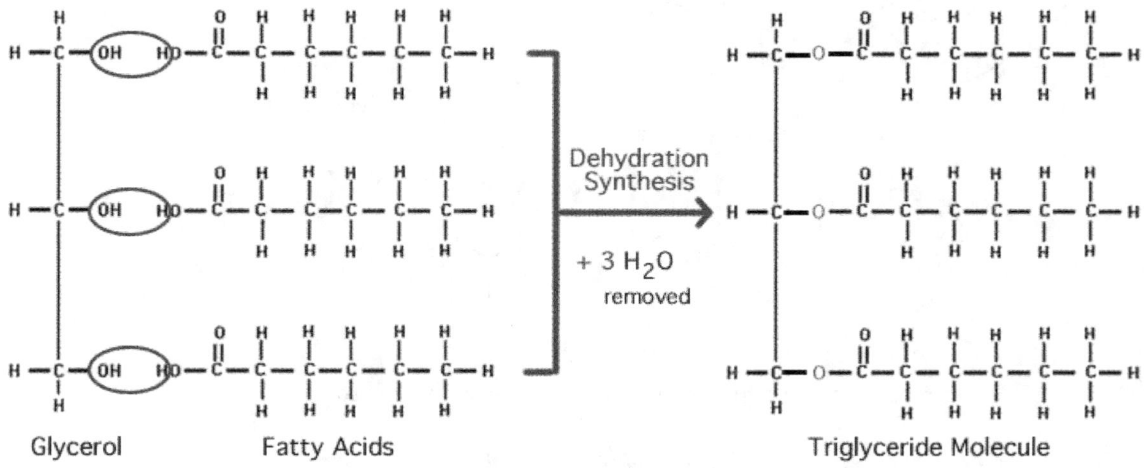

Glycerol Fatty Acids → Triglyceride Molecule

Dehydration Synthesis + 3 H$_2$O removed

The formation, via dehydration (removal of H$_2$O), of triglyceride from glycerol and 3 fatty acids.

47. D is correct.

Fat can be produced from glucose, but glucose is not produced from animal fat.

Excess glucose consumption can lead to increased levels of fat in the body.

48. D is correct.

There are two overall categories of lipids: long-chain lipids (e.g., triglycerides) and smaller, polycyclic lipids, such as steroids (e.g., cholesterol and its derivatives such as estrogen and testosterone).

Terpenes are small alkene-containing hydrocarbon building blocks that can combine and cyclize to form steroids.

Terpenes are simple lipids.

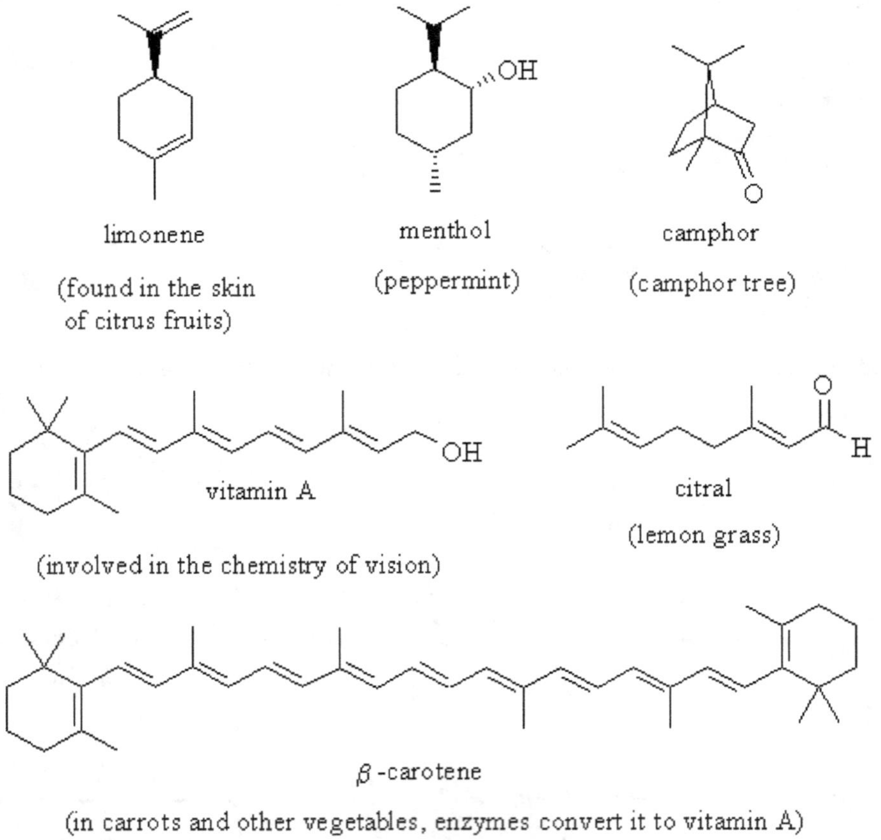
OH
O
limonene
(found in the skin of citrus fruits)
menthol
(peppermint)
camphor
(camphor tree)
OH
vitamin A
(involved in the chemistry of vision)
O
H
citral
(lemon grass)
β-carotene
(in carrots and other vegetables, enzymes convert it to vitamin A)

Examples of terpenes

Carbohydrates – Detailed Explanations

1. C is correct.

The conversion of one cyclic glucose stereoisomer to another involves the formation of an acyclic intermediate.

When the molecule cyclizes again, the opposite face of the aldehyde is accessed to generate the other cyclic diastereomer.

2. D is correct.

Glucose is an example of reducing sugar because it contains an aldehyde that can undergo reduction. The reduction of its aldehyde produces a sugar known as sorbitol. This is a way for the body to reduce blood glucose levels and is typically observed in diabetic patients with high blood sugar levels.

3. C is correct.

Sugar molecules typically contain hydrogen, carbon, and oxygen atoms.

Lipids, such as phospholipids, may contain other heteroatom groups (e.g., phosphorus atoms).

Aminosaccharides are sugars with nitrogen atoms in their structure.

4. C is correct.

It may help to use a molecular model. If the hemiacetal is converted to the aldehyde, then the acyclic form of this molecule is represented by the indicated structure.

5. D is correct.

Because amylose is a polysaccharide that does not have a β(1→4) glycosidic linkage, this polysaccharide can be digested by humans.

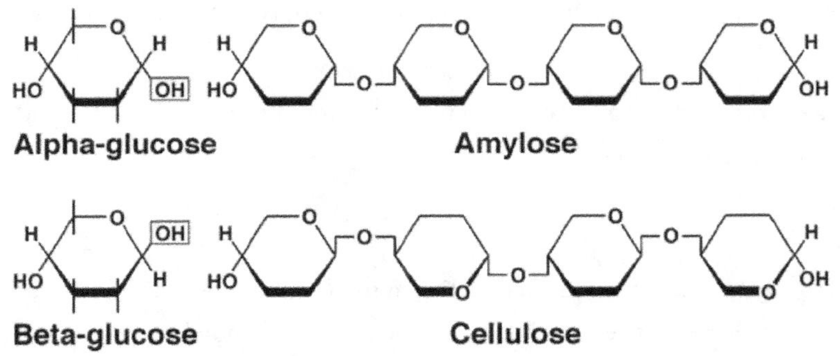

Polysaccharides (cellulose) with β(1→4) glycosidic linkages cannot be digested by humans because they lack the cellulase enzyme to cleave the β(1→4) glycosidic linkages.

6. E is correct.

Ketohexose, a monosaccharide, is one of the components of sucrose, which is a disaccharide made from fructose and glucose.

```
    Aldopentose                  Ketohexose
1  H—C=O                    1  H—C—OH
                                   |
                                   H
2  H—C—OH                   2  C=O
3  H—C—OH                   3  H—C—OH
4  H—C—OH                   4  H—C—OH
5  H—C—OH                   5  H—C—OH
      |                     6  H—C—OH
      H                            |
                                   H
(5-carbon aldehyde)         (6-carbon ketone)
```

Fructose is one of the sweetest tasting natural sugars.

```
       Glucose                    Fructose
   (an aldohexose)            (a ketohexose)

      H   O                      CH₂OH
       \ //                        |
        C                          C=O
        |                          |
   H—C—OH                     HO—C—H
        |                          |
   HO—C—H                      H—C—OH
        |                          |
   H—C—OH                      H—C—OH
        |                          |
   H—C—OH                       CH₂OH
        |
      CH₂OH
```

7. C is correct.

Tautomeric forms of a molecule change the connectivity of a functional group, typically observed for aldehydes, ketones, and imines.

The two cyclic forms of monosaccharides have the same connectivity, but the configuration of the carbons is different. These forms are, therefore, diastereomers.

8. C is correct.

One exception to the tendency for sugars to have oxygen atoms linked to every carbon is deoxyribose. It is similar to ribose; however, one of the alcohols is replaced with a carbon-hydrogen bond.

9. A is correct.

Common monosaccharides are five (pentose) or six (hexose) carbon atoms, hydrogen, and oxygen.

This molecule contains a hemiacetal group, which can open reversibly to form an aldehyde.

The presence of the aldehyde makes this carbohydrate a reducing sugar.

10. B is correct.

The ratio of these atoms that compose sugars is typically 1:2:1 (carbon : hydrogen : oxygen).

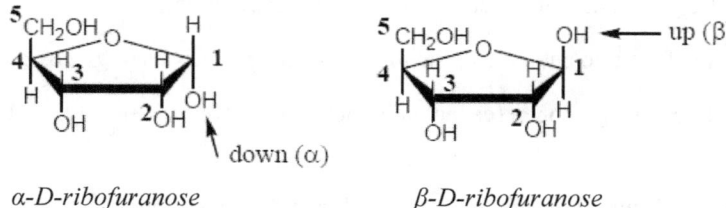

D-ribose: Fisher projections of an example ribose sugar

11. C is correct.

During the formation of disaccharides, the only configuration that can vary is at the anomeric carbon.

α-D-ribofuranose   β-D-ribofuranose

The other alcohols of the carbohydrate are primary or secondary alcohols not adjacent (i.e., allylic) to an oxygen atom (as in anomeric alcohols).

The neighboring oxygen atom in the ring facilitates the ring opening of the anomeric C-1 alcohol, which may form the other anomer (α ↔ β).

12. B is correct.

Any sugar molecule that lacks the aldehyde or the hemiacetal functional group is a non-reducing sugar. The sugars are not active during chemical tests like the Tollens test. When reducing sugars react, they are typically oxidized to carboxylic acid molecules.

Monosaccharides are reducing sugars, while some disaccharides (e.g., lactose and maltose), oligosaccharides, and polysaccharides are reducing sugars.

Reducing sugars (and *alpha* hydroxyl ketones) give a positive Benedict's test: a brown precipitate forms (as does for Fehling's solution).

Tollens' reagent forms silver ions (mirror) as a positive test for reducing sugars.

13. A is correct.

When a six-carbon sugar cyclizes (yielding a pyranose), a hemiacetal bond is formed (–linkage between the ~OH at carbon number 5 and the anomeric carbon (C_1)) and the carbonyl of C_1 is converted to alcohol.

The resulting C_1 hydroxyl may be in the axial or equatorial positions on the ring with the two called *anomers*.

continued…

α-D-glucopyranose ⇌ (mutarotation) ⇌ β-D-glucopyranose

If the ~OH at carbon 1 is in the down position (axial for D-glucopyranose or glucose), the ring is the α anomer.

When the ~OH is in the axial (up) position, it is the β anomer.

A monosaccharide in solution freely converts between the open-chain form and the two anomeric forms (mutarotation). It is necessary to specify the anomeric form because the acetal linkage (glycosidic bond) between monosaccharides in di- and polysaccharides prevent mutarotation.

Enzymes accept only one of the two anomers in the formation and hydrolysis of glycosidic linkages.

Hence, the linkages in starch and glycogen (except at branch points) are α(1→4) glycosidic, whereas the bonds in cellulose are β(1→4).

Even though these macromolecules are simple glucose polymers, humans can digest only the former two since humans lack enzymes for hydrolysis of glycosidic linkages (humans can digest starch but not grass or wood).

14. D is correct.

Glycogen (below) contains α(1→6) glycosidic linkages (below):

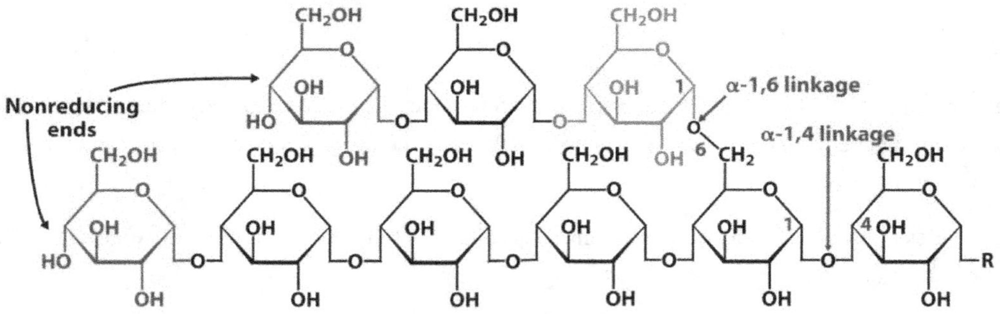

Cellulose (below) contains β(1→4) glycosidic linkages (below):

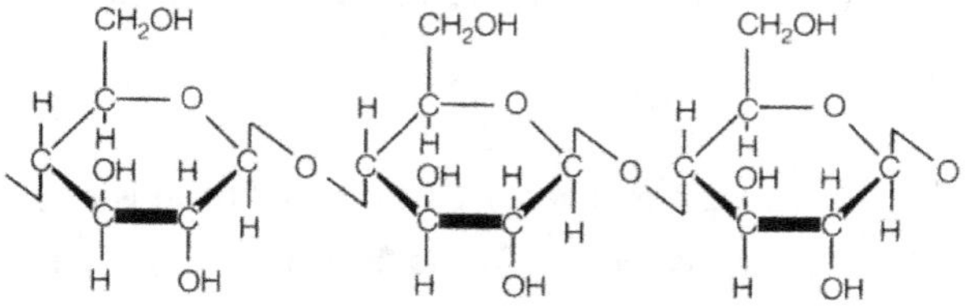

continued...

Sucrose (below) contains α (1→2) glycosidic linkages (below):

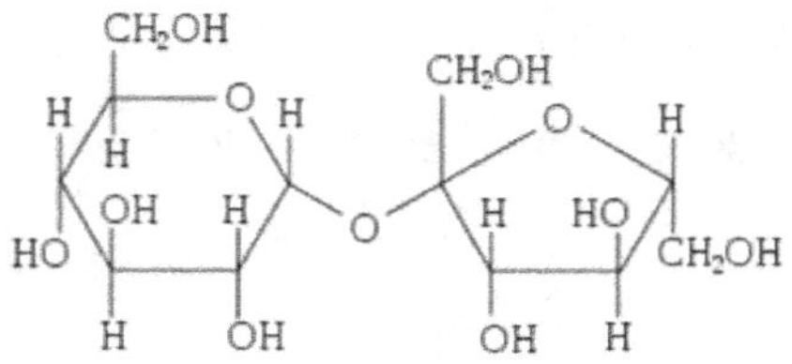

Amylose (below) contains α(1→4) glycosidic linkages (below):

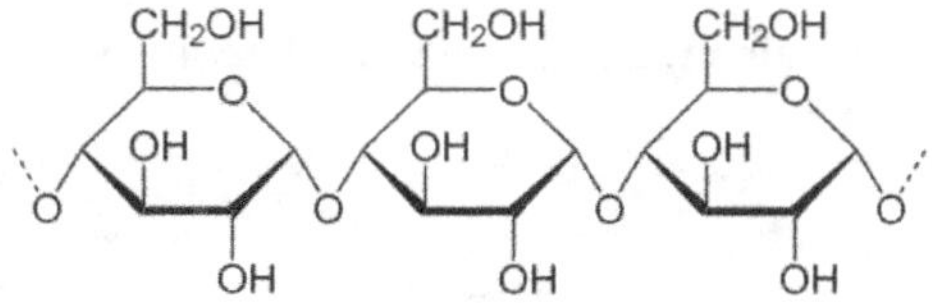

Maltose (below) contains α(1→4) glycosidic linkages (below):

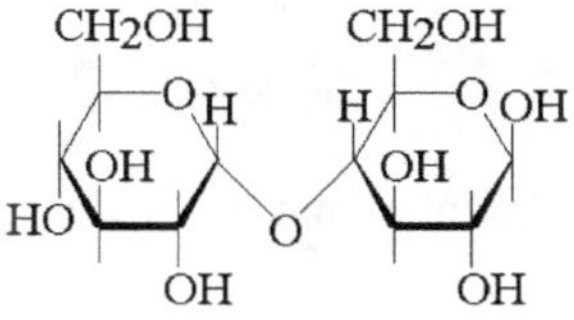

15. D is correct.

Humans cannot digest cellulose because the appropriate enzymes to break down the beta acetal linkages are lacking. The enzyme responsible for cleaving this glycosidic linkage is cellul*ase*. Cellulase is produced by plants, bacteria, fungi, and protozoans.

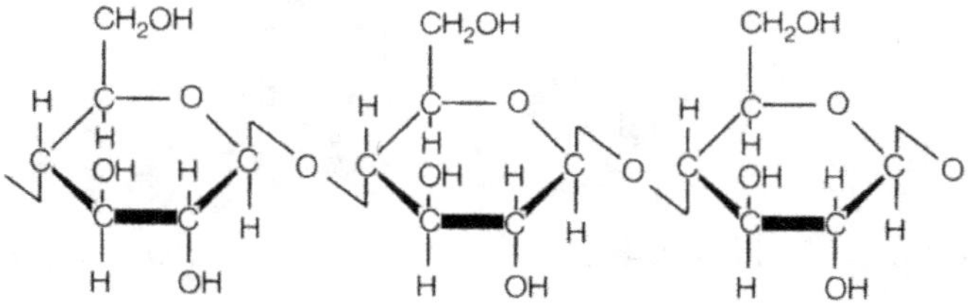

16. A is correct.

Mutarotation is possible because the anomeric carbon is a hemiacetal. When dissolved in water, hemiacetals are known to reversibly open to the acyclic form, allowing the formation of the cyclic diastereomer.

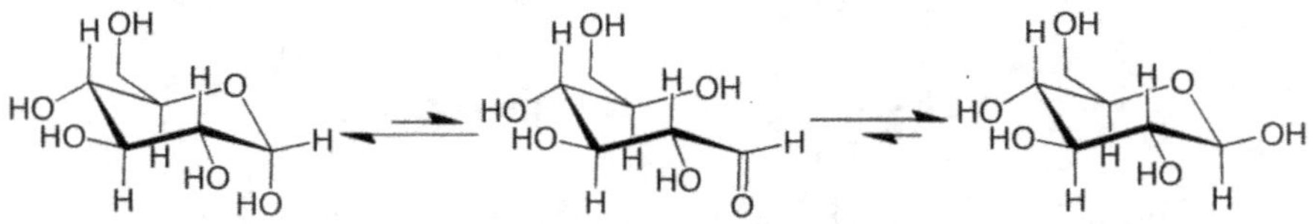

Mutarotation (open chain in the center) with the less stable α-glucose anomer (left) and more stable β-glucose anomer on the right.

The process is accelerated from the addition of acid or base.

17. A is correct.

Monosaccharides cannot be broken down into simpler sugar subunits.

However, monosaccharides can undergo *oxidative degradation* to produce carbon dioxide and carbon monoxide when treated with nitric acid.

18. C is correct.

This is how monosaccharide sugars exist as two different diastereomers.

These diastereomers are classified as α (alpha) and β (beta) diastereomers.

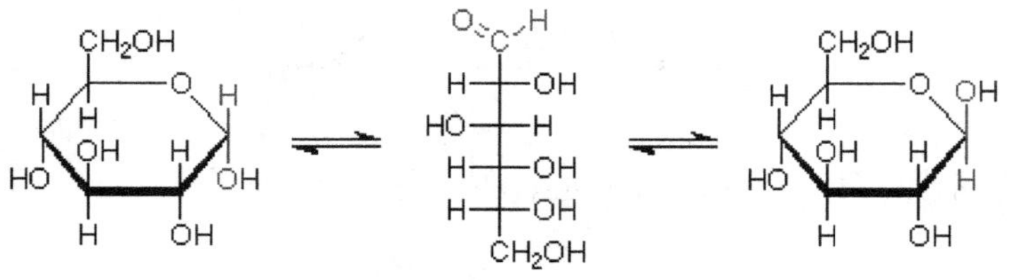

α-D-glucose ⇌ D-glucose ⇌ β-D-glucose

Two diastereomers are possible because the alcohol that cyclizes (i.e., mutarotation) to form the ring may add to either face of the aldehyde.

Epimers are isomers that differ at a single chiral center (shown below):

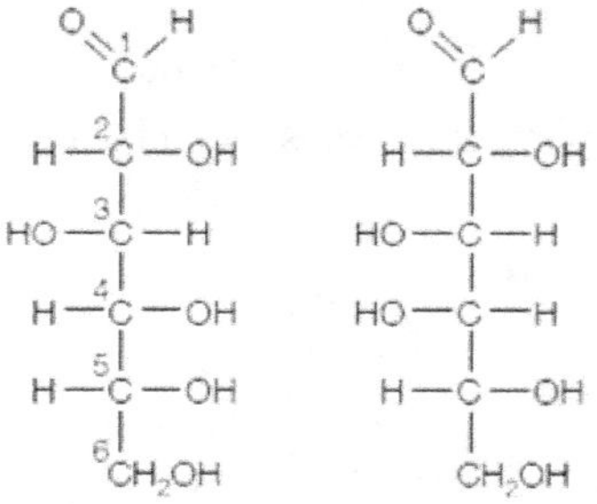

Epimers of D-glucose (left) and D-galactose (right) with inversion at the 4th chiral carbon

19. B is correct.

Reducing sugars possess a free aldehyde or a free ketone group. Monosaccharides are reducing sugars, along with some disaccharides, oligosaccharides, and polysaccharides.

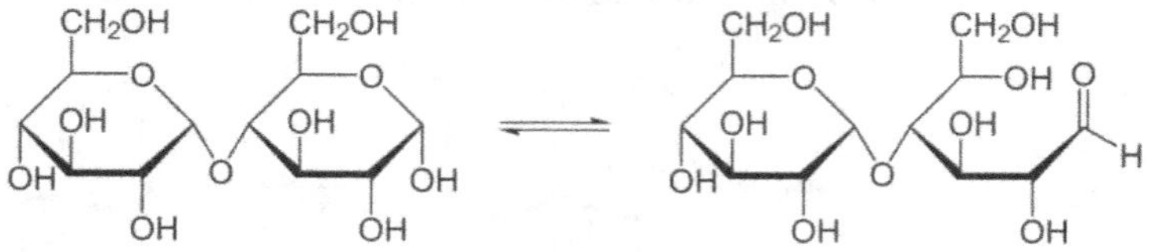

Maltose exists in equilibrium between the closed ring (left) and open form

Open form (with the aldehyde on the far) is the reducing sugar.

If a cyclic sugar possesses an acetal instead of a hemiacetal at the anomeric carbon, the sugar will not be a reducing sugar.

20. D is correct.

People who cannot produce the lactase enzyme are known to be lactose intolerant. This protein is responsible for the catalytic breakdown of lactose.

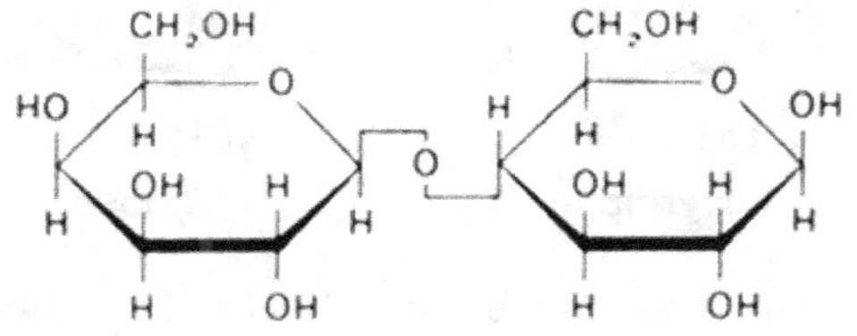

Lactose is a β(1→4) disaccharide comprised of galactose (left) and glucose.

Lactose and cellulose are two forms of sugar that possess the difficult-to-digest β(1→4) glycosidic linkage.

21. A is correct.

An aldotetrose is a four-carbon carbohydrate with an aldehyde.

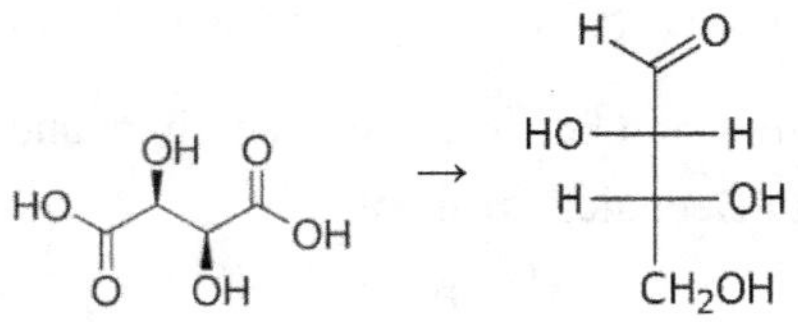

D-(–)-tartaric acid converts to aldotetrose

22. C is correct.

The key to determining a monosaccharide's category involves counting the number of carbon atoms in the structure. There are four carbon atoms in the five-membered ring, and there is a methylene-containing (~CH_2~) substituent bonded to the ring. Because there are five carbon atoms present, the molecule is a pentose.

23. D is correct.

If the chiral carbon farthest from the carbonyl points to the right, then it is a D-sugar.

If the chiral carbon farthest from the carbonyl points to the left, then it is an L-sugar.

The terminal alcohol is not an asymmetric carbon atom because it is typically bonded to two hydrogen atoms.

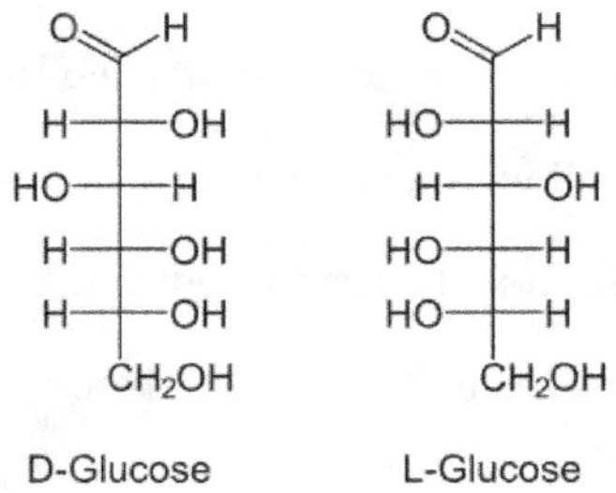

D-glucose and L-glucose are enantiomers (i.e., non-superimposable mirror images)

24. B is correct.

Deoxyribose   **Ribose**

Deoxyribose is the sugar of DNA, while ribose is the sugar of RNA. The difference is the absence of a 2'-hydroxyl (DNA) or the 2'-hydroxyl (RNA). DNA and RNA have a 3'-hydroxyl necessary for chain elongation.

Deoxyribose is a monosaccharide that contains one fewer hydroxyl group than ribose.

Deoxyribose is the sugar in DNA, and *ribose* is the sugar contained in RNA.

25. A is correct.

In an α-1,1 linkage, both anomeric carbons are in an acetal linkage.

Therefore, *mutarotation* (i.e., interconversion between the open-chain and ring-form) is not possible, and no free aldehyde is available to react with Benedict's reagent.

Mutarotation of glucose (α-glucose on the left; β-glucose on the right)

Monosaccharides are reducing sugars, while some disaccharides (e.g., lactose and maltose), oligosaccharides, and polysaccharides are reducing sugars.

B: β-1,4-glucose-glucose is a disaccharide, whereby one glucose unit has its C-1 bound as an acetal and therefore cannot mutarotate, existing only as a ring.

Therefore, the aldehyde is unavailable for reaction with Benedict's reagent. However, the other glucose unit is bonded at its C-4 hydroxyl.

C: *glucose* has a hemiacetal linkage which allows mutarotation.

Therefore, the aldehyde group is free to react in the test for reducing sugars.

D: *fructose* has a hemiacetal linkage that allows mutarotation.

Therefore, the aldehyde group is free to react in the test for reducing sugars.

E: *maltose* is a disaccharide formed from two units of glucose. Through mutarotation, the aldehyde group is free to react in the test for reducing sugars.

26. A is correct.

For a chiral molecule with n chiral centers, the number of stereoisomers is calculated by 2^n, where n is the number of chiral centers.

For example, a molecule with 3 chiral centers has 8 ($2 \times 2 \times 2$) stereoisomers. The relationship (relative to the original molecule) is 1 enantiomer and 6 diastereomers.

If there were *meso* (i.e., internal plane of symmetry) carbohydrates, the number of *meso* forms would be subtracted from the number of stereoisomers.

27. E is correct.

Lactose is a disaccharide composed of glucose and galactose.

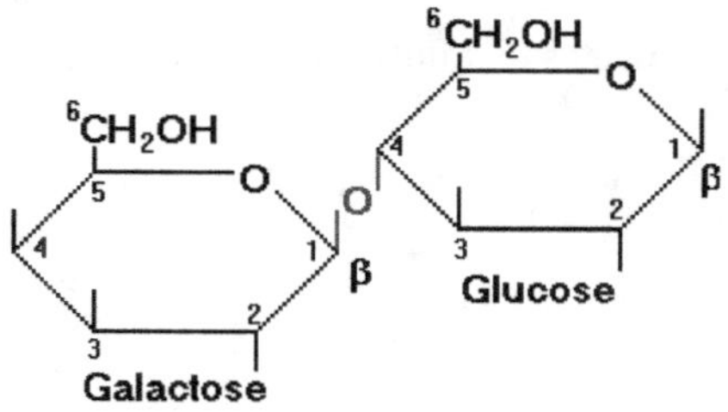

The glycosidic linkage in lactose is a β(1→4) linkage.

28. B is correct.

Ribose is an important monosaccharide that contains an aldehyde and five carbon atoms. This sugar can be in the structure of mRNA.

The 2'-deoxygenated form of ribose (i.e., deoxyribose) is found in the structure of DNA.

29. C is correct.

Cyclic carbohydrates typically have ether and alcohol functional groups.

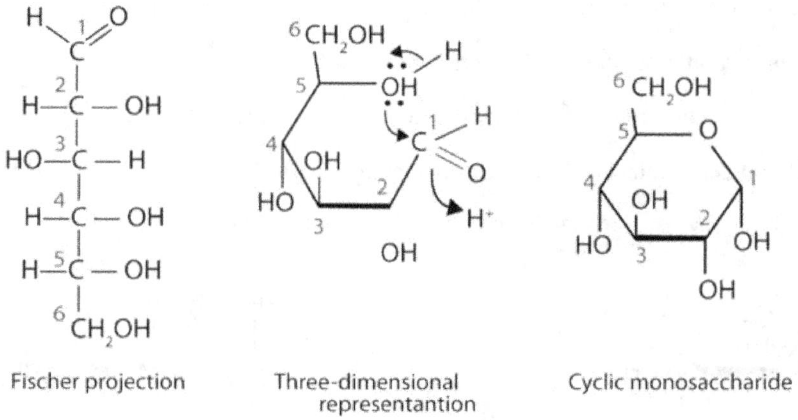

Fischer projection — Three-dimensional representantion — Cyclic monosaccharide

The hemiacetal and aldehyde forms of these compounds can equilibrate.

30. D is correct.

The numbering of carbohydrates begins at the terminal carbon closest to the most oxidized carbon (i.e., anomeric carbon).

It is an α-linkage because the linking oxygen is pointing down (axial) in the Haworth (i.e., ring) representation.

31. E is correct.

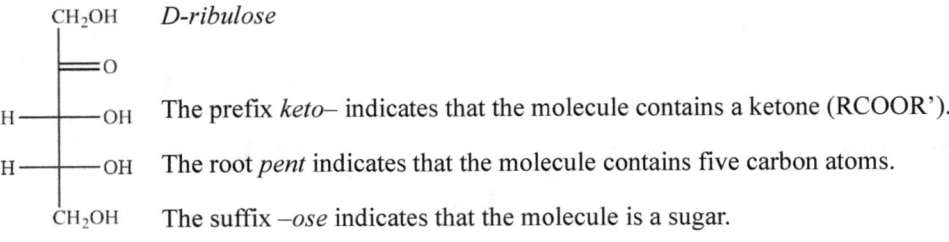

D-ribulose

The prefix *keto–* indicates that the molecule contains a ketone (RCOOR').

The root *pent* indicates that the molecule contains five carbon atoms.

The suffix *–ose* indicates that the molecule is a sugar.

32. A is correct.

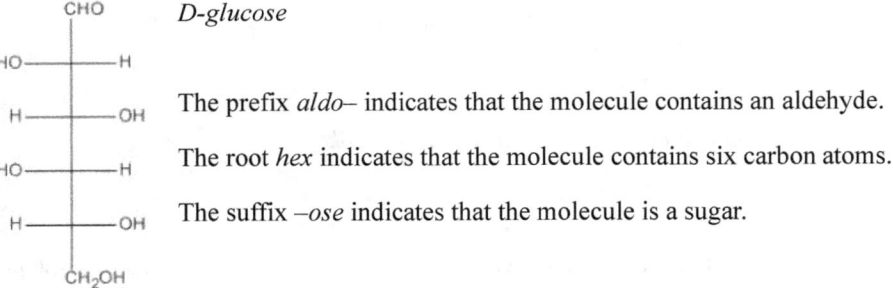

D-glucose

The prefix *aldo–* indicates that the molecule contains an aldehyde.

The root *hex* indicates that the molecule contains six carbon atoms.

The suffix *–ose* indicates that the molecule is a sugar.

33. A is correct.

Glycosidic bonds join carbohydrates and are formed between the *hemiacetal group* of a saccharide and the hydroxyl group of some organic compound, such as alcohol.

α is the designation when the hydroxyl attached to the anomeric carbon points down.

β is the designation when the hydroxyl attached to the anomeric carbon points up.

C: *acetal* is when carbon is attached to two ethers (RO–C–OR').

D: *hemiacetal* is when carbon is attached to ether and a hydroxyl (RO–C–OH).

E: *ester* is when oxygen is attached to two carbon chains (R–O–R).

Formation of a glycosidic bond: glucose and ethanol combine to form ethyl glucoside and water

continued...

The reaction often favors forming the α-glycosidic bond (as shown) due to the *anomeric effect*.

The relative size of the equilibrium is indicated by the size of the arrow

The anomeric effect describes the tendency of an element with lone pairs of electrons adjacent to a heteroatom (e.g., oxygen within a ring) of a cyclohexane ring to prefer the *axial* orientation instead of the less-hindered *equatorial* orientation expected from steric considerations.

The axial orientation permits molecular orbital overlap that increases the overall stability of the molecule.

34. C is correct.

Glycogen is a polymer of glucose that functions as the energy store of carbohydrates in animal cells (plant cells use starch). Glycogen is common in the liver, muscle, and red blood cells.

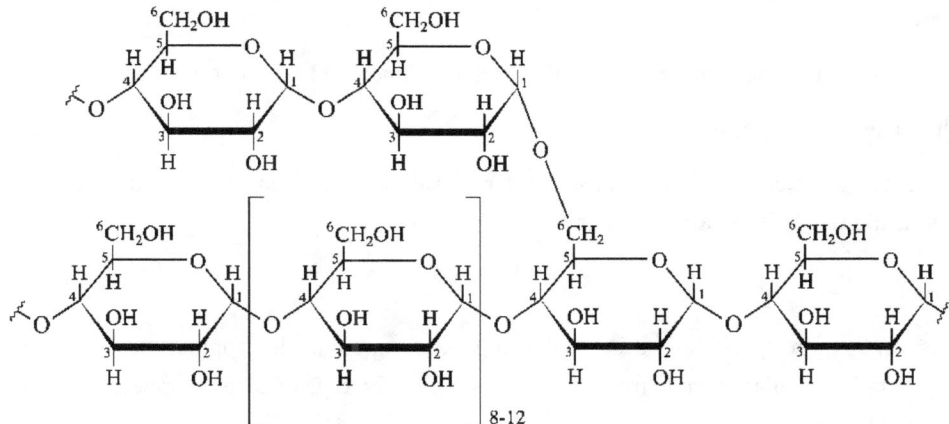

Glycogen is a large biomolecule consisting of repeating glucose subunits.

35. C is correct.

If a compound has only one chiral center, it must be chiral. A *meso* compound has an internal symmetry requiring the presence of at least two chiral centers, and this internal symmetry yields an achiral molecule, even though it contains chiral centers.

36. D is correct.

Disaccharides contain a glycosidic linkage that is an ether group. The ether can be protonated with Brønsted acids and hydrolyzed in the presence of water.

All polysaccharides can be hydrolyzed to produce monosaccharides (i.e., individual monomers of the polymer).

37. D is correct.

Multiple bonds and rings introduce *degrees of unsaturation*.

Using a subscript of n for the number of carbons, the *degree of unsaturation* is determined by the formulae:

Alkane: C_nH_{2n+2} = 0 degree of unsaturation

Alkene: C_nH_{2n} = 1 degree of unsaturation

Alkyne: C_nH_{2n-2} = 2 degrees of unsaturation

Rings = 1 degree of unsaturation

Double bonds = 1 degree of unsaturation

Acarbose has one double bond and four rings and therefore has five degrees of unsaturation.

38. D is correct.

Monosaccharides are the basic unit of carbohydrates.

Subjecting these compounds to acids or bases will not hydrolyze them further. However, they can undergo oxidative decomposition by treating them with periodic acid to form formaldehyde and formic acid.

39. B is correct.

The suffix ~*ose* is used to denote sugars. The highest priority functional group in the molecule is a ketone.

Therefore, the sugar is a ketose sugar.

The sugar ($C_nH_{2n}O_n$) has a carbon chain of 5, so it is a pentose. If the molecule had an aldehyde instead of a ketone, the molecule would be an aldose sugar.

40. D is correct.

Carbohydrates can be more specifically described as organic compounds that contain carbon, hydrogen, and oxygen. The general molecular formula may vary depending on the type of carbohydrates, but many examples have the formula of $C_nH_{2n}O_n$.

41. B is correct.

Mutarotation occurs when cyclic hemiacetals form from monosaccharides with different configurations around the anomeric carbon (i.e., carbonyl carbon in the straight chain).

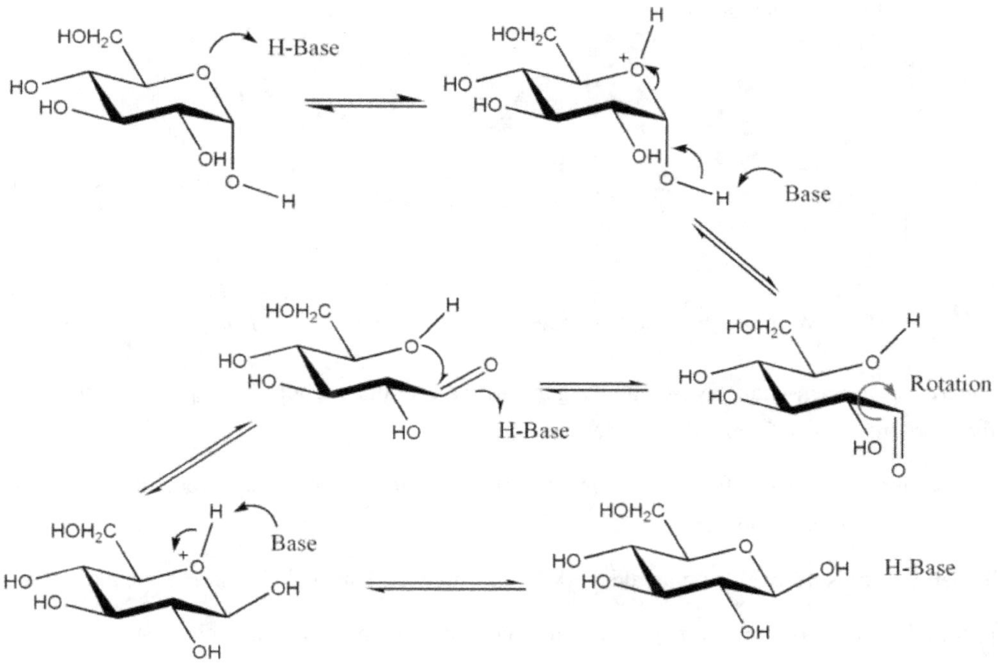

The reaction mechanism for the interconversion of α and β anomers

The bond to the *anomeric carbon* is easily broken in aqueous solutions, as either α (hydroxyl points downward) or β (hydroxyl points upward) anomer becomes an open chain.

This open chain is easily recyclized in an aqueous solution, especially slightly acidic, forming a mixture containing both anomers (α or β) in their equilibrium concentrations.

Thus, the chain's initial opening and subsequent closing result in a mixture of anomers, known as *mutarotation*.

A: *reduction* decreases the number of bonds to oxygen (or gain of electrons in inorganic chemistry).

Examples of organic chemistry reduction include converting an aldehyde to primary alcohol, a carboxylic acid to an aldehyde or primary alcohol, or a ketone to a secondary alcohol.

C: *hemiacetals* form from nucleophilic addition of hydroxyl oxygen to a carbonyl (aldehyde or ketone).

D: the *open-chain form* has a carbonyl group, and therefore an aldehyde is formed.

E: *oxidation* increases the number of bonds to oxygen (or loss of electrons in inorganic chemistry).

Examples of oxidation in organic chemistry include converting primary alcohol to an aldehyde or carboxylic acid and converting secondary alcohol to a ketone.

42. D is correct.

The cyclic hemiacetal sugars exist as either five-membered (furanose) or six-membered (pyranose) rings.

A *furanose* is a carbohydrate with a chemical structure with a five-membered ring system consisting of four carbon atoms and one oxygen atom.

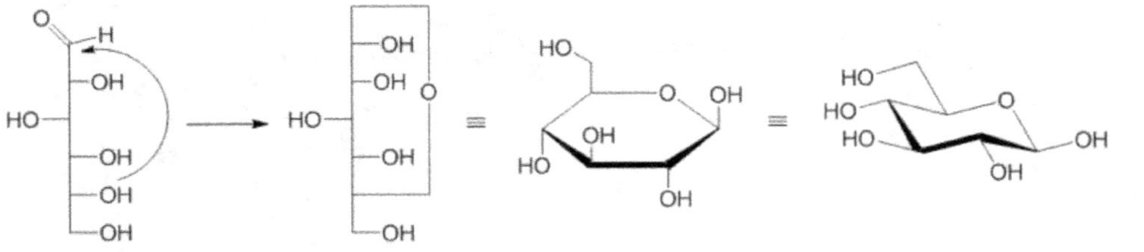

Formation of pyranose hemiacetal and representations of β-D-glucopyranose

A *ketopentose* is an open-chain five-carbon sugar that has a ketone carbonyl group. An aldopentose is an open-chain five-carbon sugar with an aldehyde carbonyl group.

A *cyclic hemiacetal* is derived from an aldehyde; if the anomeric carbon (i.e., carbon attached to two oxygen) has an H attached – it is an aldose.

A cyclic hemiacetal is derived from an aldehyde if the anomeric carbon lacks an H attached – it is a *ketose*.

The structure is a furanose form of sugar that has a ketone carbonyl in its open-chain structure.

A: the *furanose* (five-membered ring) form of an aldopentose is a structure that lacks the H on the anomeric carbon necessary in an aldose.

B: the *pyranose* (six-membered ring) form of an aldopentose is a structure that lacks the H on the anomeric carbon necessary in an aldose.

D: the pyranose form of a ketopentose is a *six-membered ring*.

43. B is correct.

The cyclic and acyclic isomers of glucose exist as an equilibrium mixture in aqueous solutions.

Because monosaccharide cyclization is reversible, cyclic isomers are a mixture of diastereomers; α and β isomers.

44. C is correct.

Isomers have identical atoms in different arrangements.

The double bond and hydroxyl groups shift positions, which indicates that one carbon was oxidized, and another was reduced.

In glycolysis, the phosphoglucose isomerase assists in the removal of a hydrogen anion (i.e., hydride ion), which attacks a carbonyl group to form a hydroxyl group.

The carbon where the hydride ion is removed – now a cation – is attacked by a water molecule.

A: $CH_3CH_2COCl + H_2O$ reaction is hydration.

B: the number of carbons in the products is one greater than in the reactants; therefore, this is *not* an example isomerization reaction that requires the number of atoms to remain constant.

D: $CH_3CH_2CH_2CHOHCH_3$ reaction is a transesterification reaction.

45. C is correct.

The "di" prefix in the name suggests that there are two smaller subunits.

Monosaccharides are linked through glycosidic (i.e., oxygen bonded to two ethers) functional groups.

Lactose is a disaccharide formed by a β(1→4) linkage between galactose and glucose.

46. C is correct.

Benedict's test (or Tollens' reagent) is used to detect reducing sugars. An oxidized copper reagent is reduced by a sugar's aldehyde, and the aldehyde is oxidized to a carboxylic acid in the process.

Reducing sugars (and alpha hydroxyl ketones) give a positive Benedict's test: a red-brown precipitate forms.

Fehling's solution gives a positive test for reducing sugars by changing from blue to clear and forming a red-brown precipitate.

Tollens' reagent forms silver ions (mirror) as a positive test for reducing sugars.

Tartaric acid has no aldehyde or ketone to react to because its first and last carbons have carboxylic functional groups.

Therefore, it cannot reduce the reagent and yields a negative result (Tollens remains clear, while Benedict's and Fehling's remain blue).

47. D is correct.

Maltose is a disaccharide composed of two glucose molecules, and the glycosidic linkage is an α(1→4) linkage.

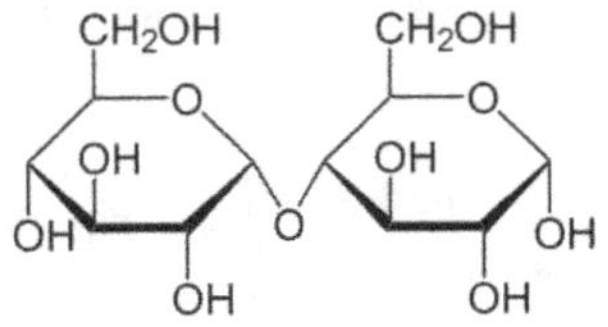

48. A is correct.

For D-sugar monosaccharides, the hydroxyl group of the last asymmetric carbon atom is oriented to the right.

Notes for active learning

Notes for active learning

Nucleic Acids – Detailed Explanations

1. D is correct.

A *complementary base pair* is used for RNA molecules during transcription and translation.

Errors are sometimes introduced to the sequence, and these errors can lead to mutations in the proteins assembled from them.

2. A is correct.

The *dipole interactions* that join the strands of DNA are specifically hydrogen bonds.

These bonds form from the acid protons between the amides and imide functional groups and the carbonyl and amide Lewis basic sites of the matched nitrogen base pairs.

3. C is correct.

Deoxyribose sugars are carbohydrates utilized in DNA synthesis, while ribose sugars are used to synthesize RNA molecules.

4. B is correct.

NADH is the biochemical equivalent to a hydride (H^-) reduction reagent for carbonyl groups.

Alternatively, NAD^+ is an oxidizing agent that can accept a hydride equivalent.

5. C is correct.

In double-stranded DNA molecules, adenine nucleotides pair with thymine (A=T), and cytosine pairs with guanine (C≡G) nucleotides.

Adenine forms two hydrogen bonds with thymine (A=T).

Cytosine forms three hydrogen bonds with guanine (C≡G).

The DNA strands are antiparallel with the 5' written on the top left side of the molecule.

    5'–ATATGGTC–3'

    3'–TATACCAG–5'

6. D is correct.

The DNA molecule has a deoxyribose sugar-phosphate backbone with bases (A, C, G, T) projecting into the center to join the antiparallel strand of DNA (i.e., double helix).

The deoxyribose sugar-phosphate backbone is negatively charged due to the formal charge of the oxygen attached to the phosphate group.

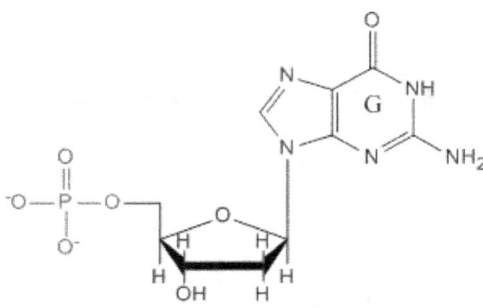

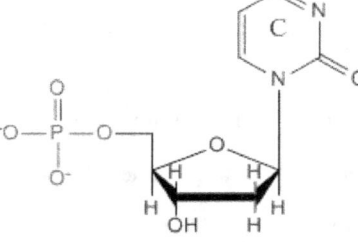

The purines (adenine and guanine) are double-ringed nitrogenous bases, while the pyrimidines (cytosine and thymine) are single-ringed nitrogenous bases.

A nucleotide has a deoxyribose sugar, base, and phosphate, while a nucleoside is a deoxyribose sugar and base without the phosphate group.

7. B is correct.

Transcription is a biomolecular event that takes place in the nucleus of the cell. During transcription, RNA molecules are synthesized using complementary base pairs of DNA single strands.

8. D is correct.

Amino acids are the monomers that comprise proteins.

Nucleotides are comprised of a nitrogenous base (i.e., adenosine, cytosine, guanine, and thymine or uracil), a phosphate group, and a five-carbon sugar (i.e., ribose for RNA or deoxyribose for DNA).

9. B is correct.

Four common nucleotides are in DNA molecules:

adenine (A), cytosine (C), guanine (G) and thymine (T)

Four common nucleotides are in RNA molecules:

adenine (A), cytosine (C), guanine (G) and uracil (U)

10. A is correct.

When DNA is replicated in the cell, one strand of DNA acts as a template for a new strand synthesized as the replication fork opens. This continuously synthesized strand is the *leading strand* and, when combined with one of the original strands of DNA, makes up one new DNA molecule.

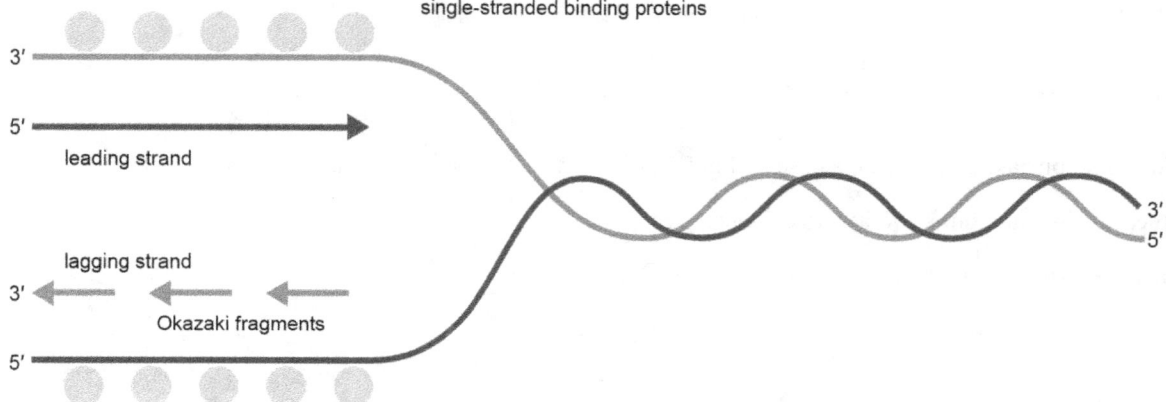

Furthermore, an additional DNA molecule is created in the process, as the other parent strand acts as a template for its complementary strand. This lagging strand comprises smaller segments called *Okazaki fragments* (about 150-200 nucleotides long). Okazaki fragments are combined with the enzyme DNA ligase.

11. E is correct.

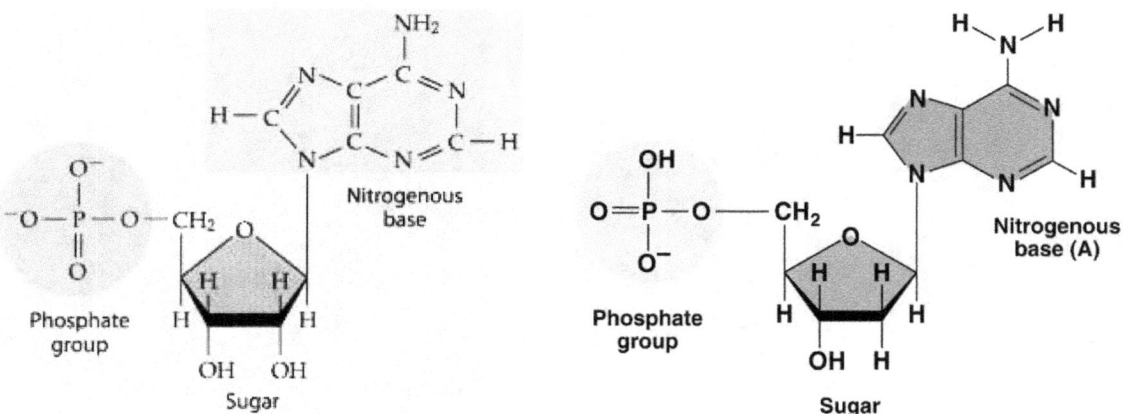

RNA molecules contain ribose as the carbohydrate component of the backbone.

DNA contains deoxyribose (lacking a 2'-hydroxy) as part of the backbone.

12. A is correct.

A nucleoside is a nitrogenous base linked to a sugar, while a nucleotide is a nucleoside and 1 or more phosphate groups attached to the ribose or deoxyribose sugar.

The uracil base is directly bonded to the 1' position of 2'-ribofuranose (i.e., RNA sugar) moiety, rather than the 2'-deoxyribofuranose (i.e., DNA sugar) ring.

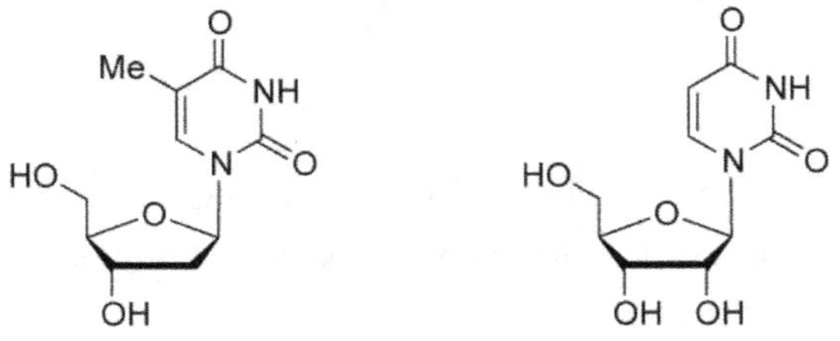

thymidine (DNA)   uridine (RNA)

13. D is correct.

The common base pairing that is observed in DNA is adenine-thymine and guanine-cytosine.

Adenine-thymine forms two hydrogen bonds (A=T).

Guanine-cytosine forms three hydrogen bonds (G≡C).

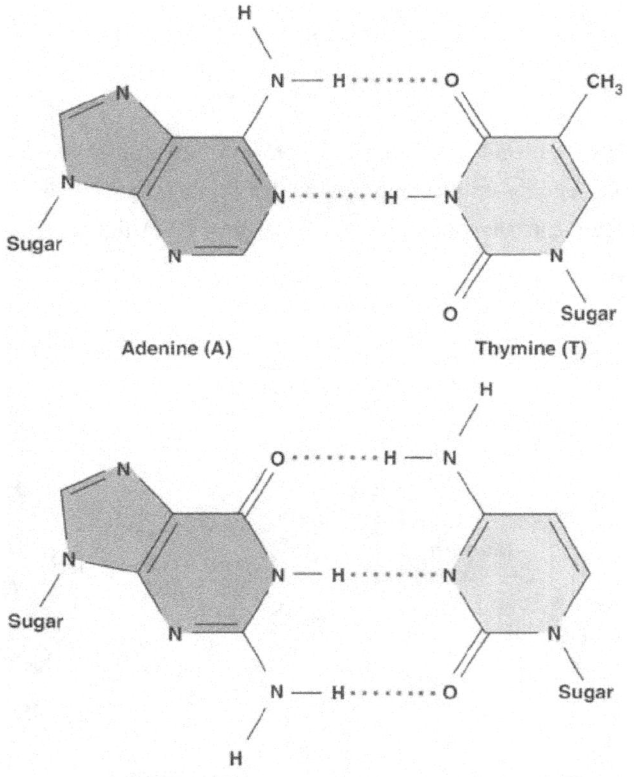

Adenine (A)   Thymine (T)

Guanine (G)   Cytosine (C)

This pairing is consistent with DNA structure, and complementary pairing guides the formation of daughter strands of DNA.

14. D is correct.

Because adenine forms two hydrogen bonds with thymine, and cytosine forms three hydrogen bonds with guanine. DNA polymerase (in the S phase of interphase) uses the complementary nitrogen base to synthesize the new (daughter) strand.

15. C is correct.

DNA molecules hold the genetic information of organisms.

RNA molecules are synthesized from the DNA strand to make proteins for the cell.

Genes are the sections of DNA responsible for the synthesis of proteins in cells.

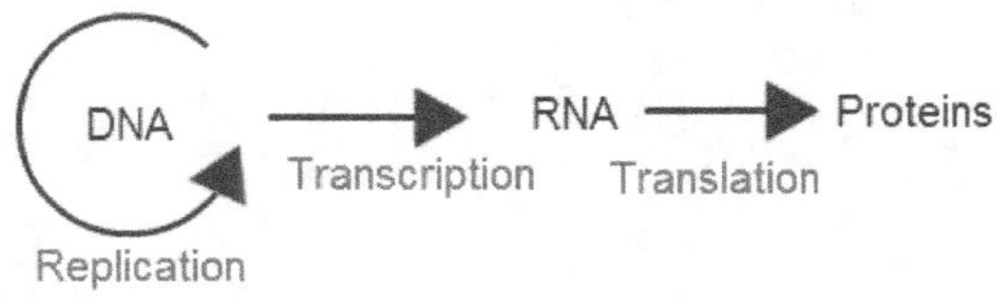

The central dogma of molecular biology (designates information flow)

16. D is correct.

There are five nucleotides, four of which appear in DNA.

These nucleotides are cytosine, guanine, adenine, and thymine.

In RNA molecules, the thymine is replaced with another pyrimidine nucleotide known as uracil.

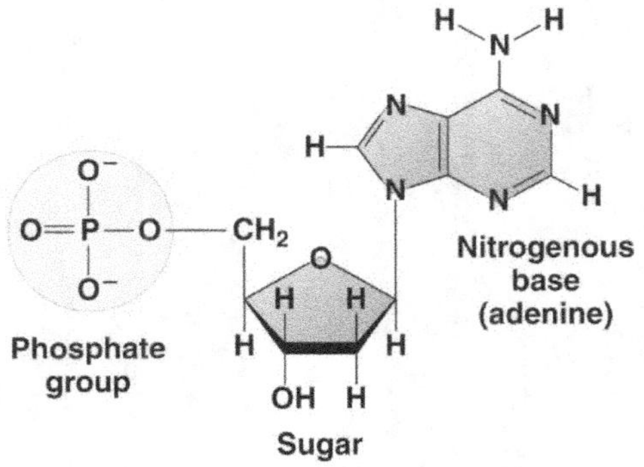

Nucleotides consist of a phosphate group (note the negative charge on oxygens), deoxyribose sugar (lack a 2'-hydroxyl), and a nitrogenous base (adenine, cytosine, guanine, or thymine)

17. D is correct.

DNA	DNA	mRNA	tRNA
A	T	A	U
C	G	C	G
G	C	G	C
T	A	U	A

Complementary base pairing for nucleotides

DNA → DNA (replication); DNA → RNA (transcription); RNA → protein (translation)

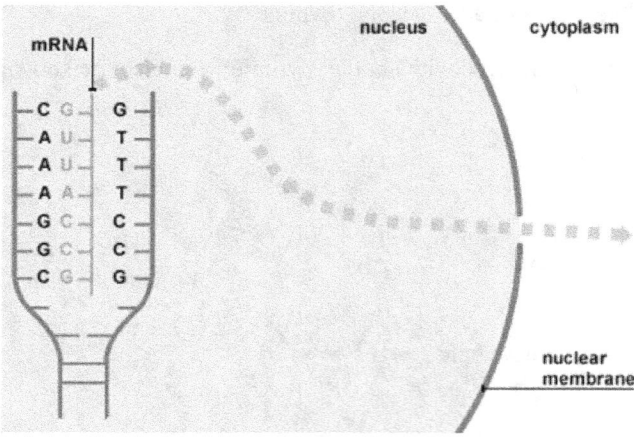

Adenosine (left) and Uracil (replaces thymine in RNA)

A single strand of DNA is the template for RNA synthesis during transcription.

The mRNA (after processing) is translocated to the cytoplasm for translation to proteins.

18. B is correct.

Cellular respiration involves the breakdown of sugar molecules to produce energy. When this occurs, water and carbon dioxide are given off as byproducts.

A campfire (i.e., wood is mainly the sugar of cellulose) gives off the same byproducts of carbon dioxide and water because it is a combustion reaction.

$$\text{wood} + O_2 \rightarrow CO_2 + H_2O$$

19. A is correct.

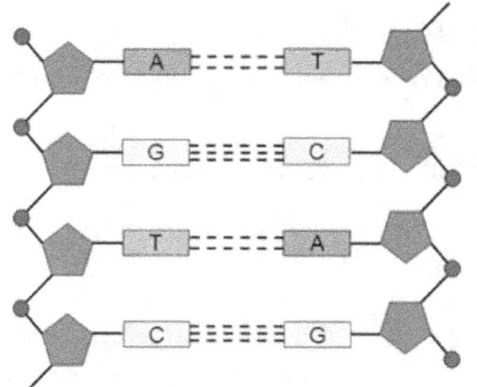

Double-stranded DNA with two hydrogen bonds between A=T and three hydrogen bonds between C≡G. The backbone comprises deoxyribose sugar and phosphates (shown as circles between the sugars).

20. D is correct.

Codons are three-nucleotide sequences located on the mRNA, while anticodons are three nucleotides sequences on the tRNA. The codon-anticodon sequences hybridize by forming hydrogen bonds to the complementary base pair.

The relationship between codon (on mRNA), anticodon (on tRNA), and the resulting amino acid:

Codon (mRNA):
5'–AUG–CAA–CCC–GAC–UCC–AGC–3'

Anticodon (tRNA):
3'–UAC–GUU–GGG–CUG–AGG–UAG–5'

Amino acids:
Met–Gln–Pro–Asp–Phe–Ser

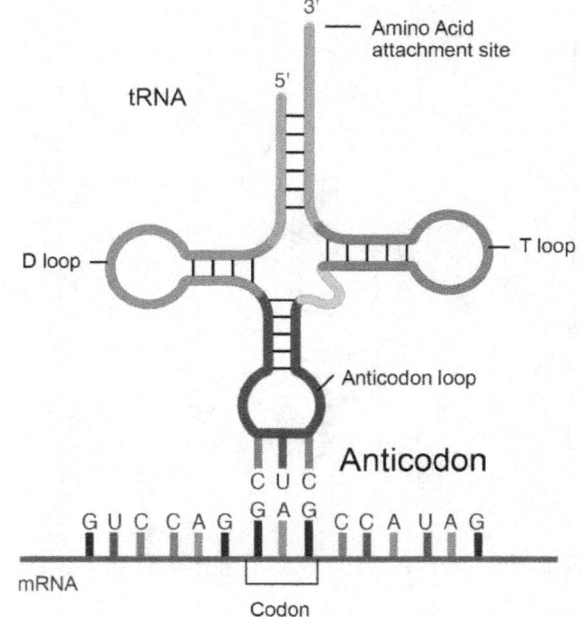

The genetic code is the nucleotide sequence of the codon (mRNA) that complementary base pairs with the nucleotide sequence of the anticodon (tRNA). From the genetic code, the identity of the amino acid encoded by a gene (synthesized during translation into mRNA) can be deduced.

21. A is correct.

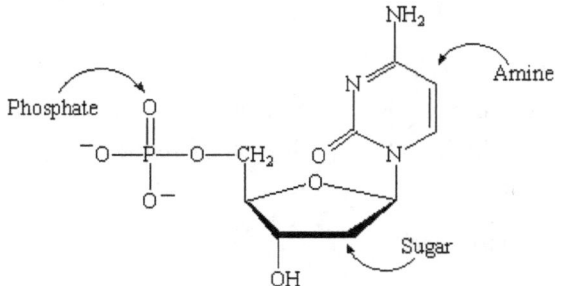

The hexose sugar, nitrogen base, and phosphoric acid group make up nucleotides, and these nucleotides are used to make larger molecules of nucleic acids.

Nucleotides consist of sugar, phosphate, and a nitrogenous base.

22. D is correct.

Identical copies of DNA are necessary for cell division; these cells are daughter cells.

When the daughter strand is synthesized, a complementary nitrogenous base containing nucleotides (A↔T and C↔G) is incorporated into the growing strand.

Replication (i.e., synthesis of DNA) occurs during the S phase (i.e., within interphase) of the cell cycle.

23. B is correct.

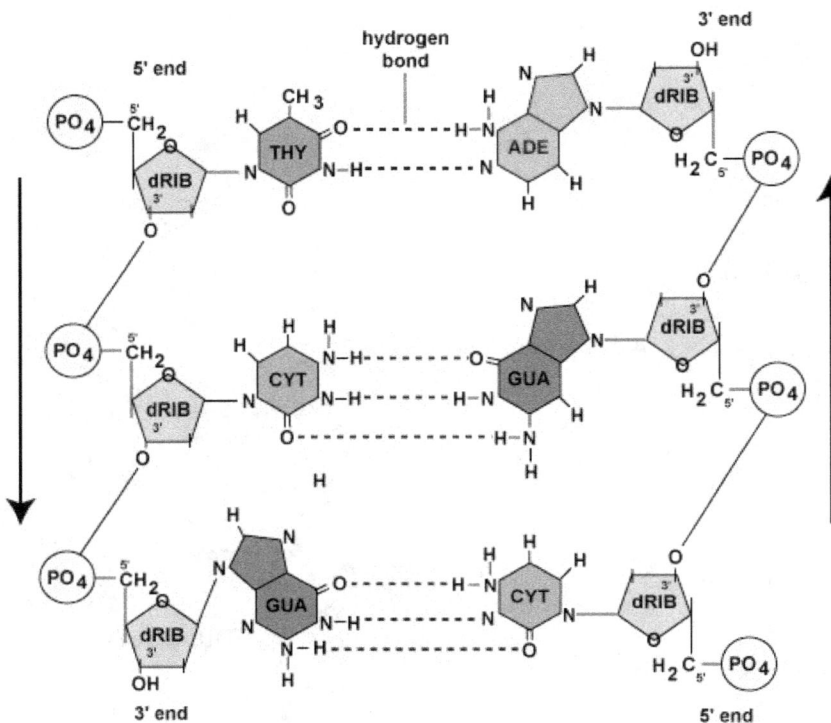

Two antiparallel strands of DNA with the vertical arrows indicating the 3'-hydroxyl (~OH) of the sugar (i.e., from point of chain elongation). The nucleotide is the DNA building block and comprises sugar (i.e., deoxyribose), phosphate, and base (adenosine, cytosine, guanine, and thymine).

Two hydrogen bonds (A=T) hold the base pairs together for nucleotide pairs with adenine and thymine. Cytosine and guanine nitrogen base pairs use three hydrogen bonds (C≡G) to support the nucleic acid structure.

24. B is correct.

The *central dogma of molecule biology*: DNA → RNA → protein

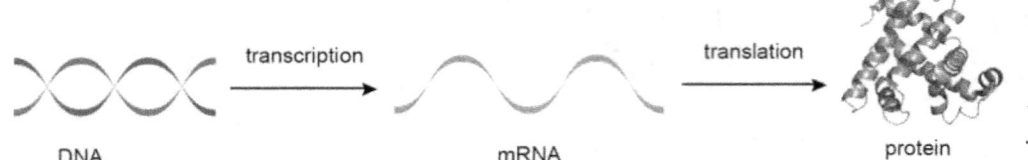

tRNA and ribosomes are used to make peptide chains from mRNA.

Nucleotide triplets known as codons are combined with a complementary tRNA (i.e., containing the anticodon that is complementary in base pairing with the codon of the mRNA). Each tRNA brings an appropriate (i.e., anticodon ↔ amino acid) to the growing polypeptide during translation.

The amino acid residues are combined in the order of the codon sequence.

25. C is correct.

After transcription (DNA → mRNA), the synthesized mRNA is transported out of the nucleus and to the ribosome so that proteins can be synthesized during translation (mRNA → protein).

RNA contains ribose, a sugar similar to deoxyribose of DNA molecules, except it has a hydroxyl group (~OH) at the 2' position on the sugar.

Ribonucleotide (left) and deoxyribonucleotide (right): note the 2' position of the sugars

Nucleic acids (DNA and RNA) bind *via* hydrogen bonds between bases (A=T/U and C≡G). The sugar and phosphate groups form the backbone and do not affect (to any appreciable degree) the hydrogen bonding between bases.

The hydroxyl group (~OH) in RNA causes the backbone to experience steric and electrostatic repulsion.

Therefore, the grooves formed in helixes or hairpin loops between chains are larger. The larger groove permits nucleases (i.e., enzymes that digest DNA or RNA) to bind to the RNA chain more easily and digest the covalent bonds between the alternating sugar-phosphate monomers.

26. E is correct.

The nitrogen bases of nucleotides are responsible for the hydrogen bonding that keeps one nucleic acid strand attracted to the other.

The number of hydrogen bonds that exist between the pairs varies between two and three bonds.

Adenine and thymine nitrogenous base pairs bond with two hydrogen bonds (A=T).

Cytosine and guanine nitrogenous base pairs bond with three hydrogen bonds (C≡G).

27. A is correct.

Three nucleotides are combined to make a codon or anticodon required for translation.

A larger segment of transcribed DNA is a gene, and sections of different genes make up nucleic acid strands.

Two single strands of DNA are used to make one DNA double helix.

28. B is correct.

In any given molecule of DNA, each of the thymine nitrogen bases forms two hydrogen bonds with adenine.

Because each thymine pairs with an adenine residue, there are equal nitrogen bases in the molecule.

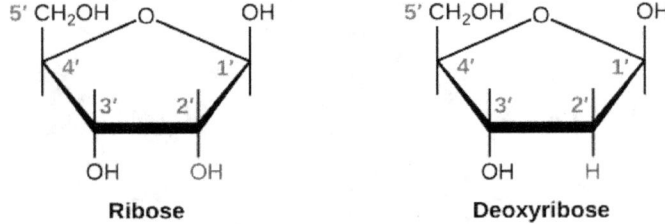

29. D is correct.

Ribose differs from deoxyribose sugars in that ribose lacks one alcohol (i.e., hydroxyl) group.

Ribose sugar (RNA) and deoxyribose sugar (DNA) are used to synthesize nucleic acid polymers.

30. B is correct.

Replication: DNA → DNA during the S phase of the cell cycle.

A: translation is the process of synthesizing proteins from mRNA.

C: transcription is the process of synthesizing mRNA from DNA.

D: complementation is observed in genetics when two organisms with different homozygous recessive mutations produce the same mutant phenotype (e.g., thorax differences in *Drosophila* flies), when mated or crossed, produce offspring with the wild-type phenotype.

Complementation only occurs if the mutations are in different genes. Each organism's genome supplies the wild-type allele to *complement* the mutated allele of the other. Since the mutations are recessive, the offspring display the wild-type phenotype.

Complementation (i.e., *cis/trans*) tests can be used to evaluate whether the mutations are in different genes.

E: restriction digestion is a process used in molecular biology for cleaving (i.e., cutting) DNA for analysis (e.g., restriction fragment analysis) or processing (e.g., PCR amplification).

31. C is correct.

When NADH is oxidized, it loses an equivalent of hydrogen.

NADH is a neutral compound; a loss of hydride (H^-) means that the substrate develops a positive charge.

32. D is correct.

The *central dogma of molecular biology*:

DNA → RNA → protein

DNA → RNA is transcription.

RNA → protein is translation.

DNA is the nucleic acid biomolecule that can give rise to other nucleic acids and proteins.

DNA strands are synthesized from parental DNA strands during replication.

33. B is correct.

The intermolecular forces among the nitrogen base pairs are hydrogen bonds.

Adenine forms 2 hydrogen bonds with thymine (A=T).

Cytosine forms 3 hydrogen bonds with guanine (C≡G).

34. C is correct.

The peptide chain is assembled depending on the amino acid residue order, dictated by the mRNA sequence.

rRNA is the nucleic acid that comprises the ribosome used during translation (converting the codon into a corresponding amino acid in the growing polypeptide chains of the nascent protein).

Each codon of RNA has a corresponding anticodon located on the tRNA.

tRNA molecules have the 3-nucleotide sequence of the anticodon and the appropriate amino acid at its 3' end that corresponds to the anticodon.

The genetic code is the language for converting DNA (i.e., nucleotides) to proteins (i.e., amino acids).

DNA → mRNA → protein

DNA to mRNA is transcription.

mRNA to protein is translation.

There are 20 naturally occurring amino acids with one start codon (methionine) and three stop codons (containing releasing factors that dissociate the ribosome).

35. E is correct.

Because RNA utilizes a uracil (U) nitrogenous base instead of thymine (T) nitrogenous base, thymine should not appear in the codon.

36. A is correct.

Thymine is a pyrimidine nitrogenous base pair that forms two hydrogen bonds with adenine (purine) in the base-paired structure of DNA.

In RNA molecules, the nitrogenous base thymine is replaced by uracil.

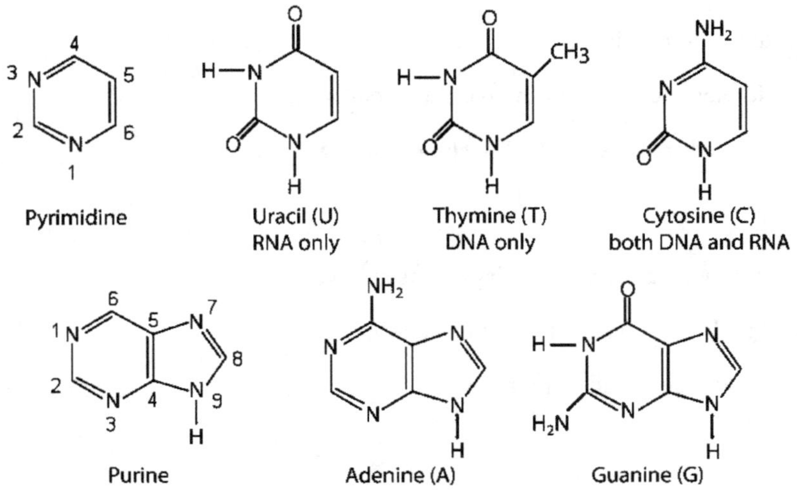

Pyrimidine

Uracil (U)
RNA only

Thymine (T)
DNA only

Cytosine (C)
both DNA and RNA

Purine

Adenine (A)

Guanine (G)

Purines (A, G) are single-ring structures, while pyrimidines (C, T, U) are double-ring structures.

37. B is correct.

There are three general components of nucleotides: a phosphate group, a cyclic five-carbon sugar, and a nitrogenous base.

Fat molecules are biomolecules for phospholipid membranes, storage fat molecules, and lipid molecules.

38. A is correct.

There are three major components of a nucleotide, the subunit that makes up nucleic acids.

All nucleic acids have a nitrogen base used for hydrogen bonding, a hexose sugar (ribose or deoxyribose), and a phosphate group that contains a phosphate linkage with the sugar.

Ester linkages are found in fats, glycosidic linkages are in sugars, and peptide linkages are in proteins.

39. B is correct.

One of the two complementary codes important for constructing peptide chains is the codon made of three RNA nucleotides complementary to the anticodon of tRNA molecules.

40. C is correct.

Because RNA contains uracil, this nitrogenous base forms hydrogen bonds with adenine. It is important to note that RNA molecules are single-stranded, and DNA molecules are double-stranded.

In DNA:

> Adenosine (A) forms two hydrogen bonds with thymine (T).

> Cytosine (C) forms three hydrogen bonds with guanine (G).

41. C is correct.

Thymine is a pyrimidine nucleotide base that occurs in DNA but does not occur in RNA. Instead, RNA has uracil.

42. C is correct.

In DNA, thymine (T) hydrogen bonds with adenine (A).

However, in RNA, the thymine is exchanged for uracil (U).

43. C is correct.

Nucleic acids determine the sequences of amino acids because of groupings of nucleotides along a sequence corresponding to a particular amino acid.

The information of certain nucleic acids (i.e., mRNA) is translated on ribosomes with the help of tRNA.

Prions are infectious, disease-causing agents of misfolded proteins.

44. A is correct.

The sugar component of nucleic acid (ribose or deoxyribose sugar) has many alcohol groups.

The nitrogenous base contains a basic nitrogen atom (i.e., amino group) hydrogen bonds with complementary nitrogen bases.

There are two hydrogen bonds between the nitrogenous bases adenine and thymine (A=T) and three hydrogen bonds between cytosine and guanine (C≡G).

45. C is correct.

DNA (deoxyribonucleic acid) is a long biological molecule composed of smaller units called nucleotides (i.e., sugar, phosphate, and base).

The sugar is deoxyribose (compared to ribose for RNA), and the bases are adenine, cytosine, guanine, and thymine (with thymine replaced by uracil in RNA).

The strands that these nucleotides make up are *nucleic acids*.

46. D is correct.

Ribose is the structural sugar of RNA, while deoxyribose is the sugar for DNA.

Uracil is a nucleotide (i.e., sugar, phosphate, and base) that contains ribose sugar.

This sugar is similar to deoxyribose; however, one difference is that deoxyribose has one fewer alcohol group (2'~position of the sugar) than ribose.

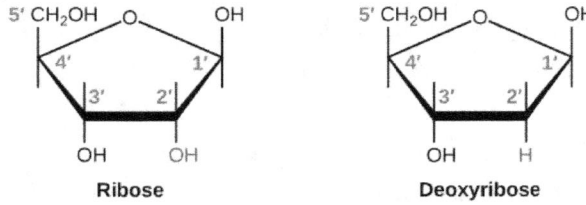

Uracil is a base of RNA and does not appear in DNA strands.

47. E is correct.

During transcription, the two nucleic acid strands of DNA dissociate, and mRNA is synthesized (transcription) using the nitrogen bases of DNA as a template.

Therefore, the mRNA strand is a complementary strand to the DNA strand.

The mRNA then exits the nucleus to be used as a template for synthesizing proteins (translation).

48. C is correct.

The bonds that make up every single strand of DNA are covalent, and the bonds linking the antiparallel strands of DNA are hydrogen bonds.

Two hydrogen bonds form between adenine and thymine (A=T).

Three hydrogen bonds form between cytosine and guanine (C≡G).

Notes for active learning

Notes for active learning

Glossary of Organic Chemistry Terms

A

Acetal – product formed by the reaction of an aldehyde with *alcohol*; the general structure of an acetal is:

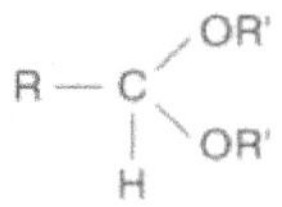

$$R-\underset{H}{\overset{|}{C}}(OR')(OR')$$

Achiral (or non-chiral) – the opposite of *chiral*; can be superimposed on its mirror image (e.g., CH_4); do not rotate plane-polarized light.

Acid – an agent able to produce positively charged hydrogen ions (H^+); since the hydrogen ion is a bare proton, it usually exists in a solvated form (such as H_3O^+); a proton donor or an electron pair acceptor; see *Brønsted-Lowry theory of acids and bases*, *Lewis acid* and *Lewis base*.

Acid-base reaction – a neutralization reaction in which the products are salt and water.

Activated complex – molecules at an unstable intermediate stage in a reaction.

Activating group – a substituent that increases the rate of electrophilic aromatic substitution (EAS) when bonded to an aromatic ring.

Activation energy – the minimum energy which reacting species must possess in order to be able to form an "activated complex," or "transition state," before proceeding to the products; the difference in potential energy between the ground state and the transition state of molecules; molecules of reactants must have this amount of energy to proceed to the product state; the activation energy (E_a) may be derived from the temperature dependence of the reaction rate using the *Arrhenius equation*.

Acyl group – a substituent with the following structure, where R can be an alkyl or aryl group:

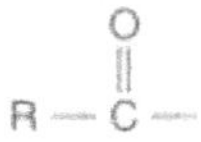

$$R-\overset{O}{\underset{}{\overset{\|}{C}}}-$$

Acyl halide – a compound with the general structural formula:

$$R-\overset{O}{\underset{}{\overset{\|}{C}}}-X$$

Acylation – a reaction in which an acyl group is added to a molecule.

Acylium ion – the resonance stabilized cation:

$$R\overset{+}{C}=\ddot{O}: \longleftrightarrow RC\equiv\overset{+}{O}:$$

Addition – a reaction that produces a new compound by combining the elements of the original reactants.

Addition elimination mechanism – the two-stage mechanism by which nucleophilic *aromatic* substitution (NAS) occurs; in the first stage, the addition of the nucleophile to the carbon bearing the *leaving group* occurs; an elimination follows in which the leaving group is expelled.

Addition reactions – an unsaturated system is saturated or partly saturated by adding a molecule across the multiple bonds (e.g., adding bromine to ethene to form 1,2-dibromoethane).

Adduct – the product of an addition reaction.

Alcohol – a molecule containing a hydroxyl (~OH) group; also, a functional group.

Aldehyde – a molecule containing a terminal carbonyl (~CHO) group; also, a functional group.

Alicyclic compound – an *aliphatic cyclic* hydrocarbon; a compound contains a ring but not an *aromatic* ring; see *aliphatic compound* and *cyclization*.

Aliphatic compound – a straight-chain or branched-chain hydrocarbon, an *alkane*, *alkene* or *alkyne*.

Alkaloid – organic substances occurring naturally, which are basic, forming salts with acids; the basic group is usually an amino function.

Alkane – a hydrocarbon that contains only single covalent bonds (i.e., containing only C–H and C–C single bonds); the general *alkane* formula is C_nH_{2n+2}.

Alkene – a molecule containing one or more carbon-carbon double bonds; also, a *functional group*.

Alkoxide ion – an anion formed by removing a proton from an *alcohol*; the RO⁻ ion.

Alkoxy free radical – formed by homolytic cleavage of an *alcohol* ~OH bond; the RO· radical.

Alkyl group – an *alkane* molecule from which a hydrogen atom has been removed; abbreviated as "R" in structural formulas.

Alkyl halide – a hydrocarbon that contains a halogen substituent, such as fluorine, chlorine, bromine, or iodine.

Alkyl-substituted cycloalkane – a cyclic hydrocarbon to which one or more *alkyl groups* are bonded; compare with *cycloalkyl alkane*.

Alkylation – a reaction in which an *alkyl group* is added to a molecule.

Alkyne – a molecule containing one or more carbon-carbon triple bonds; also, a functional group; the general formula is C_nH_{2n-2}.

Allyl group – the H₂C=CHCH₂ group contains 3 carbon atoms and a double bond, $C_1=C_2-C_3$, where C_3 is the "allylic position" or "*allylic carbon* atom."

Allylic carbon – a sp^3 carbon adjacent to a double bond.

Allylic carbocation – the $H_2C=CHCH_2^+$ ion.

Allylic rearrangement – the migration of a double bond in a 3-carbon system from carbon atoms one and two to carbon atoms two and three (e.g., $C_1=C_2-C_3-X$ to $X-C_1-C_2=C_3$).

Analogue – in organic chemistry, chemicals that are similar but not identical (e.g., the hydrocarbons are similar, but an *alkane* is different from *alkenes* and *alkynes* because of the types of bonds they contain; therefore, an *alkane* and an *alkene* are analogs).

Angle of rotation (α) – in a polarimeter, the angle right or left in which plane-polarized light is turned after passing through an optically active compound in solution.

Amide – a molecule containing a carbonyl group attached to nitrogen (~$CONR_2$); also, a functional group.

Amine – a molecule containing isolated nitrogen (NR_3); also, a functional group.

Anion – a negatively charged ion.

Anomers – the specific term used to describe carbohydrate stereoisomers differing only in configuration at the hemiacetal carbon atom.

***Anti*-addition** – a reaction in which the two groups of a reagent X–Y add on the opposite faces of a carbon-carbon bond.

***Anti*-aromatic** – a highly unstable planar ring system with 4*n pi* electrons.

Antibonding molecular orbital – contains more energy than the atomic orbitals (AO) from which it was formed; an electron is less stable in an antibonding orbital than in its original atomic orbital.

***Anti*-conformation** – a type of staggered conformation in which the two big groups are opposite each other in a *Newman projection*.

***Anti*-Markovnikov addition** – a reaction in which the hydrogen atom of a hydrogen halide bonds to the carbon of a double bond that is bonded to *fewer* hydrogen atoms; the addition takes place *via* a free-radical intermediate rather than a carbocation; compare with *Markovnikov rule*.

***Anti*-periplanar (anti-coplanar)** – the conformation in which hydrogen and the leaving group are in the same plane and on opposite sides of a carbon-carbon single bond; also, the conformation required for E_2 elimination.

Aprotic solvents – solvents that do not contain O–H or N–H bonds.

Arene – an *Aromatic* hydrocarbon.

Aromatic – possesses aromaticity; aromaticity is the property of planar (or nearly planar) cyclic, conjugated systems having (4n + 2) conjugated *pi* electrons; the delocalization of the (4n + 2) *pi* electrons give them exceptional stability (aromatic compounds are unusually stable compounds); for benzene, the most common aromatic system (n = 1, therefore 6 *pi* electrons), the aromaticity confers the characteristic reactivity of electrophilic substitution.

Aromatic compound – possesses a closed-shell electron configuration and resonance; obeys *Hückel's rule*.

Aryl – an *aromatic* group as a substituent.

Aryl group – a group, produced by the removal of a proton from an *aromatic* molecule.

Aryl halide – a compound in which a halogen atom is attached to an *aromatic* ring.

Association – a term applied to the combination of molecules of a substance with one another to form more complex systems; see *dissociation* and *dissociation constant*.

Asymmetry – a term applied to an object or molecule that does not possess symmetry.

Asymmetric induction – a term applied to the selective synthesis of one diastereomeric form of a compound resulting from the influence of an existing *chiral* center adjacent to the developing asymmetric carbon atom; this usually arises because, for steric reasons, the incoming atom or group does not have equal access to both sides of the molecule.

Atom – the smallest amount of an element; a nucleus surrounded by electrons.

Atomic mass (A) – the sum of the weights of the protons and neutrons in an atom; a proton and neutron each have a mass of 1 atomic mass unit.

Atomic number (Z) – the number of protons or electrons in an atom.

Atomic 1s orbital – the spherical orbital nearest the nucleus of an atom.

Atomic orbital – a region in space around the nucleus of an atom where the probability of finding an electron is high, which may be described in terms of the four quantum numbers.

Atomic p orbital – an hourglass-shaped orbital, oriented on x, y and z-axes in three-dimensional space.

Atomic s orbital – a spherical orbital.

Avogadro's constant – number of particles (atoms or molecules) in one mole of a pure substance: 6.022×10^{23}.

Axial bond – a bond perpendicular to the equator of the ring (up or down), typically in a chair cyclohexane.

B

Baeyer reagent – cold, dilute potassium permanganate; used to oxidize *alkenes* and *alkynes*.

Base – a substance that can combine with a proton (i.e., a proton acceptor or an electron-pair donor); see *Brønsted-Lowry theory of acids and bases*, *Lewis acid* and *Lewis base*.

Benzene (Benzenoid) ring – an *Aromatic* ring with a benzene-like structure.

Benzyl group – a *benzene ring* plus a methylene (~CH$_2$~) unit: $C_6H_5CH_2$.

Benzylic position – the position of carbon attached to a *benzene ring*.

Benzyne – an unstable intermediate consisting of a benzene ring with an additional (triple) bond created by the side-to-side overlap of sp^2 orbitals on adjacent carbons of the ring.

Bicyclic – a molecule with two rings that share at least two carbons.

Bimolecular reaction – two species (e.g., molecules, ions, radicals) react to form new chemical species; most reactions are bimolecular or proceed through a series of bimolecular steps.

Bond angle – the angle formed between two adjacent bonds on the same atom.

Bond-dissociation energy – the amount of energy needed to homolytically fracture a bond.

Bond energy – the energy required to break a particular bond by a homolytic process.

Bond length – the equilibrium distance between the nuclei of two atoms or groups bonded to each other.

Bond strength – see *bond-dissociation energy*.

Bonding electron – see *valence electrons*.

Bonding molecular orbital – formed by the overlap of adjacent atomic orbitals.

Branched-chain *Alkane* – an *alkane* with *alkyl groups* bonded to the central carbon chain.

Brønsted-Lowry theory of acids and bases – a compound capable of donating a proton (a hydrogen ion); a Brønsted-Lowry base can accept a hydrogen ion; in *neutralization,* an acid donates a proton to a base, creating a conjugate acid and a conjugate base.

Buffer solution – a solution of definite pH made so that the pH alters only gradually by adding an acid or base.

C

Canonical structures – any of two or more hypothetical structures of *resonance* theory which can be written for a molecule simply by rearranging the valence electrons of the molecule (e.g., the two Kekule structures of benzene); sometimes called "valence bond isomers."

Carbanion – a carbon atom bearing a negative charge; a carbon anion.

Carbene – a reactive intermediate, characterized by a neutral, electron-deficient carbon center with two substituents - two single bonds and just six electrons in its valence shell (R_2C:).

Carbenoid – a chemical that resembles a carbene in its chemical reactions.

Carbocation – a carbon *Cation*; a carbon atom bearing a positive charge (sometimes referred to as a "carbonium ion").

Carbonyl group – a carbon double bonded to oxygen (C=O).

Carboxylic acid – a molecule containing a carboxyl (~COOH) group; also, a functional group.

$$-\overset{\overset{\displaystyle}{\|}}{\underset{\displaystyle O}{C}}-OH$$

Catalyst – a substance that, when added to a reaction mixture, changes (speeds up) the rate of attainment of equilibrium in the system without itself undergoing a permanent chemical change.

Catalytic cracking – the method for producing gasoline from heavy petroleum distillates; generally, the catalysts are mixtures of silica and alumina or synthetic conjugates, such as zeolites.

Catalytic reforming – the process of improving the octane number of straight-run gasoline by increasing the proportion of *aromatic* and *branched-chain alkanes*; catalysts employed are either molybdenum-aluminum oxides or platinum-based.

Cation – a positively charged ion.

Cationic polymerization – occurs *via* a cation intermediate and is less efficient than *free-radical polymerization*.

Chain reaction – once started, produces enough energy to keep the reaction running; proceed by a series of steps which produce intermediates, energy, and products (e.g., the free radical addition of hydrogen bromide (HBr) to an *alkene*).

Chair conformation – typically, the most stable cyclohexane conformation; looks like a chair.

Chemical shift – the location of an NMR peak relative to the standard tetramethylsilane (TMS), given in units of parts per million (ppm).

Chiral – describes a molecule that is not superimposable on its mirror image, like the relationship of a left hand to a right hand; see *chiral molecule*.

Chiral center – a carbon or other atom with four non-identical substituents.

Chiral molecule – a molecule that is not superimposable on its mirror image; rotate plane-polarized light.

Chirality – any asymmetric object or molecule; the property of non-identity of an object with its mirror image.

Chromatography – a series of related techniques for separating a mixture of compounds by their distribution between two phases; in gas-liquid chromatography, the distribution is between a gaseous and a liquid phase; in column chromatography, the distribution is between a liquid and a solid phase.

Cis – two identical substituents on the same side of a double bond or ring.

Closed-shell electron configuration – a stable electron configuration in which the electrons are in the lowest energy orbitals available.

Competing reactions – two reactions that start with the same reactants but form different products.

Compound – a term used generally to indicate a definite combination of elements into a more complex structure (a molecule); also applied to systems with non-stoichiometric proportions of elements.

Concerted – taking place at the same time without forming an intermediate.

Condensation reaction – a reaction in which two molecules join with the liberation of a small stable molecule.

Configuration – the order and relative three-dimensional orientation of the atoms in a molecule; given the designation R or S; "absolute configuration" is when the relative three-dimensional arrangement in the space of atoms in a *chiral* molecule has been correlated with an absolute standard.

Configurational isomers – a series of compounds that have the same constitution and bonding of atoms, but which differ in their atomic spatial arrangement (e.g., glucose and mannose).

Conformation – the spatial arrangement of a molecule in space at any moment; most molecules can adopt an infinite number of conformations because of rotation about single covalent bonds; of these possibilities, most compounds spend the most time in one or a few *preferred conformations*.

Conformer – a conformation of a molecule; generally, these will be at energy minima.

Conjugate acid – the acid that results when a Brønsted-Lowry base accepts a hydrogen ion.

Conjugate base – the base that results when a Brønsted-Lowry acid loses a hydrogen ion.

Conjugated double bonds – double bonds separated by a carbon-carbon single bond; alternating double bonds.

Conjugation – the overlapping in all directions of a series of *p* orbitals; a sequence of alternating double (or triple) and single bonds (e.g., C=C–C=C and C=C–C=O); can be relayed by the participation of lone pairs of electrons or vacant orbitals.

Conjugation energy – see *resonance energy*.

Constitution – the number and type of atoms in a molecule.

Constitutional isomers – molecules with the same molecular formula but with atoms attached differently.

Coordinate bond – the linkage of two atoms by a pair of electrons, both electrons provided by one of the atoms (the donor); covalent bonds.

Coupling constant (*J*) – the distance between two neighboring lines in an NMR peak (given in units of Hz); the separation in frequency units between multiple peaks in one chemical shift; this separation results from the spin-spin coupling.

Coupling protons – protons interacting and splitting the NMR peak into some lines following the *n*+1 rule.

Covalent bond – a bond formed by the sharing of electrons between atoms.

Cyano group – the ~C≡N group.

Cyanohydrin – a compound with the general formula:

$$R-\underset{\underset{R'}{|}}{\overset{\overset{OH}{|}}{C}}-C\equiv N$$

Cyclization – the forming of ring structures.

Cycloaddition – a reaction that forms a ring.

Cycloalkane – a ring hydrocarbon made of carbon and hydrogen atoms joined by single bonds.

Cycloalkyl *Alkane* – an *alkane* to which a ring structure is bonded.

Cyclohydrocarbon – an *alkane*, *alkene* or *alkyne* formed in a ring structure rather than a straight or branched chain; the general formula is C_nH_{2n} (n must be a whole number of 3 or greater).

D

Deactivating group – causes an *Aromatic* ring to become less reactive toward electrophilic aromatic substitution.

Debye unit (D) – the unit of measure for a dipole moment; one debye equals 1.0×10^{-18} electrostatic units (esu · cm); see *dipole moment*.

Decarboxylation – a reaction in which carbon dioxide is expelled from a carboxylic acid.

Dehalogenation – the elimination reaction in which two halogen atoms are removed from adjacent carbon atoms to form a double bond.

Dehydration – the elimination reaction in which water is removed from a molecule.

Dehydrohalogenation – the elimination reaction in which a hydrogen atom and a halogen atom (a hydrohalic acid, like HBr, HCl) are removed from a molecule to form a double bond.

Delocalization – electron systems in which bonding electrons are not localized between two atoms as for a single bond but are spread (delocalized) over the whole group (e.g., *pi-bond* electrons the delocalized pi-electrons associated with *aromatic* molecules).

Delocalization energy – see *resonance energy*.

Delta value (δ value) – the chemical shift; the location of an NMR peak relative to the reference standard tetramethylsilane (TMS), given in units of parts per million (ppm).

Deprotonation – the loss of a proton (hydrogen ion) from a molecule.

Deshielding – an effect in NMR spectroscopy that the movement of *sigma* and *pi* electrons within the molecule causes; causes chemical shifts to appear at lower magnetic fields (downfield).

Dextrorotatory – the phenomenon in which plane-polarized light is turned in a clockwise direction.

Diastereomers (diastereoisomers) – stereoisomeric structures which are not enantiomers (mirror images); often applied to systems that differ in the configuration at one carbon (e.g., *meso-* and *d-* or *l-*tartaric acids).

Diels-Alder reaction – a cycloaddition between a conjugated diene and an *alkene* that produces a 1,4-addition product.

Diene – a molecule that contains two alternating double bonds; also, a reactant in the *Diels-Alder reaction*.

Dienophile – a reactant in the *Diels-Alder reaction* that contains a double bond; often substituted with electron-withdrawing groups.

Dienophile – the *alkene* that adds to the *diene* in a Diels-Alder reaction.

Dihalide – a compound that contains two halogen atoms; also called "*a dihaloalkane.*"

Dihedral angle – the angle between groups attached on adjacent carbons when viewed in a *Newman projection*.

Diol – a compound that contains two hydroxyls (~OH) groups; also called a "*dihydroxy alkane.*"

Dipole moment – a measure of the polarity of a molecule; the mathematical product of the charge in electrostatic units (ESU), and the distance that separates the two charges in centimeters (cm) (e.g., substituted *Alkyne*s have dipole moments caused by differences in electronegativity between the triple-bonded and single-bonded carbon atoms).

Disproportionation – a process in which a compound of one oxidation state changes to compounds of two or more oxidation states (e.g., $2\ Cu^+ \rightarrow Cu + Cu^{2+}$).

Dissociation – the process whereby a molecule splits into simpler fragments that may be smaller molecules, atoms, free radicals, or ions.

Dissociation constant – the measure of the extent of dissociation, measured by the dissociation constant K. For the process: $AB = A + B$

$$K = ([A] \cdot [B]) / [AB]$$

Dissymmetric – see *chiral*.

Distillation – the separation of components of a liquid mixture based on differences in boiling points.

Double bond – some atoms can share two pairs of electrons to form a double bond (two covalent bonds); formally, the second (double) bond arises from the overlap of *p* orbitals from two atoms, already united by a *sigma* bond, to form a *pi bond*; hydrocarbons that contain one double bond are *alkenes*, and hydrocarbons with two double bonds are *dienes*.

Doublet – describes an NMR signal split into two peaks.

Dyestuffs – intensely colored compounds applied to a substrate; colors are due to the absorption of light to give electronic transitions.

E

E₁ elimination reaction – a reaction that eliminates a hydrohalic acid (e.g., HCl, HBr) to form an *alkene*; a first-order reaction that goes through a *carbocation* mechanism.

E₂ elimination reaction – a reaction that eliminates a hydrohalic acid (e.g., HCl, HBr) to form an *alkene*; a second-order reaction that occurs in a single step in which the double bond is formed as the hydrohalic acid is eliminated.

E **isomer** – *stereoisomer* with the two highest priority groups on opposite sides of a ring or double bond.

Eclipsed – a *conformation* in which substituents on two attached saturated carbon atoms overlap when viewed as a *Newman projection*.

Eclipsed conformation – *conformation* about a carbon-carbon single bond in which the bonds off two adjacent carbons are aligned (0° apart when viewed in a *Newman projection*).

Electron – negatively charged particles of little weight that exist in quantized probability areas around the atomic nucleus.

Electron affinity – the amount of energy liberated when an electron is added to an atom in the gaseous state.

Electronegativity – measures an atom's ability to attract electrons toward itself in a covalent bond; the halogen fluorine is the most electronegative element; measured by the highest occupied molecular orbital (*HOMO*) and the lowest unoccupied molecular orbital (*LUMO*) energy levels.

Electronegativity scale – an arbitrary reference by which the electronegativity of elements can be compared.

Electronic configuration – the order in which electrons are arranged in an atom or molecule; used in a distinct and different sense from stereochemical *configuration*; see *stereochemistry*.

Electronic transition – in an atom or molecule, the electrons only have specific allowed energies (orbitals); if an electron passes from one orbital to another, an electronic transition occurs, and the emission or absorption of energy corresponds to the difference in energy of the two orbitals.

Electrophile – an "electron seeker;" an atom, molecule, or ion able to accept an electron pair to stabilize itself; a *Lewis acid*.

Electrophilic addition – a reaction in which the addition of an *electrophile* to an unsaturated molecule forms a saturated molecule.

Electrophilic substitution – an overall reaction in which an *electrophile* binds to a substrate with the expulsion of another electrophile (e.g., the electrophilic substitution of a proton by another electrophile, such as a nitronium ion, on an *Aromatic* substrate, such as benzene).

Electrostatic attraction – the attraction of a positive ion for a negative ion.

Electrovalent (ionic) bond – bonding by *electrostatic attraction*.

Element – a substance which cannot be further subdivided by chemical methods.

Element of unsaturation – a *pi bond*; a multiple bond or ring in a molecule.

Enantiomers – a pair of isomers related as mirror images of one another (e.g., isomers differing only in the configuration of the *chiral* atoms).

Enantiomorphic pair – in optically active molecules with more than one stereogenic center, the two structures are mirror images.

Endothermic – a reaction in which heat is absorbed.

Energy diagram (or *reaction energy diagram*) – a graph of the energy against the progress of the reaction.

Energy of reaction – the difference between the total energy content of the reactants and the total energy content of the products; the greater the energy of reaction, the more stable the products.

Enol – an unstable compound (e.g., *vinyl alcohol*) in which a hydroxide group is attached to a carbon in a carbon-carbon double bond; these compounds *tautomerize* to form more stable *ketones*.

Enolate ion – the resonance stabilized ion formed when an *aldehyde* or *ketone* loses an α hydrogen:

$$H-\overset{H}{\underset{H}{C}}-\overset{O}{\overset{\|}{C}}-H \longleftrightarrow H-\overset{H}{\underset{H}{C}}=\overset{O^-}{\underset{}{C}}-H$$

Enthalpy (*H*) – a thermodynamic state function, generally measured in kilojoules per mole; in chemical reactions, the enthalpy change (ΔH) is related to changes in the free energy (ΔG) and *entropy* (ΔS) by the equation: $\Delta G = \Delta H - T\Delta S$.

Entropy (*S*) – a thermodynamic quantity that is a measure of the degree of disorder within a system. The greater the degree of order, the higher the entropy; for an increase in entropy, *S* is positive; has the units of joules per degree K per mole.

Enzyme – a naturally occurring substance able to catalyze a chemical reaction.

Epimerization – a process in which the configuration about one *chiral* center of a compound, containing more than one *chiral* atom, is inverted to give the opposite configuration; the term "epimers" describes two related compounds that differ only in the configuration about one chiral atom.

Epoxide – a three-membered ring that contains oxygen.

Epoxidation – the addition of an oxygen bridge across a double bond to give an oxirane; achieved by use of a peracid or, in a few cases, by use of a *Catalyst* and oxygen.

Equatorial – the bonds in a chair cyclohexane oriented along the equator of the ring.

Equilibrium constant – according to the law of mass action, for a reversible chemical reaction, aA + bB = cC + dD, the equilibrium constant (*K*) is defined as: $K = ([C]^c[D]^d) / ([A]^a[B]^b)$.

Ester – a functional group; a molecule containing a carbonyl group adjacent to an oxygen (RCOOR').

$$-\overset{\overset{\displaystyle O}{\|}}{C}-OR$$

Ether – a molecule containing oxygen singly-bonded to two carbon atoms; also, a functional group; the general formula is R—O—R'; epoxyethane, an epoxide, is a cyclic ether; often refers to diethyl ether.

Excited state – the state of an atom, molecule, or group when it has absorbed energy and becomes excited to a higher energy state than the ground state; may be electronic, vibrational, rotational, etc.

F

Fischer projection – a convention for drawing carbon chains so that the relative three-dimensional stereochemistry of the carbon atoms is relatively easy to portray as a 2-dimensional drawing.

Fingerprint region – an IR spectrum below 1,500 cm^{-1}; often complex and difficult to interpret.

Free energy (ΔG) – a thermodynamic state function; the free energy change (ΔG) in any reaction is related to the *enthalpy* and *entropy*: $\Delta G = \Delta H - T\Delta S$.

Free radicals – molecules or ions with unpaired electrons; generally, extremely reactive; "stable" free radicals include molecular oxygen, NO, and NO$_2$; organic free radicals range from those of transient existence only to very long-lived species; alkyl free radicals tend to be very reactive and short-lived.

Free-radical chain reaction – a reaction that proceeds by a free-radical intermediate in a chain mechanism (a series of self-propagating, interconnected steps); compare with a *free-radical reaction*.

Free-radical polymerization – a polymerization initiated by a *free radical*.

Free-radical reaction – a reaction in which a covalent bond is formed by the union of two radicals; compare with a *free-radical chain reaction*.

Frontier orbital symmetry – the theory that the site and rates of reaction depend on the geometries, the sign of the wave function and relative energies of the highest occupied molecular orbital (*HOMO*) of one molecule and the lowest unoccupied molecular orbital (*LUMO*) of the other.

Functional group – a set of bonded atoms that displays a specific molecular structure and chemical reactivity when bonded to a carbon atom in the place of a hydrogen atom.

G

Gauche – a conformational isomer in which the groups are neither eclipsed nor trans to one another; often taken as the conformation where the dihedral angle between the groups is 60°.

Gauche conformation – a type of staggered conformation in which two bulky groups are next to each other.

Geometrical isomerism – isomerism from the restricted rotation about a bond (e.g., (*Z*) and (*E*) isomers of unsymmetrically substituted *alkene*s).

Grignard reagent – an organometallic reagent in which magnesium metal inserts between an *alkyl group* and a halogen (e.g., CH_3MgBr).

Ground state – the lowest energy state of an atom, molecule, or ion.

H

Half-life, $t^{1/2}$ – the time taken for the concentration of a substance in a reaction to reduce to half its original value; used in first-order reactions and as a measure of the rate of radioactive decay.

Halide – a member of the VIIA column of the periodic table (e. g., F, Cl, Br, I) or a molecule that contains one of these atoms; also, a functional group.

Haloalkane – an *alkane* that contains one or more halogen atoms; also called an *alkyl halide*.

Halogen – an electronegative, nonmetallic element in Group VII of the periodic table, including fluorine, chlorine, bromine, and iodine; often represented in structural formulas by an "X."

Halogenation – a reaction in which halogen atoms are bonded to an *Alkene* at the double bond.

Halonium ion – a halogen atom that bears a positive charge; highly unstable.

Hard and soft acids and bases – a classification of acids and bases depending on their polarizability; hard bases include fluoride ions; soft bases include triphenylphosphine; hard acids include Na^+, whilst an example of a soft, polarizable acid is Pt^{2+}; hard-hard and soft-soft interactions are favored; hardness and softness can be described in terms of the *HOMO* and *LUMO* interactions.

Heat of reaction – the amount of heat absorbed or evolved when specified amounts of compounds react under constant pressure; expressed as kilojoules per mole; for exothermic reactions, the convention is that the *Enthalpy* (*H*) change (heat of reaction) is negative.

Hemiacetal – a functional group with a hydroxyl and ether attached to the same carbon; the structure:

```
      OR
   —C<
    |  OH
    H
```

Hemiketal – a functional group with a hydroxyl and ether attached to the same carbon; the structure:

$$-\underset{R}{\underset{|}{C}}\begin{matrix}OR\\OH\end{matrix}$$

Hertz – a measure of a wave's frequency; equals the number of waves that passes a specific point per second.

Heteroatom – in organic chemistry, an atom other than carbon.

Heterocyclic compound – a class of cyclic compounds in which one of the ring atoms is not carbon (e.g., epoxyethane).

Heterogeneous reaction – occurs between substances mainly present in different phases (e.g., between a gas and a liquid).

Heterogenic bond formation – a type of bond formed by the overlap of orbitals on adjacent atoms. One orbital of the pair donates both electrons to the bond.

Heterolytic cleavage – the fracture of a bond so that one of the atoms receives both electrons; in reactions, this asymmetrical bond rupture generates carbocation and carbanion mechanism.

Heterolytic reaction – a reaction with a covalent bond broken by unequal sharing of bonding electrons.

HOMO – the highest occupied molecular orbital of a molecule, ion, or atom.

Homologous series – compounds with common compositions (e.g., *alkane*s, *alkene*s and *alkyne*s).

Homolog – one of a series of compounds in which each member differs by a constant unit.

Homolytic cleavage – the fracture of a bond in such a manner that both atoms receive one of the bond's electrons; this symmetrical bond rupture forms free radicals; in reactions, it generates *free-radical* mechanisms.

Homolytic reaction – a covalent bond is broken with equal sharing of the electrons from the bond.

Hückel's rule – a compound with $4n + 2$ π electrons has a closed-shell electron configuration and is *aromatic*.

Hybrid orbitals – formed from mixing atomic orbitals (AO), like the sp^x orbitals, which result from mixing *s* and *p* orbitals.

Hydration – the addition of the elements of water to a molecule.

Hydride shift – the movement of a hydride ion (a hydrogen atom with a negative charge; H^-) to form a more inductively-stabilized carbocation.

Hybridization – the process whereby atomic orbitals of different types but similar energies are combined to form a set of equivalent hybrid orbitals; these hybrid orbitals do not exist in the atoms but only by forming molecular orbitals by combining atomic orbitals from different atoms.

Hydroboration – the *cis*-addition of B–H bonds across double (or triple) carbon-carbon bonds.

Hydroboration-oxidation – the addition of borane (BH_3) or an alkyl borane to an *alkene* and its subsequent oxidation to produce the *anti-Markovnikov* indirect addition of water.

Hydrocarbon – a molecule that exclusively contains carbon and hydrogen atoms; the central bond may be a single, double, or triple covalent bond, and it forms the molecule's backbone.

Hydrogenation – the addition of hydrogen to a multiple bond.

Hydrogenolysis – the cleaving of a chemical bond by hydrogen, generally with a hydrogenation *catalyst*.

Hydrohalogenation – a reaction in which a hydrogen atom and a halogen atom are added to a double bond to form a saturated compound.

Hydrolysis – the addition water to a substance, often with the partition of the substance into two parts (e.g., the hydrolysis of an *ester* to an acid and an *alcohol*).

Hydrolyze – to cleave a bond *via* the elements of water.

Hyperconjugation – weak interaction (electron donation) between *sigma* bonds with *p* orbitals; explains why alkyl substituents stabilize carbocations.

I

Inductive effect – an electronic effect transmitted through bonds in an organic compound due to the electronegativity of substituents and the permanent polarization thereof; the substituent either induces charges towards or away from itself with the formation of a dipole.

Infrared spectroscopy (IR) – the study of the absorption of infrared light by substances; corresponding to vibrational (and some rotational) changes, infrared spectroscopy provides valuable information about the molecule's structure; detailed correlation tables exist relating infrared bands (absorbances) to functional groups.

Inhibitor – a general term for a compound that inhibits (slows down) a reaction; can be used to slow or stop free radical chain reactions.

Initiation step – the first step in the mechanism of a reaction.

Initiator – a material capable of being fragmented into free radicals, which initiates a *free-radical reaction*.

Insertion – placing between two atoms.

Intermediate – a species that form in one step of a multistep mechanism; unstable and cannot be isolated.

Ion – an atom or group of atoms that has lost or gained one or more electrons to become a charged species.

Ionic bond – a bond formed by the transfer of electrons between atoms, resulting in forming ions of opposite charge; the electrostatic attraction between these ions.

Ionization energy – the energy needed to remove an electron from an atom.

IR spectroscopy – an instrumental technique that measures IR (infrared) light absorption by molecules and can determine functional groups in an unknown molecule.

Isolated double bond – a double bond more than one single bond away from another double bond in a *diene*.

Isomers – compounds having the same atomic composition (*constitution*) but differing in their chemical structure; includes structural isomers (chain or positional), tautomeric isomers, and stereoisomers (including geometrical isomers, optical isomers, and conformational isomers).

IUPAC nomenclature – a systematic method for naming molecules based on a series of rules developed by the International Union of Pure and Applied Chemistry, not the only reference body, but the most common.

J

***J* value** – the coupling constant between two peaks in an NMR signal; given in units of Hz.

K

Kekulé structure – the structure for benzene in which there are three alternating double and single bonds in a six-membered ring of carbon atoms.

Ketal – the product formed by the reaction of a *ketone* with *alcohol*; the general structure is:

$$R-C(OR')(OR')R$$

Keto-enol tautomerization – the process by which an *enol* equilibrates with its corresponding *aldehyde* or *ketone*.

Ketone – a compound in which an oxygen atom is bonded *via* a double bond to a carbon atom, bonded to two more carbon atoms.

Kinetic product – the product that forms the fastest; has the lowest *activation energy*.

Kinetics – the study of the rate of reactions.

Kinetically controlled product – the product formed from the fastest reaction in competing reaction pathways.

L

Levorotatory – the phenomenon that turns plane-polarized light in a counterclockwise direction.

LCAO – a method for calculating molecular orbitals from a "Linear Combination of Atomic Orbitals."

Leaving group – the negatively charged group that departs from a molecule undergoing a *nucleophilic substitution* reaction.

Lewis acid – an agent capable of accepting a pair of electrons to form a *coordinate bond*.

Lewis base – an agent capable of donating a pair of electrons to form a *coordinate bond*.

Lone pair – a pair of electrons in a molecule not shared by two of the constituent atoms.

Linear – the shape of a molecule with *sp* hybrid orbitals; an *alkyne*.

LUMO – the lowest unoccupied molecular orbital in a molecule or ion.

M

Markovnikov rule – the positive part of a reagent (e.g., a hydrogen atom) adds to the carbon of the double bond that already has more hydrogen atoms attached; the negative part adds to the other carbon of the double bond; leading to the more stable *carbocation* over other less-stable intermediates; useful to predict the major product; *free radical reactions* proceed in the opposite sense, giving rise to *anti-Markovnikov addition*.

Mass number – the total number of protons and neutrons in an atom.

Mass spectrometry – a form of spectrometry in which, generally, high energy electrons are bombarded onto a sample, generating charged fragments of the parent substance; electrostatic and magnetic fields then focus these ions for a spectrum of the charged fragments.

Mechanism – the series of steps that reactants go through during their conversion into products.

Meso compounds – molecules with *chiral* centers but are *achiral* due to one or more planes of symmetry.

Mesomerism – see *resonance*.

Meta – describes the positions of two substituents on a *benzene ring* separated by one carbon.

Meta-directing substituent – groups on an *aromatic* ring directing incoming *electrophile*s to meta position.

Methylene group – a ~CH_2~ group.

Microwave spectroscopy – the interaction of electromagnetic waves with wavelengths in the range 10^{-2} to 1 meter; this energy range corresponds to rotational frequencies; helpful in studying the structure of materials (generally gases) and their characterization.

Molecular ion – the fragment in a mass spectrum corresponding to the *cation* radical (M+) of the molecule; gives the molecular mass of the molecule.

Molecular orbitals – the electron orbitals belonging to a group of atoms forming a molecule.

Molecular orbital theory – a model for depicting the location of electrons that allows electrons to delocalize across the entire molecule, a more accurate but less user-friendly theory than the *valence bond theory*.

Molecule – a covalently-bonded collection of atoms with no electrostatic charge; the smallest particle of matter that can exist in a free state; in the case of ionic substances, such as sodium chloride, the molecule is considered as a pair of ions (e.g., NaCl).

Multiple bonds – a double or triple bond; atomic *p* orbitals in side-to-side overlap, preventing rotation.

Multistep synthesis – synthesis of a compound that takes several steps to achieve.

N

n+1 rule – rule for predicting the coupling for a proton in ^{1}H NMR spectroscopy; an NMR signal splits into n+1 peak, where n is the number of equivalent adjacent protons.

Natural product – a compound produced by a living organism.

Neutralization – the reaction of an acid and a base; the acid and base reaction products are salt and water.

Neutron – an uncharged particle in the atomic nucleus with the same weight as a proton; additional neutrons do not change an element but convert it to one of its isotopic forms.

Newman projection – a projection obtained by viewing along a carbon-carbon single (double) bond.

Nitrile – a compound with a cyano group (a carbon triply-bonded to nitrogen (~C≡N)); a functional group.

NMR – nuclear magnetic resonance spectroscopy; a technique that measures radiofrequency light absorption by molecules; a powerful structure-determining method; see *nuclear magnetic resonance spectroscopy*.

Node – a region of zero electron density in an orbital; a point of zero amplitude in a wave.

Nonbenzenoid aromatic ring – an *aromatic* ring system that does not contain a *Benzene ring*.

Nonbonding electrons – *valence electrons* not used for covalent bond formation.

Nonterminal Alkyne – an *alkyne* in which the triple bond is somewhere other than the 1 position.

Nuclear magnetic resonance (NMR) spectroscopy – a form of spectroscopy that depends on the absorption and emission of energy arising from changes in the spin states of the nucleus of an atom; for aggregates of atoms, as in molecules, minor variations in these energy changes are caused by the local chemical environment; the energy changes used are in the radiofrequency range of the electromagnetic spectrum and depend upon the magnitude of an applied magnetic field.

Nucleofuge – see *leaving group*.

Nucleophile – a "nucleus lover;" a molecule with the ability to donate a lone pair of electrons (a *Lewis base*).

Nucleophilic substitution – an overall reaction in which a *nucleophile* reacts with a compound displacing another nucleophile; such reactions commonly occur in aliphatic chemistry; if the reaction is unimolecular, they are *S_N1 reactions*; for bimolecular reactions, they are *S_N2 reactions*.

Nucleophilicity – a measure of the reactivity of a *nucleophile* in a *nucleophilic substitution* reaction.

Nucleus – the central core of an atom; the location of the protons and neutrons.

O

Optical activity – the property of certain substances to rotate plane-polarized light; associated with asymmetry; compounds that possess a *chiral* carbon atom of the same "handedness" rotate plane-polarized light; isomers that rotate light in equal but opposite directions are "optical isomers," although the better term is *enantiomers*.

Orbit – an area around an atomic nucleus with a high probability of finding an electron; also called a *shell*, divided into *orbitals* or *subshells*.

Orbital – an area in an *orbit* with a high probability of finding an electron; a "subshell;" the orbitals have the same principal and angular quantum numbers.

Organic compound – carbon-containing compound.

Ortho – describes the positions of two substituents on a *benzene ring* on adjacent carbons.

Ortho-para director – an *aromatic* substituent that directs incoming *electrophile*s to *ortho* or *para* positions.

Outer-shell electron – see *valence electrons*.

Overlap region – the region in space where atomic or molecular orbitals overlap, creating an area of high electron density.

Oxidation – a chemical process in which the proportion of electronegative substituents in a compound is increased (the loss of electrons by an atom in a covalent bond), the charge is made more positive, or the oxidation number is increased; in organic reactions, when a compound accepts additional oxygen atoms.

Oxonium ion – a positively-charged oxygen atom.

Ozonide – a compound formed by the addition of ozone to a double bond.

Ozonolysis – the cleavage of double and triple bonds by ozone, O_3.

P

Paired spin – the spinning in opposite directions of the two electrons in a bonding orbital.

Para – describes the positions of two substituents on a *benzene ring* separated by two carbons.

Parent name – the root name of a molecule according to the *IUPAC nomenclature* rules (e.g., hexane is the parent name in *trans*-1,2-dibromocyclohexane).

Peroxide – a compound that contains an oxygen-oxygen single covalent bond.

Peroxyacid – an acid of the general form:

$$R-\overset{\overset{\displaystyle O}{\|}}{C}-O-OH$$

Phenyl ring – a *benzene ring* as a substituent, abbreviated Ph.

Photochemical reaction – a chemical reaction brought about by the action of light.

Pi (π) bond – formed by the side-to-side overlap of atomic *p* orbitals (with electron density above and below the two atoms, but not directly between the two atoms); weaker than a *sigma bond* because of poor orbital overlap caused by nuclear repulsion; create unsaturated molecules; found in double and triple bonds.

Pi (π) complex – an intermediate formed when a *cation* is attracted to the high electron density of a *pi bond*.

Pi (π) molecular orbital – a molecular orbital created by the side-to-side overlap of atomic *p* orbitals.

pKa – the scale for defining a molecule's acidity (pKa = –log K_a).

Plane-polarized light – light that oscillates in a single plane.

Plane of symmetry – a plane cutting through a molecule in which both halves are mirror images of each other.

Polar covalent bond – a bond in which the shared electrons are not equally available in the overlap region, forming partially positive and partially negative ends on the molecule.

Polarimeter – measures rotation of plane-polarized light by a compound, generally prepared in a solution.

Polarity – the asymmetrical distribution of electrons in a molecule, positive and negative ends on the molecule.

Precursor – the substance from which another compound is formed.

Preparation – a reaction in which the desired chemical is produced (e.g., the dehydration of an *alcohol* is a preparation for an *alkene*).

Primary carbocation – a *carbocation* to which one *alkyl group* is bonded.

Primary (1°) carbon – a carbon atom that is attached to one other carbon atom.

Product – the substance that forms when reactants combine in a reaction.

Propagation step – the event in a free radical reaction in which both a product and energy are produced; the energy keeps the reaction going.

Protecting group – formed on a molecule by the reaction of a reagent with a substituent on the molecule; the resulting group is less sensitive to further reaction than the original group, but it must be easy to be reconverted to the original group.

Protic solvent – a solvent that contains O–H or N–H bonds.

Proton – an H^+ ion; also, a positively-charged nuclear particle.

Protonation – the addition of a proton (a hydrogen ion) to a molecule.

Pure covalent bond – in which the shared electrons are equally available to the bonded atoms.

Pyrolysis – the application of high temperatures to a compound.

R

R group – abbreviation given to an unimportant part of a molecule; indicates rest of molecule; see *alkyl group*.

Racemate – a 50:50 mixture of two enantiomers; another name for *racemic mixture*.

Racemic mixture – an equimolar mixture of the two enantiomeric isomers of a compound; because of the equal numbers of *levo-* and *dextro-*rotatory molecules present in a *racemate*, there is no net rotation of plane-polarized light (i.e., they are optically inactive).

Radical – an atom or molecule having one or more free valences; see *free radicals*.

Rate-determining step – the step in a reaction's mechanism that requires the highest activation energy and is, therefore, the slowest.

Rate of reaction – the speed with which a reaction proceeds.

Reactant – a starting material.

Reaction energy – the difference between the energy of the reactants and that of the products.

Reagent – the chemicals that ordinarily produce reaction products.

Rearrangement reaction – a reaction that causes the skeletal structure of the reactant to change in converting to product.

Reduction – chemical processes in which the proportion of more electronegative substituents is decreased, the charge is made more negative, or the oxidation number is lowered.

Resolution – the separation of a *racemate* into its two enantiomers using some *chiral* agency.

Resonance – 1) the representation of a compound by two or more canonical structures in which the *valence electrons* are rearranged to give structures of similar probability; the actual structure is a hybrid of the resonance forms; 2) the process by which a substituent removes electrons from or gives electrons to a *pi bond* in a molecule; a delocalization of electrical charge in a molecule.

Resonance energy – the difference in energy between the calculated energy content of a *resonance structure* and the actual energy content of the hybrid structure.

Resonance hybrid – the actual structure of a molecule that shows resonance; possesses the characteristics of possible structures (and consequently cannot be drawn); lower in energy than any structure for the molecule and is more stable.

Resonance structures – intermediate structures of one molecule that differ only in the positions of their electrons; used to depict the location of *pi* and nonbonding electrons on a molecule; a molecule looks like a hybrid of all resonance structures; none of the drawn resonance structures are correct, and the best representation is a hybrid of the drawn structures.

Reversible process – the forward reaction can reach an equilibrium with the reverse reaction.

Ring structure – a molecule in which the end atoms have bonded, forming a ring rather than a straight chain.

Rotamers – isomers formed by restricted rotation.

Rotation – the ability of carbon atoms attached by single bonds to freely turn, which gives the molecule an infinite number of *conformations*.

R/S convention – a formal non-ambiguous nomenclature system for assigning absolute configuration of structure to *chiral* atoms using the Cahn, Ingold and Prelog priority rules.

S

s-*cis* conformation – a relationship in which the two double bonds of a conjugated *diene* are on the same side (*cis*) of the carbon-carbon single bond connecting them; the required conformation for the *Diels-Alder reaction*.

s-*trans* conformation – the *conformation* in which the two double bonds of a conjugated *diene* are on opposite (*trans*) sides of the carbon-carbon single bond that connects them.

Saturated – the term given to organic molecules that contain no multiple bonds.

Saturated compound – a compound containing single bonds.

Saturation – the condition of a molecule containing the most atoms possible; a molecule with single bonds.

Sawhorse projection – the sideways projection of a carbon-carbon single bond and the attached substituents; gives a clearer representation of stereochemistry than the *Fischer projection*; see *Newman projection*.

Secondary carbocation – a positively charged intermediate to which two *alkyl group*s are bonded.

Secondary (2°) carbon – a carbon atom that is directly attached to two other carbon atoms.

Separation technique – a process by which products are isolated from each other and impurities.

Shielding – an effect, in NMR spectroscopy, caused by the movement of *sigma* and *pi* electrons within the molecule; causes chemical shifts to appear at higher magnetic fields (upfield).

***Sigma* (σ) antibonding molecular orbital** – in which one or more electrons are less stable than when localized in the isolated atomic orbitals from which the molecular orbital (MO) was formed.

***Sigma* (σ) bond** – formed by the linear combination of orbitals so that the maximum electron density is along a line joining the two nuclei of the atoms.

***Sigma* (σ) bonding molecular orbital** – when electrons are more stable than localized in the isolated atomic orbitals from which the molecular orbital was formed.

Singlet – describes an NMR signal consisting of only one peak.

Skeletal structure – the carbon backbone of a molecule.

S_N1 – a *substitution reaction* mechanism in which the slow step is a self-ionization of a molecule to form a *carbocation*; the rate-controlling step is unimolecular.

S_N1 reaction – a first-order substitution reaction that goes through a *carbocation* intermediate.

S_N2 – a *substitution reaction* mechanism in which the rate-controlling step is a simultaneous attack by a *Nucleophile* and a departure of a *leaving group* from a molecule; the rate-controlling step is bimolecular.

S_N2 reaction – a second-order *substitution reaction* that takes place in one step and has no intermediates during the reaction pathway.

***sp* hybrid orbital** – a molecular orbital (MO) created by combining wave functions of an *s* and a *p* orbital.

***sp^2* hybrid orbital** – a molecular orbital (MO) created by combining wave functions of an *s* and two *p* orbitals.

***sp^3* hybrid orbital** – a molecular orbital (MO) created by combining wave functions of an *s* and three *p* orbitals.

Spectrometer – an instrument that measures the spectrum of a sample (e.g., a *mass spectrometer*).

Spectrophotometer – an instrument that measures the degree of absorption (or emission) of electromagnetic radiation by a substance. The measuring system generally includes a photomultiplier UV., IR, visible and microwave regions of the electromagnetic spectrum.

Spin-spin splitting – NMR (nuclear magnetic resonance) signals caused by the coupling of nuclear spins on neighboring nonequivalent hydrogens.

Stability constant – when a complex is formed between a metal ion and a ligand in solution, the equilibrium may be expressed by a constant related to the free energy change for the process: $M + A = MA : \Delta G = -RT\ln K$.

Staggered conformation – the orientation about a carbon-carbon single bond in which bonds off one carbon are at a maximum distance apart from bonds coming from an adjacent carbon (60° apart when viewed in a *Newman projection*).

Stereochemistry – the study of the spatial arrangements of atoms in molecules and complexes.

Stereoisomers – molecules that have the same atom connectivity but different orientations of those atoms in three-dimensional space. Another name for *configurational isomer*.

Stereospecific reactions – bonds are broken and made at a particular carbon atom and lead to a single stereoisomer; if the configuration is altered in the process, the reaction undergoes inversion of configuration; if the configuration remains the same, the transformation occurs with retention of configuration.

Steric hindrance – a physical blockage of a site within a molecule by the presence of local atoms or groups of atoms; therefore, a reaction at a particular site will be impeded.

Straight-chain alkane – a saturated hydrocarbon that has no carbon-containing side chains.

Structural isomer (or *constitutional isomer*) – have the same molecular formula but different bonding among their atoms (e.g., C_4H_{10} can be butane or 2-methylpropane, and C_4H_8 can be 1-butene or 2-butene).

Subatomic particles – a component of an atom; a proton, neutron, or electron.

Substituent – a piece that sticks off the main carbon chain or ring.

Substituent group – an atom or group that replaces a hydrogen atom on a hydrocarbon.

Substitution – the replacement of an atom or group bonded to a carbon atom with a second atom or group.

Substitution reactions – an atom or group of atoms replace one atom or group of atoms; see *electrophilic substitutions* and *nucleophilic substitutions*.

Syn addition – a reaction in which two groups of a reagent X–Y add on the same face of a carbon-carbon double bond.

T

Tautomerism – a form of structural isomerism where the two structures are interconvertible using the migration of a proton.

Tautomers – molecules that differ in the placement of hydrogen and double bonds and are easily interconvertible; keto and *enol* forms are tautomers; see *keto-enol tautomerization*.

Terminal alkyne – an *alkyne* whose triple bond is between the first and second carbon atoms of the chain.

Terminal carbon – the carbon atom on the end of a carbon chain.

Termination step – the step in a reaction mechanism ending the reaction, often between two free radicals.

Tertiary carbocation – a *carbocation* to which three *alkyl group*s are bonded.

Tertiary (3°) carbon – a carbon atom that is directly attached to three other carbon atoms.

Tetrahaloalkane – an *alkane* that contains four halogen atoms on the carbon chain; the halogen atoms can be on vicinal or non-vicinal carbon atoms.

Thermodynamic product – the reaction product with the lowest energy.

Thermodynamically controlled reaction – conditions permit two or more products to form; the products are in an equilibrium condition, allowing the more stable product to predominate.

Thermodynamics – the study of the energies of molecules.

Thiol – a molecule containing an –SH group; also, a functional group.

Tosyl group – a *p*-toluenesulfonate group:

Tosylation – a reaction that introduces the toluene-4-sulphonyl group into a molecule, generally by reacting an *alcohol* with tosyl chloride to give the tosylate *ester*.

Transition state – the point of highest energy on an energy against reaction coordinate curve; the least stable point (peak) on a reaction path; a reaction path may involve more than one transition state.

Trigonal planar – the shape of a molecule with an sp^2 hybrid orbital; in this arrangement, the *sigma bonds* are in a single plane separated by 60° angles.

Triple bond – a multiple bond composed of one *sigma bond* and two *pi bond*s; rotation is not possible around a triple bond; hydrocarbons containing triple bonds are *alkyne*s.

Triplet – describes an NMR signal split into three peaks.

U

Ultraviolet light (UV) – radiation of a higher energy range than visible light but lower than that of ionizing radiations (e.g., X-rays); many substances absorb ultraviolet light, leading to electronic excitation; useful both for characterizing materials and stimulating chemical reactions (*photochemical reactions*).

Ultraviolet spectroscopy – spectroscopy that measures how much energy a molecule absorbs in the ultraviolet region of the spectrum.

Unsaturated – an organic compound containing multiple bonds.

Unsaturated compound – contain one or more multiple bonds (e.g., *alkene*s and *alkyne*s).

Unsaturation – a molecule containing less than the maximum number of single bonds due to multiple bonds.

V

Valence bond theory – the mechanical wave basis of *resonance* theory.

Valence electrons – the outermost electrons of an atom (e.g., the valence electrons of the carbon atom occupy the $2s$, $2p_x$ and $2p_y$ orbitals).

Valence isomerization – the isomerization of molecules that involve structural changes resulting only from a relocation of single and double bonds; if a dynamic equilibrium is established between the two isomers, it is referred to as "valence *tautomerism*" (e.g., the valence tautomerism of cyclo-octa-1,3,5-triene).

Valence shell – the outermost electron orbit.

Vinyl alcohol – ~CH_2=CH–OH

Vinyl group – the ethenyl group: ~CH=CH_2.

W

Walden inversion – a Walden inversion occurs at a tetrahedral carbon atom during an S_N2 *reaction* when the entry of the reagent and the departure of the leaving group are synchronous; the result is an inversion of configuration at the center under attack.

Wurtz reaction – the coupling of two *alkyl halide* molecules to form an *alkane*.

X

X group – "X" is the abbreviation for a halogen substituent in the structural formula of an organic molecule.

Y

Ylide – a neutral molecule in which two oppositely charged atoms are bonded to each other.

Z

Z isomer – the two highest-priority substituents are on the same side of a double bond or ring.

Zaitsev's rule – the major product in the formation of *alkene*s by elimination reactions will be the more highly substituted alkene or the alkene with more substituents on the carbon atoms of the double bond.

Customer Satisfaction Guarantee

Your feedback is important because we strive to provide the highest quality prep materials. Email us comments or suggestions.

info@sterling–prep.com

We reply to emails – check your spam folder

Highest quality guarantee

Be the first to report a content error for a $10 reward
or a grammatical mistake to receive a $5 reward.

Thank you for choosing our book!

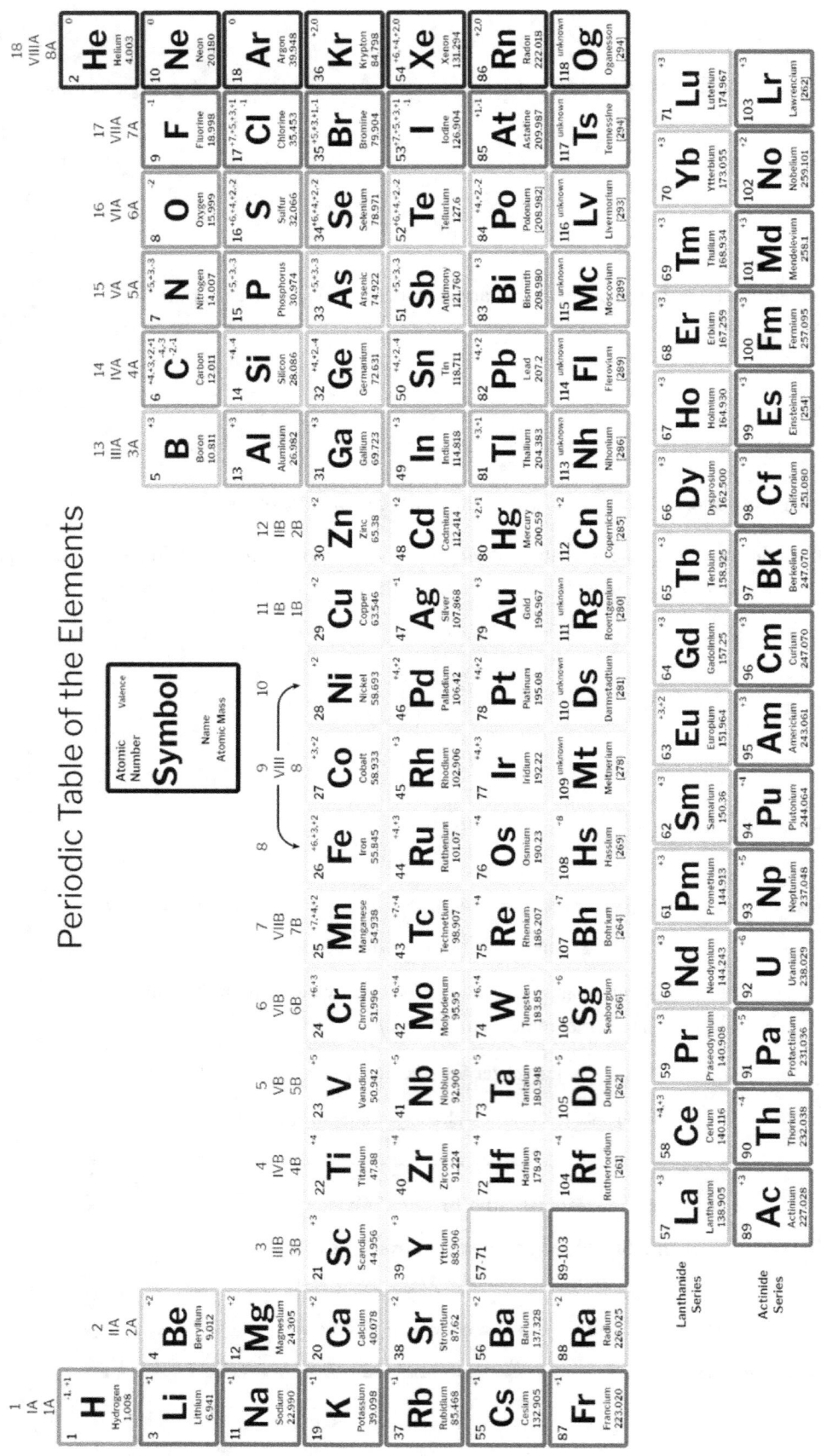

Isomer Classification

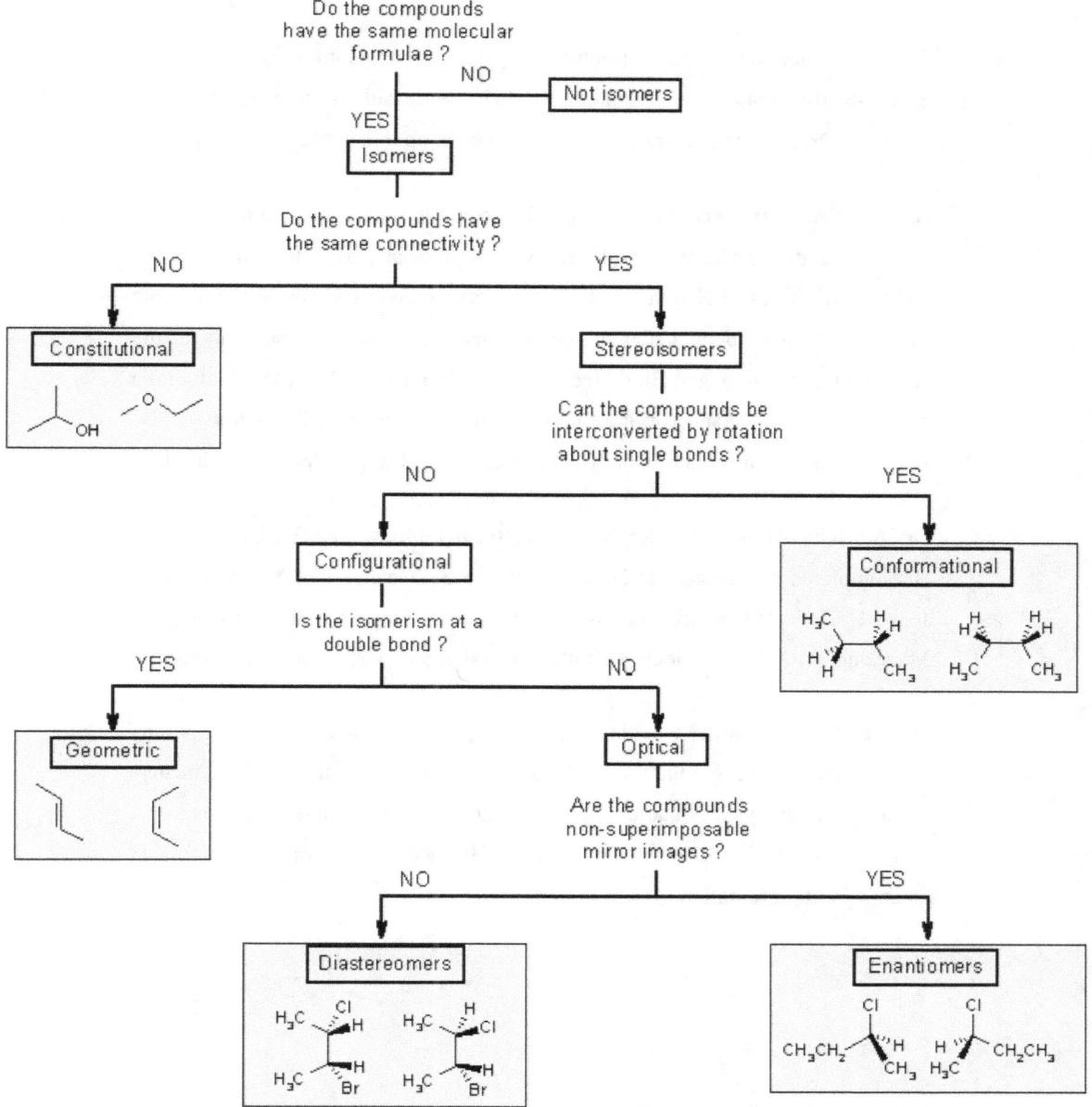

Frank J. Addivinola, Ph.D.

The lead author and chief editor of this study guide is Dr. Frank Addivinola. With his outstanding education, laboratory research, and decades of university science teaching, Dr. Addivinola lent his expertise to develop this book.

Dr. Frank Addivinola conducted original research in developmental biology as a doctoral candidate and pre-IRTA fellow in Molecular and Cell Biology at the National Institutes of Health (NIH). His dissertation advisor was Nobel laureate Marshall W. Nirenberg, Chief of the Biochemical Genetics Laboratory at the National Heart, Lung, and Blood Institute (NHLBI). Before NIH, Dr. Addivinola researched prostate cancer in the Cell Growth and Regulation Laboratory of Dr. Arthur Pardee at the Dana Farber Cancer Institute of Harvard Medical School.

Dr. Addivinola holds an undergraduate degree in biology from Williams College. He completed his Masters at Harvard University, Masters in Biotechnology at Johns Hopkins University, and five other graduate degrees at the University of Maryland University College, Suffolk University, and Northeastern University.

During his extensive teaching career, Dr. Addivinola taught numerous undergraduate and graduate-level courses, including biology, biochemistry, organic chemistry, inorganic chemistry, anatomy and physiology, medical terminology, nutrition, and medical ethics. He received several awards for his research and presentations.

If you benefited from this book, we would appreciate if you left a review on Amazon, so others can learn from your input. Reviews help us understand our customers' needs and experiences while keeping our commitment to quality.

College study aids

Cell and Molecular Biology Review

Organismal Biology Review

Cell and Molecular Biology Practice Questions

Organismal Biology Practice Questions

Physics Review

Physics Practice Questions

Organic Chemistry Practice Questions

United States History 101

American Government and Politics 101

Environmental Science 101

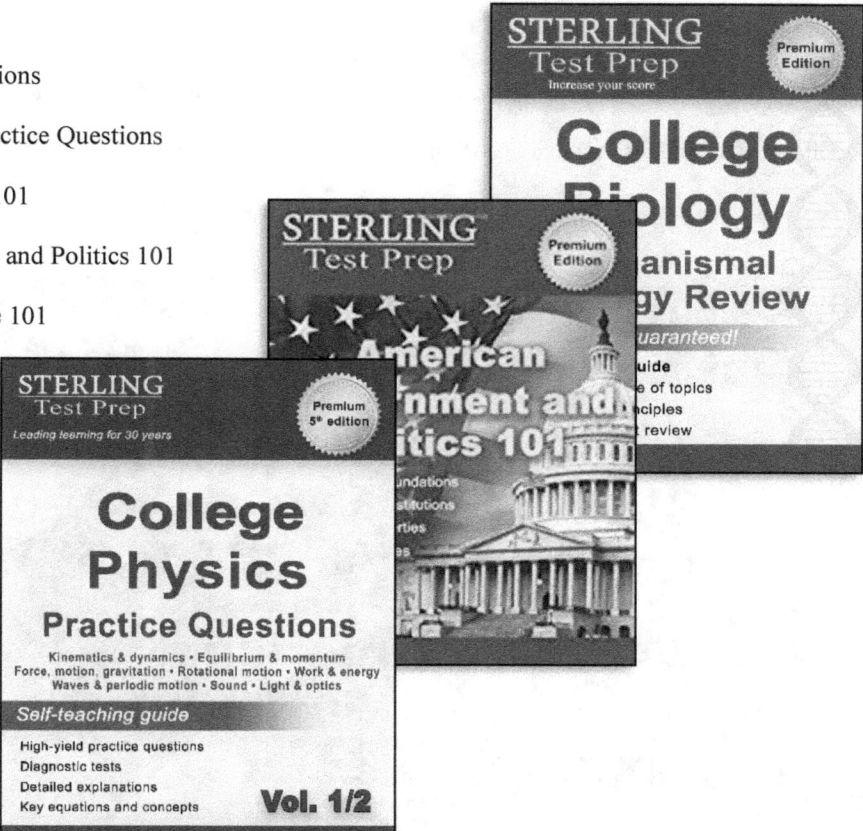

Visit our Amazon store

College Level Examination Program (CLEP)

Biology Review

Biology Practice Questions

Chemistry Review

Chemistry Practice Questions

Introductory Business Law Review

College Algebra Practice Questions

College Mathematics Practice Questions

History of the United States I Review

History of the United States II Review

Western Civilization I Review

Western Civilization II Review

Social Sciences and History Review

American Government Review

Introductory Psychology Review

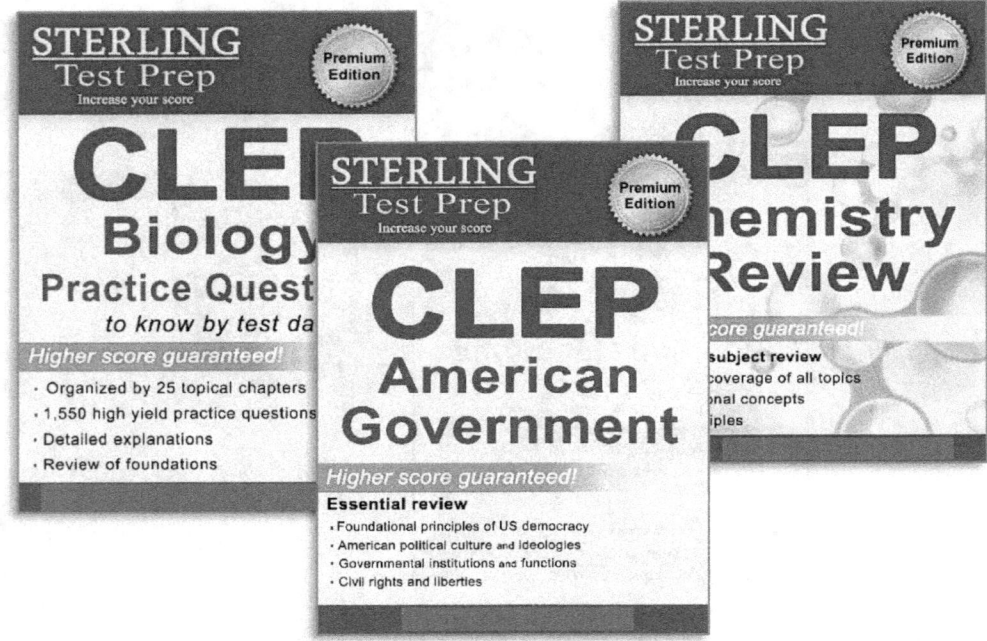

Visit our Amazon store

www.ingramcontent.com/pod-product-compliance
Lightning Source LLC
Chambersburg PA
CBHW081342070526
44578CB00005B/696